Evolution

W9-AHG-609

EVOLUTION
The History of an Idea

THIRD EDITION,
COMPLETELY REVISED AND EXPANDED

PETER J. BOWLER

University of California Press
BERKELEY   LOS ANGELES   LONDON

University of California Press
Berkeley and Los Angeles, California

University of California Press, Ltd.
London, England

© 1983, 1989, 2003 by The Regents of the University of California

Library of Congress Cataloging-in-Publication Data

Bowler, Peter J.
   Evolution : the history of an idea / Peter J. Bowler.—3rd ed.,
completely rev. and expanded.
      p.   cm.
   Includes bibliographical references (p.   ) and index.
   ISBN 0–520–23693–9 (pbk. : alk. paper).
   1. Evolution (Biology)—History.   I. Title.

QH361.B69 2003
576.8—dc21   2002007569

Manufactured in the United States of America
13   12   11   10   09   08   07   06   05   04
10   9   8   7   6   5   4   3   2

The paper used in this publication is both acid-free and totally chlorine-
free (TCF). It meets the minimum requirements of ANSI/NISO
Z39.48–1992 (R 1997) (*Permanence of Paper*).

Contents

List of Illustrations   ix

Preface to the Third Edition   xi

Preface to the First Edition   xv

1. THE IDEA OF EVOLUTION: ITS SCOPE AND IMPLICATIONS   1
    The Old Worldview and the New   3
    The Possibilities of Change   7
    The Nature of Science   13
    The Historian's Problems   18

2. THE PRE-EVOLUTIONARY WORLDVIEW   27
    Human History   29
    Theories of the Earth   31
    The Meaning of Fossils   35
    Natural Theology   38
    The New Natural History   41
    The Problem of Generation   45

3. EVOLUTION IN THE ENLIGHTENMENT   48
    Human Nature   50
    The Origin of Society   54
    The History of the Earth   57
    The Chain of Being   62
    The New Classification   66
    The New Theories of Generation   71
    The Materialists   81
    The First Transmutationists   84

4. NATURE AND SOCIETY, 1800–1859   96
    The Invention of Progress   99
    The Framework of Science   106
    Georges Cuvier: Fossils and the History of Life   108
    Catastrophism and Natural Theology in Britain   115
    The Philosophical Naturalists   120
    Radical Science   126
    The Principle of Uniformity   129
    The Vestiges of Creation   134

5. THE DEVELOPMENT OF DARWIN'S THEORY   141
    Darwin's Early Career   148
    The Crucial Years: 1836–1839   155
    Development of the Theory, 1840–1859   164
    Wallace and Publication of the Theory   173

6. THE RECEPTION OF DARWIN'S THEORY   177
    The Foundations of Darwinism   180
    The Scientific Debate   188
    Darwinism and Design   202
    Human Origins   207
    Evolution and Progress   216

7. THE ECLIPSE OF DARWINISM:
    SCIENTIFIC EVOLUTIONISM, 1875–1925   224
    Reconstructing the History of Life   227
    The Age of the Earth   234
    Neo-Lamarckism   236
    Orthogenesis   247
    Neo-Darwinism   251
    Mendelism and the Mutation Theory   260

8. EVOLUTION, SOCIETY, AND CULTURE, 1875–1925   274
    The Missing Link   278
    The Origins of Culture and Society   284
    Evolution and Race   292
    Social Evolutionism   297
    Biological Determinism   307
    Neo-Lamarckism and Society   315

Evolution and Philosophy 317
Evolution and Religion 322

9.   THE EVOLUTIONARY SYNTHESIS 325
Population Genetics 327
The Modern Synthesis 333
The Origin of Life 339
Wider Implications of the Synthesis 340

10.   MODERN DEBATES AND DEVELOPMENTS 347
The History of Life 349
Human Origins 353
Sociobiology and Ultra-Darwinism 356
Opponents of Ultra-Darwinism 361
Anti-Darwinians 366
Darwinism not Scientific? 369
Creationism 375

Bibliography 383
Index 451

Illustrations

1. Frontispiece to Thomas Burnet's *Sacred Theory of the Earth* 33
2. Ammonites, from Robert Hooke's *Posthumous Works* 38
3. Petrus Camper's facial line and facial angle 53
4. Nebular hypothesis of planetary origins 60
5. The chain of being 64
6. Classification and the binomial nomenclature 69
7. Linnaeus's system of hybridization 71
8. Buffon's theory of degeneration 78
9. Lamarck's theory of organic progress 90
10. Lamarck's branching chain of being 91
11. Cuvier's four types of animal organization 111
12. Cuvier's reconstruction of the mammoth 112
13. Examples of fossils described by Cuvier and their geologic relations 114
14. Sequence of geological formations 117
15. Teeth of the *Iguanodon* 119
16. The law of parallelism and von Baer's law 122
17. Chambers's system of linear development 137
18. The Galápagos finches 154
19. Charles Darwin 165
20. The relationship between varieties and species 167
21. The branching tree from Darwin's *Origin of Species* 171
22. Variation and the recapitulation theory 192
23. Ernst Haeckel's tree of life 193

24. Evolution of the horse family — 195

25. Frontispiece to T. H. Huxley's *Man's Place in Nature* — 209

26. Pangenesis and the germ plasm — 254

27. Continuous variation and selection — 258

28. First hybrid generation — 265

29. Second hybrid generation — 265

30. Elliot Smith's tree of human evolution — 279

31. The skull of *Pithecanthropus* — 281

32. De Mortillet's sequence of cultures — 286

33. Darwinian and developmental views of race — 295

34. Patterns of evolution — 365

35. Cladism and evolutionary classification — 373

36. Cladograms and evolutionary trees — 374

Preface to the Third Edition

Evolution: The History of an Idea has been used as a textbook to teach introductory courses in this area of the history of science for many years now. To judge by the reports which reach me, many instructors find it extremely useful, and I have been gratified at the number who have expressed their enthusiasm for the book. I have also had good reports from members of the reading public who have found it useful as an introduction to what is still a massively controversial area of science, especially in the United States. The chief reason for preparing a completely revised edition is to incorporate the considerable amount of work that has been done by scholars since the book appeared in 1983. The revisions made in the edition of 1989 were fairly minor: although they allowed for the insertion of particular points of new information or reinterpretation, there was no chance to reflect the fairly large-scale changes that were taking place in the way we view some aspects of the rise of evolutionism. I decided to rewrite the book completely, thus avoiding the risk of being trapped by the existing text (the fact that the old text was written before the advent of word processors made this decision easier). I have also come to realize that much of the original wording was open to improvement and in some cases condensation. The new text thus reflects my current thinking both on the actual historical material presented and also on how best to make it accessible to students and to the general reader. One benefit arising from the process of condensation was that I was able to add a little more illustrative material, especially in the form of quotations.

To help teachers who have been using the original version of the text, I provide here a brief overview of the major changes and my reasons for making them.

I have eliminated the chapter "Early Theories of the Earth," although some of the material originally covered in that chapter now reappears in chapters 2 and 3. I still think that developments in geology form an indispensable background for understanding the rise of evolutionism, but more literature has now become available in this area, and I felt that it was no longer necessary to devote a whole chapter to it.

A new chapter entitled "The Pre-evolutionary Worldview" tries to capture the spirit of how people looked at both the earth and its inhabitants in the period immediately before the challenges of geologists and naturalists began to show the weaknesses of a system based on a literal reading of Genesis. Given that many people in the United States still accept the biblical literalist viewpoint, this chapter may help to highlight the many steps that had to be taken to replace it.

I have dismantled the old chapter titled "Changing Views of Man and Nature" and incorporated the material into the existing chapters on the Enlightenment and the early nineteenth century. Originally I thought it would be useful to have the broader developments of the pre-Darwinian period covered in a chapter separate from the science. But I have since come to feel that a separate chapter straddling this very long period was difficult to use for teaching because the students had to be redirected to it several times if the instructor followed a chronological format. The chapters on the Enlightenment, on the early nineteenth century, and on the origins of Darwin's theory remain substantially the same in overall coverage, except for the additions made necessary by the above-mentioned changes.

A new chapter sums up the initial debate sparked by the *Origin of Species* both in science and in society and culture—the original "Darwinian revolution." In some respects this is a controversial change, given the revisionist historiography which is now playing down Darwin's role, at least in terms of the cultural impact of evolutionism. But I think a lot of instructors still treat the *Origin* debate as a unit, and in my original format they had to use material from two different chapters to present this kind of overview. This new chapter stresses that Darwin succeeded in converting most people to evolutionism but found it impossible to gain wide acceptance for his theory of natural selection. The initial Darwinian revolution thus cannot be explained by appealing to the success of the selection theory in the twentieth century (which might be called the second Darwinian revolution, except for the continuing resistance to the theory in many quarters).

As a consequence of this, there are now two chapters covering the period from 1875 to 1925, one for science and the other for wider issues. Although this period is often neglected by historians, I think we have become increasingly aware of its importance and I want to make sure that the material is there for those who decide to study it in detail. In science, this period covers the "eclipse of Darwinism" and the emergence of genetics in its original form, which was *not* seen as compatible with Darwinism. In Western culture as a whole, this was a period when the idea of evolution was extremely influential but was routinely interpreted in the context of the ideology of progress, often expressed through various forms of so-called social Darwinism.

The final two chapters on the emergence of the Darwinian synthesis in the mid–twentieth century and the subsequent debates over its consequences have been retained, though in an updated form.

The Queen's University of Belfast
August 2001

Preface to the First Edition

Because the structure of this book is rather unconventional, most of this preface will be taken up with an explanation of why it has been written this way. The book is a survey of the history of evolutionism, presented not as an academic monograph but as an introduction accessible to someone with no background in either biology or history. It is not, however, a popularization in the sense of turning history into bedtime reading. It is aimed at those with a serious interest in the theory of evolution and its implications who have not tackled the subject in any detail before. Primarily, in fact, it is intended as a textbook for university courses in the history of science, although I hope that it will prove of value to professional biologists and historians looking for a compact guide to the field. Because the potential readership is so diverse, specialists from one side or the other will have to bear with the text when it describes what seems to them a trivially obvious point—it may not be so obvious to someone on the other side of the fence.

A number of books surveying the history of evolutionism are already in print, some of them of excellent quality. But most are a couple of decades old, and none seems to have been written as a simple guide to how historians are tackling the issues involved. Let me set out what I believe to be the essential characteristics of a book written for nonspecialists, including university students.

First, the book must be organized to allow a systematic introduction to the issues, taking nothing for granted. This is particularly the case in the history of science, where many of those who become interested do not have scientific backgrounds. A nonscientist may not understand, for instance, the principles of the Linnaean nomenclature—and unless he is told exactly what is involved, this will remain a puzzle to him whenever he encounters it. In addition, the historical issues must be presented so that the nonspecialist (in this case the scientist himself) can appreciate the different ways of looking at them. I have not hesitated to express my opinion on disputed issues, but I have tried to describe the alternatives and suggest where further information about them can be found.

Second, an introductory book must be comprehensive. It must cover all topics that could be of interest to anyone studying the field. This is especially true when the book is to be used for introductory level university courses, where something like the traditional idea of a textbook is required. So this is not a book on the Darwinian revolution alone. It is a history of "evolution," using that term in its widest possible sense to denote any theory postulating a natural process for the development of life on earth. Much important work was done in geology and natural history before the *Origin of Species* appeared, and one cannot properly understand the impact of Darwin's theory without some grounding in the earlier developments. Conversely, the post-Darwinian situation must be dealt with as well. Trying to teach nonscientists about the impact of evolutionism without introducing them to the events leading up to the "modern synthesis" is ridiculous. At all stages, the relationship between the scientific theories and the culture in which they appeared needs to be discussed.

The penalty for trying to provide comprehensive coverage of so wide a field is that the amount of space available for each topic is limited. This is why I have abandoned the normal academic practice of giving extensive quotations, footnotes, and so forth, to flesh out the narrative. I also have pared down to a minimum information on the lives and backgrounds of the scientists involved. The main purpose of this book is to introduce the ideas themselves, in all their complexity, in as straightforward a manner as possible. There are many books that provide such background material on the individual personalities, including the *Dictionary of Scientific Biography* (Gillispie, ed., 1970–80). I do not believe, in any case, that students at the introductory level are interested in such background or in the academic conventions. For them, the important thing is to get over the basic ideas as clearly as possible. In a teaching situation, the lecturer can supplement the textbook with extra material in those areas where he or she wishes to specialize.

This brings me to a third point: the book must serve as a guide to further reading. I anticipate that many professionals will find this book useful precisely because of its bibliography. For students preparing term papers, the bibliography will also prove invaluable. It might be objected that the material cited is too advanced for new students, but we have a responsibility to guide such students toward further reading. I hope that the references in the text will allow them to find the items in the bibliography most relevant to their needs. It would be easy for lecturers to provide their students with an additional guide to the bibliography. More advanced students approaching the subject for the first time also should find the bibliography useful. In the history of science, courses up to and including the graduate level are often taught to students with no previous background. Such students are placed in a difficult position, because they must absorb the basic points and then pass straight on to more advanced reading. I hope this book will ease their burden.

My original intention was to provide each chapter with its own bibliography, which might have seemed less intimidating to the uninitiated. This plan was abandoned because the amount of duplication would have significantly increased the cost of typesetting. I also had intended to list primary and secondary sources separately, but this proved impossible for a single bibliography, because the distinction becomes unworkable when dealing with twentieth-century science. Wherever possible, I have tried to cite translations of works published originally in a foreign language. For primary sources printed before 1900, I have concentrated on editions that have been recently reprinted and should be more widely available. The secondary sources concentrate on classics in the field, plus the specialized literature of the last few decades. I am grateful to my departmental secretary, Mrs. Hilary Joiner, who did the original typing for the bibliography.

As an introduction to the development of perhaps the most controversial of all scientific theories, this book is meant to help bridge the gap between the "two cultures" that still divide our society. The history of science is one of the few areas in which students from both the humanities and the sciences come face to face with each other, and can appreciate that there is a genuine relationship between what each is doing. Professional historians of science have a responsibility to ensure that suitable reading material is available to help in this reconciliation, without trivializing the issues through overpopularization. Yet despite constant calls for more to be written for the nonspecialist audience, most of us continue to produce our highly technical articles and monographs. This is my own effort to relate the issues I am familiar with at a level comprehensible to the nonspecialist. Whether or not the book is a success, the goal of providing this kind of literature is vital if the history of science is to serve its true function.

In its last two chapters, the book gives some background on the debates that currently are raging over the mechanism of evolution and the teaching of evolution in the schools. Scientists may be interested to see how some aspects of the modern debates raise issues that have been controversial throughout the growth of evolution theory. A few years ago, it would have been difficult to write a history of evolutionism except from the perspective that the modern form of Darwinism represents a triumphant climax to the process. Now, we see that the basic issues were not settled quite so conclusively. I do not expect scientists to derive any technical insights from reading about the work of their predecessors, but they may gain a better insight into the nature of the fundamental issues raised by what they are doing. On the question of creationism, I have expressed myself rather more forcefully than elsewhere in the book. The historian has as much right as anyone to comment with authority on a system that would, in effect, return us to a theoretical position last taken seriously by working naturalists in the early eighteenth century. Yet I have suggested also that scientists themselves may gain a better understanding of their own position when they acknowledge the complex status of scientific theories in general and evolution theory in particular.

My qualifications for writing this kind of book are twofold. First, I have spent a number of years teaching the history of evolutionism at various levels in universities in three different countries (Canada, Malaysia, and the United Kingdom). This, I hope, has given me some insight into the difficulties of presenting the essence of complex intellectual developments to students unfamiliar with the field. Second, I have published—originally by accident and later by design—research in most areas of the history of evolutionism from the eighteenth to the early twentieth century. Although there are others more competent than myself in any one area, few historians of science will have had direct experience in so wide a section of the time period that must be covered. The one area I am not directly familiar with is modern biology, including the material of the last two chapters. I have done my best to present a layman's introduction to modern Darwinism and its opponents. I hope that I have not oversimplified or misconstrued any important points, or misrepresented the views of those who are currently engaged in the debates.

My own intellectual debts are too numerous to record here in detail. It was Robert Young who first aroused my interest in the origins of modern evolutionism, but since then I have benefited from the thoughts of a host of scholars who are represented in the bibliography. Particular mention must be made of the two referees who read the original draft of the manuscript: John C. Greene and Malcolm Kottler. From their very different perspectives, they tore the whole thing apart and advised me on how to put it back together again in a much improved form. They also advised on a host of detailed points of information and interpretation. Needless to say, where opinions are expressed, they are my own, and any remaining mistakes are my responsibility.

The Queen's University of Belfast
September 1982

1   The Idea of Evolution

Its Scope and Implications

For historians of science, the "Darwinian revolution" has always ranked alongside the "Copernican revolution" as an episode in which a new scientific theory symbolized a wholesale change in cultural values. In both cases, fundamental aspects of the traditional Christian worldview were replaced by new interpretations of nature. Medieval cosmology had pictured the earth as the center of a hierarchical cosmos stretching up through the perfect heavens to the abode of the Almighty. Because the physical universe was the stage on which the spiritual drama of humanity's creation, fall, and redemption was played out, it was natural to believe—as common sense dictated—that the earth was at the center of the whole system. Copernicus taught instead that the earth is merely the third planet orbiting around the sun. His followers realized that, by breaking down the barrier between the earth and the heavens, he had revealed a nonhierarchical universe governed by fixed laws of dynamics. The human race was not the center of everything: we are merely the inhabitants of a single planet orbiting what was soon recognized as an insignificant star lost in the immensities of space.

One could still believe that the human race had a spiritual role to play. For the Christian, humans are unique in their capacity to appreciate the moral dilemma of their existence. On the earth, at least, we are the lords of creation, the highest link in a "chain of being" that ranks all living things into a natural hierarchy. Surely no natural process could explain the existence of such an orderly system of life or of humanity's spiritual faculties. The Book of Genesis assures us that everything—including the founders of the human race, Adam and Eve—was created by divine will in a period of six days. The Darwinian revolution—which actually began before Darwin was born—undermined this belief in humanity's innate superiority over nature. Geologists showed that Genesis was not a good guide to earth history, because there were vast periods before humans appeared on the scene, when the earth was inhabited by bizarre creatures unlike any known today. Evolutionists saw the possibility that this sequence of living things might have been developed by natural laws rather than by divine miracles. And if so, the human race itself became merely another animal species, no longer the lords of creation but only superior apes.

Such changes were not accomplished without resentment. The Catholic Church resisted the new cosmology, as Galileo discovered when he was brought to trial. In this case, the opposition soon died down. Darwin's theory, however, aroused stronger passions that still generate active resistance today. If the religious controversies over Darwin's *Origin of Species* seemed to decline in the late nineteenth century, they broke out again following the emergence of twentieth-century fundamentalism, with its insistence that the Book of Genesis must be taken seriously as the history of creation. Whatever the success of evolution theory in science, the cultural phase of the Darwinian revolution has not yet reached its conclusion.

The antagonism of the creationists should not blind us to the fact that science and religion have sometimes been able to work in harmony. The history of evolutionism reveals many attempts to see the development of life on earth as the unfolding of a divine plan. There have certainly been some scientists who would adopt a more militant posture, arguing that humanity simply has to come to terms with the unpleasant fact that it is the product of a purposeless sequence of natural events. Efforts to apply the resulting theories to human affairs have, however, sparked opposition from moralists and philosophers, who may not be religious in the conventional sense. They point out that what the biologists have claimed to be "natural" human behavior all too often turns out to reflect the prejudices of a particular interest group within our society. The term *social Darwinism* has been coined to denote the illegitimate use of scientific arguments to bolster harsh or discriminatory social policies. When pushed to its logical conclusion, this critique threatens the objectivity of science itself by suggesting that no interpretation of nature can be completely free from the values of those who articulate it.

Because evolution theory so directly affects our views on human nature and the relationship between humans and the natural world, it continues to provide a focus for debate. The issues raised by Darwin and his followers are still alive today, and a study of the history of evolutionism provides invaluable background for anyone interested in how the various positions—scientific, religious, or ideological—came into existence. Much can be gained from the lessons of history, especially an awareness of the dangers of oversimplification. Before we can undertake such a study, however, we must establish a framework for understanding the rise of modern evolutionism. We must identify the full range of issues over which evolutionism has challenged the traditional worldview. We must appreciate how aspects of the old view can be modernized to fit in with certain kinds of evolutionary thought. The basic idea of evolution can be developed in many different ways, each with its own broader implications. We must also look more closely at the problems facing the historian who has to seek a balance between the conventional image of science as the objective evaluation of factual data and the suspicion of many critics that scientific theories are value-laden contributions to philosophical and ideological debates.

THE OLD WORLDVIEW AND THE NEW

The theory of biological evolution is only one part of a new approach to the study of the earth's past that has been developed over the last few centuries. Long before Darwin tackled the question of the origin of species, geologists had challenged the literal interpretation of the Genesis creation story by showing that the earth and its inhabitants have changed significantly over a vast period of time. Only within this new vision of an evolving physical universe did it become possible to imagine that living things might also be subject to natural change. Once that possibility had been confronted, religious thinkers had to ask if the Creator might have structured the universe through the operations of the laws He established, rather than by supernatural fiat. Some scientists began to doubt that any kind of divine providence was visible in the operations of nature. Thus we can distinguish several steps in the process by which the old worldview was challenged.

The existence of a series of intermediate positions between biblical creationism and materialistic evolutionism forces us to recognize the complexity of the relationship between science and religion. All too often this relationship has been depicted as a state of open warfare. But the scholars who first wrote on the war between science and religion, such as J. W. Draper (1875) and Andrew White (in 1896), were part of a secularist movement that rejoiced in sweeping away what was regarded as outdated superstition. Modern creationists also have much to gain from depicting all evolutionism as an outgrowth of militant atheism. To construct a more balanced picture, historians have had to adopt a more flexible view of the interaction between science and religion (for recent surveys, see Brooke 1991; and Lindberg and Numbers 1986). Moore (1979) has criticized Draper's and White's analyses of the Darwinian debates, pointing out that some religious thinkers were able to exploit the idea of evolution as part of their campaign to liberalize Christian theology. Many scientists too were deeply concerned over the religious implications of what they were doing. Both scientists and religious thinkers had to accommodate themselves to the new ideas being advanced to explain the history of life on earth.

The Expansion of the Time Scale

In the Judeo-Christian worldview, the universe was extremely short-lived, since the six days of creation were supposed to have occurred only a few thousand years ago. James Ussher, archbishop of Armagh in the mid–seventeenth century, has become notorious for his estimate (based on a sophisticated analysis of the generations of Hebrew patriarchs) that the creation took place in 4004 B.C. By 1800, geologists had shown that the earth must be far older than this estimate, and, by the time Darwin published his book, all educated persons accepted that the biblical timescale was untenable (Toulmin and Goodfield 1965). There were still debates over exactly how old the planet is, some of which affected evolutionary theory, but by the early twentieth century, radioactive dating techniques were beginning to suggest that the earth had been formed several billions of years ago. Ironically, Darwin himself did not have to respond to claims that the true age is only a few thousand years: not until the mid–twentieth century did creationists begin to revive Ussher's timescale.

The Concept of a Changing Universe

Ussher saw no reason to posit a time before the human race appeared, but the geological evidence suggested that during vast periods of time the earth had been inhabited by creatures unlike those we see today. The whole of human history formed merely the most recent episode in the development of life. It became clear that a sequence of populations of animals and plants had appeared and disappeared one after the other. The idea of divine creation could be defended only by abandoning the biblical model of a single episode of creation and invoking a succession of creative acts spread over eons. But an alternative became conceivable: the possibility that the successive populations were linked by a process of gradual change leading to the eventual appearance of the human race.

The theory of a sequence of miraculous creations spread through geological time represents the first and simplest attempt to work out a compromise that would preserve at least some aspects of the traditional worldview. One aspect of ancient Greek natural philosophy that had fitted neatly into the Christian view of divine creation was the belief that each species of animal and plant is based on a fixed pattern. Casual observation suggests that we can assign each individual animal to a particular species. The simplest explanation of this fact is that species have a real existence; they are fixed groupings into which individual animals *must* fall, like prearranged pigeonholes into which mail is sorted (on the emergence of biological concepts from folk belief, see Atran 1990). The Greek philosopher Plato saw individual animals as being patterned on a fixed plan or type which exists eternally at a level of reality transcending the physical universe. The Christian could believe that the pattern existed in the mind of God. According to the evolutionist Ernst Mayr (1964, 1991), the great triumph of modern Darwinism is the elimination of this ancient typological view of nature. Darwinism denies the existence of these fixed types: the species is merely a breeding population. If the average nature of the individuals in the population changes, then the species itself has changed.

If species are miraculously created, then the design of each is established by God and remains constant throughout the generations of animals. This would be true even if the species first appeared at some point in the earth's history and then disappeared (became extinct) at a later date. While it exists, the type remains fixed, and thus the idea of a sequence of creations preserves the fixity of species. At first sight it might seem that any theory of evolution must break down this element of fixity: if species change, how can they be fixed? Fixity is certainly impossible in Darwinism and in any theory which supposes that species change gradually. But some theories have been based on the assumption that transmutation occurs suddenly by saltations (from the Latin *saltus*, a leap). For example, occasionally individuals may be born so different from their parents that they count as a new species—critics call this possibility the theory of the "hopeful monster." If it is assumed that significant change occurs only through saltations, the species remain fixed between the jumps. This theory preserves the typological view of species even while accepting a form of natural transmutation.

The Elimination of Design

In the old worldview, the pattern of each species is designed by its creator. The whole point of the theory of miraculous creations is to insist that only supernatural design can explain the complex and orderly structure of each animal's body. The laws of nature can perpetuate existing designs but cannot create new ones. The "argument from design" popular among the exponents of natural theology holds that the perfection of each design, and the adaptation of each species to a particular way of life, confirms the wisdom and benevolence of God. This offers what is known as a teleological explanation of organic structures (from the Greek *telos,* purpose, and *logos,* reason). The species is constructed that way for the purpose of enabling the organisms to function in their environment. It is also possible to argue that the relationships between species exhibit a rational order or pattern that can be explained only as part of the divine plan.

The theory of a sequence of miraculous creations can preserve the notion of a pattern of creation, even if only part of the pattern exists at any one time. But what would be the consequence of denying miracles and asserting that new species can actually be formed by the operation of natural laws that presumably act to modify existing species? The most radical extension of this belief would be to argue that, since the laws of nature act blindly, the production of new species cannot be an expression of divine providence. In the Darwinian theory of natural selection, evolution exploits individual variations that are purposeless (although not absolutely random, as is popularly supposed) by simply eliminating those that confer no benefit to the organism. Evolution becomes a process of trial and error based on massive wastage and the death of vast numbers of unfit creatures. On such a model, it is difficult to see how the structure of any particular species, or the overall distribution of species, can be the product of divine forethought. There can be no goal toward which evolution is striving, only a multitude of different species each responding to local environmental challenges in an ad hoc way.

If this were the only version of evolutionism, then a war between science and religion might be inevitable. In fact, however, there are again a series of intermediate positions which allow the naturalist to view the production of new species as an expression of God's handiwork. After all, any religious thinker must accept that the laws of nature were established or instituted by the Creator. It has even been argued that Christianity fostered the rise of science precisely because it encouraged the belief in a universe governed by a divine lawgiver (Jaki 1978a). If those laws govern processes that can change species, are not the changes therefore part of the divine plan? Many efforts were made to show that the radical materialism of Darwin's theory was not the only explanation of evolutionary change. Evolution must somehow represent the unfolding of the divine plan. Some nonreligious thinkers, frightened by the haphazard and open-ended nature of evolution according to the Darwinian system, also argue that there must be orderly patterns built into the history of life on earth.

The most obvious way of retaining an element of design is to suggest that the steady progress of evolution up to the human species is itself a sign of divine purpose, a position revived in the mid–twentieth century by Pierre Teilhard de Chardin (1959). Some biologists now argue that it is difficult to see such a progressive trajectory in evolution; progress is difficult to define except by imposing human values onto nature and even then does not occur all the time. All attempts to see evidence of purpose at a more detailed level are dismissed as illusory because the laws of nature, as understood by science, cannot be programmed to work toward the fulfillment of a future goal. Other evolutionists accept a long-range and nondeterministic progressive trend without feeling any need to ascribe this to the action of a supernatural creator.

The Inclusion of Humankind within Nature

The Bible tells us that the human race was formed in the image of the Creator and given dominion over the rest of nature. Christianity (unlike some other religions) insists that only humans have a spiritual component to their existence; the other animals are "the brutes that perish." Evolutionism states that the human race must have evolved from the lower animals and cannot have this unique status. In this view, we are part of nature, and our higher faculties cannot stem from a spiritual factor with which we alone are endowed. Even our moral or ethical feelings (our conscience) must be seen as an extension of animal social instincts produced by the laws of natural evolution. This prospect created perhaps the greatest barrier to the acceptance of Darwin's theory, and it still horrifies many people today. It was overcome in Darwin's time by invoking the idea of progress: if nature was intended to progress toward evolving these higher mental faculties, then we retain our position as the goal of creation. But critics still maintain that it is precisely in the reduction of humanity to the status of beasts that we see the true materialism of the evolutionary worldview. Social scientists also prefer to believe that some aspects of human behavior cannot be explained in purely biological terms. They warn that biologists are constantly tempted to see behavioral differences between humans as innate and hence unchangeable products of the evolutionary past. In fact, the differences may be the products of education and culture.

THE POSSIBILITIES OF CHANGE

In trying to specify the range of positions intermediate between biblical creationism and the Darwinian theory of evolution, we have already been forced to confront the fact that there are different ways of understanding how the natural world could change through time. There is no single theory of evolution, only an array of rival depictions of how new forms of life originate. Over the last century or more, biologists have tried out many different ideas, and nonscientists have expressed passionate support for some of the alternatives. In scientific biology, the Darwinian theory of natural selection has achieved dominant status following its synthesis with genetics in the mid–twentieth century. But some biologists continue to challenge the Darwinian orthodoxy, and many nonscientists advocate less materialistic theories of the kind that biologists themselves once took seriously. Modern debates often revive issues that once were thought to have been settled, suggesting that the basic alternatives have a fascination that transcends the factual evidence (the "eternal metaphors" of Gould 1977a).

The word *evolution* itself has been applied to different concepts of change (Bowler 1975). The Latin *evolutio* means "to unroll," implying no more than the unpacking of a structure already present in a more compact form. Its first use in biology was to describe the development of the embryo, a process sometimes thought to offer a speeded-up model of the history of life on earth. Early embryologists believed that the growth of the organism was no more than the expansion of a preexisting miniature, a process which fits the literal meaning of *evolution*. Biologists soon came to believe that the development of the embryo was the classic example of a purposeful process moving from the simple to the complex. When the philosopher Herbert Spencer popularized the term *evolution* to denote the natural development of life on earth, he certainly intended to convey the impression of a necessary progress toward higher states. Most people still think that evolution is progressive, but Darwin (who seldom used the term *evolution*) pioneered a rival model in which it is by no means clear that life must advance toward higher levels.

Development or Steady State?

The assumption that evolution is necessarily progressive reflects the more general view that there must be an "arrow of time" defining the direction of history. It has been argued that the idea of the universe itself having a history was impressed on Western culture by the biblical story of the creation and the threatened end of the world. A classic scientific expression of this view is big bang cosmology, in which all matter in the universe is held to be flying outward from a gigantic explosion which marks the beginning of time. Geologists take it for granted that the earth had a beginning and has changed significantly since its origin. Even those biologists who say that evolution is not *necessarily* progressive admit that life has occasionally moved onto a new plane of structural complexity.

If this assumption that there is a direction to history seems self-evident, it must be noted that an alternative position is tenable. Many non-Western cultures have no sense of universal history running from beginning to end: rather, everything repeats itself in an endlessly repetitive cycle (Eliade 1951). In modern science, an analogous view is expressed in steady state cosmology. Here, all change is seen to be self-correcting: if the galaxies move farther apart, new ones are formed in the intervening space so that the density remains constant. In such a model there can be no origin—or at least we can never hope to see any evidence of a starting point, since, however far back in time we look, more or less the same state of affairs will always obtain. Such a view was applied to the history of the earth by the geologist Charles Lyell, who exerted great influence on Darwin. Lyell's uniformitarian geology supposed that the erosion of mountains exactly compensated for uplift, so the earth's surface would have looked more or less the same at any point in its history. Again, Lyell insisted that we could never hope to see evidence of an original state from which the earth began, however ancient the rocks we study. Lyell's viewpoint was overthrown, not least by the evidence for the development of life on earth, but the modern Darwinians' insistence that evolution has no built-in direction of change may be a legacy of the steady state model.

Continuity or Catastrophe?

Why would Lyell advance such an apparently bizarre view of earth history? He began from the assumption that scientific geology can use only observable causes to explain the past. To invoke a state of affairs different from what we observe is to speculate and hence to become unscientific. Since we observe only slow, gradual changes, every structure on the earth's surface must have been built gradually over a long period—and this must apply even to the oldest rocks left intact. Lyell's steady state worldview was thus a by-product of his desire to uphold the "principle of continuity," according to which there are no breaks or sudden steps in the sequence of events, no causes outside the everyday range of experience (Gould 1987; Hooykaas 1959). This aspect of his geology impressed Darwin, and Darwin's theory of evolution is a classic expression of the principle of continuity in biology. It uses only processes that can be observed at work in modern populations to explain changes in the past. Since we see no evidence of species being formed by sudden saltations today, we may not invoke such discontinuities in the past.

Lyell developed his position because he was annoyed by the tendency of the rival catastrophist school to propose massive upheavals to explain any large-scale structure, such as a mountain range. Curiously, the catastrophists accepted the progressive development of life on earth, since they were quite willing to admit that things were very different "in the beginning." Modern evolutionism has managed to combine the elements of continuity and cumulative change that were polarized in Lyell's time. Yet discontinuity continued to play a role in biology long after Darwin tried to eliminate it. The concept of saltative evolution depends on the assumption that normal, continuous processes break down at the point when a new species is formed: the radically new individual is the product of causes not normally observed in the reproductive process. This position has been largely abandoned today, but the catastrophist form of discontinuity has reemerged in the theory that the history of life has been punctuated by mass extinctions caused by asteroid impacts.

Exponents of the typological view of species outlined above often were drawn to the idea of saltation. They distrusted the theory of continuous evolution because it seemed to blur the sharp distinctions between the specific types. Darwinism shows that distinct (but not, of course, fixed) species can exist even though all change is gradual. If a population is divided in two by a geographical barrier, the two halves evolve in different directions and end up significantly different from one another, even though every step in the separation has been gradual.

Internal and External Control

The possibility that dramatic geological changes may have profound effects on the earth's inhabitants is the most extreme manifestation of a more general belief that the chief agent driving biological evolution is the external environment. Darwin's theory of natural selection builds on the same assumption but concentrates on the steady pressure of the environment when changes are very slow. In this theory, there can be no direction imposed on evolution by factors internal to the organisms, because the variation upon which selection acts is random in the sense that it is composed of many different and apparently purposeless modifications of structure. The environment determines which shall live and reproduce, and which shall die, thus defining the direction in which the population evolves. Evolution is essentially a process by which species adapt to their environment. Any other trend, for example, a progress toward more complex structures, can at best be only a by-product of the ceaseless pressure to find a better way of coping with the demands of the external world. In such a theory, the course of evolution is open-ended and unpredictable, because each population is subject to changes in its local environment or may encounter entirely new environments through migration.

Opposed to this philosophy of environmental determinism is a range of theories based on the assumption that evolution is controlled by something other than the demands of adaptation. If one assumes that variation is not random, that something in the genetic or physical constitution of the organisms directs the production of variant characters along certain predetermined lines, then the evolution of the species would be forced to move in the direction defined by this trend. This is the theory of orthogenesis, once popular among anti-Darwinian biologists. Advocates of evolution by saltation would also argue that the production of new characters in this way is controlled by internal forces. Such theories are based on the assumption that the environment does not exert an unremitting pressure on the organisms. Nonadaptive trends can be sustained because natural selection is powerless to prevent the appearance of the predetermined characters.

Individuals and Populations

A related distinction insisted on by Richard C. Lewontin (1983) centers on the role played by the individual organism in evolution. In Darwin's theory, only populations evolve: changes which affect the individual in the course of its lifetime cannot become part of the evolutionary process. This is because, in the model of heredity confirmed by modern genetics, the only characters that can be transmitted to the next generation are those the individual was born with. Selection merely picks out the genetic characters transmitted to future generations, thus shaping the gene pool of the population. This rigid model of heredity is not accepted by what was once the great rival of Darwinism, the theory of the inheritance of acquired characteristics, usually identified with the French biologist J. B. Lamarck (see chap. 3). Lamarck believed that characters acquired by the organism in the course of its life can be transmitted to future generations. Acquired characters include the effects of use and disuse on organs: the weightlifter's muscles will be inherited by his children. On such a model, evolution is the sum total of individual acts of self-development. The individual can be said to evolve, or at least to provide some fraction of the effect that accumulates to create evolution in the population. Modern genetics denies the validity of the Lamarckian effect: acquired characters and other modifications of individual development cannot be imprinted on the genes and thus cannot participate in evolution.

Ladders and Trees

Our last important distinction relates to the way the evolutionary process is depicted visually. Stephen Jay Gould (1989) warned of the dangers implicit in different forms of diagrammatic representations of evolution. Each kind of representation has hidden assumptions built into it, and the reader must be aware of the values being projected. The simplest distinction is that between ladders and trees or bushes. The simplest idea of progress presupposes a ladderlike scale of organization, from simple to complex, which serves as the pattern of evolutionary progress. Many popular accounts of the animal kingdom are still based on the idea of an evolutionary hierarchy which concludes by presenting the human species as the pinnacle of development.

Biologists have long known that they cannot in fact arrange the various forms of life into a ladderlike pattern. There are many differences in structure that do not correspond to an increase or decrease in complexity. Evolution must be represented as a tree, with the branches diverging and rediverging in various different directions. Yet in the heyday of progressionism in the late nineteenth century, evolutionary trees were often drawn with a central trunk running to the human species at the top. Lower forms of life were depicted either as steps on the sequence leading up to humanity or as the products of side branches that split off without advancing any farther up the scale. Thus the essence of the ladder model was retained, and the human species was still seen as the goal of creation.

The idealized evolutionary tree in Darwin's *Origin of Species* has no central trunk and thus prevents the reader from imagining that the end product of one branch can somehow be the goal toward which all the others are striving. Although Darwin never completely shook off the legacy of progressionism, he realized that, in an evolutionary process governed solely by the response of the species to its environment, there was no force that could be imagined to drive species in a certain direction, and no goal toward which all life could be said to strive. Evolution is a bush rather than a tree: each branch has moved in its own direction, and we cannot present lower forms of life as merely relics of early stages in the ascent toward humankind. On this model, humans cannot have evolved from chimpanzees, because chimpanzees have evolved in their own direction at the same time that we have evolved toward our current state. The common ancestor from which we both diverged existed only in the past, and we must look for it in the fossil record, not in any living species of ape. Some modern evolutionists regard the diversity of living forms as so great that we can establish no reliable scale of progress against which to measure their degree of evolutionary development.

Given the wide range of different positions which have been explored by evolutionary biologists, it should come as no surprise that the history of the field (like the Darwinian model of evolution) cannot be represented as a step-by-step process by which the modern theory was assembled. In the last few centuries, scientists have tried out many different theoretical models in their efforts to make sense of the bewildering array of evidence presented by the natural world. Our task is to outline not only the acquisition of factual knowledge but also the factors that have led scientists to prefer one model to another. Given the fact that some issues are still debated today, we must be careful not to assume that the Darwinian orthodoxy preferred by the majority of biologists is a goal toward which biology has been striving throughout its history. Even those who do accept modern Darwinism as the best possible explanation of the evolutionary process would do well to take account of the forces which have led biologists of previous generations to prefer rival theories. This leads us to the question of the scientific method itself and historians' efforts to understand how new theories are introduced.

THE NATURE OF SCIENCE

Historians have become involved in a more general process by which the nature of science has been questioned. Traditionally, scientists have presented themselves as disinterested providers of factual information. The scientific method was supposed to guarantee total freedom from the influence of subjective factors such as religion, philosophy, and moral values. It has now become apparent even to most scientists that this was an inadequate model of how science actually functions. Historians dealing with the development of evolution theory are in an excellent position to see why the traditional image of objectivity has broken down. Here is a theory which, even in its most basic form, begins to challenge a worldview that was deeply rooted in Western culture. Any theory advanced to explain the facts of nature had implications for these deeply rooted beliefs, and history suggests that the scientists' behavior often reflected the positions they took up on the broader issues.

The Scientific Method

The simplest model of science represents it as a process of factual discovery, the piling up of brute facts one upon another. A moment's reflection shows that such a model is inadequate. Scientists are interested not in single facts but in universal generalizations abstracted from the facts, what we call laws of nature. At one time it was thought that the method of induction allowed scientists to recognize a law on the basis of a large number of factual instances of the law's operation. But it is impossible to perceive generalizations in a barrage of unclassified facts which might or might not be relevant. Philosophers of science have long recognized that any investigation starts not from mere fact collecting but from a hypothesis proposed to explain how the phenomenon under investigation *might* operate. The hypothesis is tested against the facts by observation and experiment: this is the hypothetico-deductive method (Hempel 1966). The most basic kind of hypotheses might eventually be accepted as laws of nature; more general hypotheses which link many related phenomena together yield theories. If a hypothesis is successful in passing the experimental tests to which it is subjected, we might be tempted to regard it as an established truth about how nature works. One who makes such a move would commit the logical fallacy of affirming the consequent: although false, a hypothesis might nevertheless pass some early tests by luck. Only when exposed to more rigorous testing would its inadequacy be revealed. In principle, all scientific knowledge must be treated as provisional, because we can never be sure that currently accepted laws and theories will not turn out to be false when exposed to future tests.

If scientific knowledge is only provisional, why should it be given higher status than other forms of knowledge? Scientists argue that their laws and theories are more objective because they have been formulated and tested in such a way as to rapidly expose any weaknesses. A scientific hypothesis is constructed to maximize its testability—or, since any test may potentially refute it, its degree of "falsifiability" (Popper 1959, 1974). By formulating their statements in a way that leaves them open to rigorous testing, scientists ensure that their kind of knowledge can be distinguished from nonscience and from pseudosciences, such as astrology, which offer vague generalizations that can never be falsified by any empirical test. Karl Popper argues that scientists have consistently been guided in their choice of hypotheses by the criterion of which is the more falsifiable.

Most scientists have been prepared to endorse Popper's definition of their objectivity because it preserves a line of demarcation between science and nonscience. If this position is ascribed to the scientists of previous generations as well, we could retain the conventional image of science as an essentially progressive force that steadily expands our knowledge of the external world. Yet this image of steady progress seems inadequate to deal with those episodes conventionally known as scientific revolutions, where a theory that has long been accepted as valid is suddenly exposed as inadequate and replaced by something different. In principle, revolutions are not ruled out in the hypothetico-deductive system: all hypotheses, however fundamental and well tested, may eventually turn out to be false. But the prospect of scientists having to reconstruct the foundations of the way they think about nature seems to violate any impression of continuity in the development of knowledge. New theories ought to explain more facts than the old ones they replace, so there is still a sense in which science increases our level of understanding—but the increase is not brought about by a continuous development at the level of theoretical principles.

According to Popper, as soon as a theory fails an experimental test, it should be abandoned by the scientific community. But the analysis of scientific revolutions by Thomas S. Kuhn (1962) suggests that the replacement of theories is a much more complex affair (for a comparison of the Popperian and Kuhnian views of science, see Lakatos and Musgrave 1970). Kuhn used history to show that successful theories establish themselves as the paradigm for scientific activity in the field: they define not only acceptable techniques for tackling problems but also which problems are relevant for investigation. Not surprisingly, the cards are stacked in favor of the theory, because the chance of falsification is minimized by working in "safe" areas. Science done under the influence of a dominant paradigm is what Kuhn calls "normal science." Even when anomalies begin to appear, the scientific community has become so loyal to the paradigm that older scientists refuse to admit the significance of facts that falsify it and continue as though it were still functioning smoothly. Only when the number of anomalies becomes unbearable will a crisis state emerge, as younger and more radical scientists begin to look around for a new theory. When a new theory is found which deals with the outstanding problems, it soon establishes itself as the new paradigm, and another period of unadventurous normal science begins.

Kuhn's approach treats science as a social activity: scientists develop professional loyalties to their paradigm which restrict their ability to challenge the status quo. If this interpretation is valid, there are episodes in which science is anything but objective. On the contrary, scientists will do anything to defend the theory upon which their careers were founded. Objectivity may seem to be restored at the time of a revolution, but this is soon lost. And although the new paradigm seems to expand our range of knowledge by dealing with facts that could not be fitted into the old theory, Kuhn notes that there are cases in which successful lines of investigation under the old paradigm were abandoned under the new.

Science and Society

Whether Kuhn's scheme applies to the Darwinian revolution is a topic we shall return to. But we must also consider the broader implications of the claim that science cannot be defined by a method of objective study but can be understood only as a social process subject to the same rules as any other human activity. There is one level at which even the Popperian scheme accepts that the scientist constantly must go beyond the facts. A hypothesis cannot simply emerge in the scientist's mind as a mechanical response to the facts. It goes beyond the available facts and thus represents a leap of the imagination, an act as creative as that of any artist. If we admit that any theory is only an approximate model of the real world, it is possible that at any one point in time several different hypotheses might act as useful guides to research. Only time and testing will determine which is most successful, but in the meantime we can ask why particular scientists conceive or accept the theory they choose to work with. Theories often have philosophical or ideological implications, so it is not unreasonable to suppose that scientists may be influenced in their choice of hypothesis by their feelings on these wider issues.

Science thus has to be understood in the context of the "sociology of knowledge" (Barnes, Bloor, and Henry 1996; Mulkay 1979). Sociologists accept that other areas of knowledge are shaped by social and cultural factors; what is acceptable as knowledge of God (theology) clearly has varied from society to society and from time to time. Science may have to be treated in the same way: although science increases our ability to manipulate nature, its underlying theoretical perspectives may nevertheless reflect social values. Historians of science obviously must take this possibility on board, because they must be on the lookout for the effect of these subjective factors in any past debate—especially when the theories involved are intrinsically more likely to affect sensitive issues (Barnes and Shapin 1979; Shapin 1982). Evolution theory provides a fertile field of study for historians to debate the relative significance of objective and subjective factors in determining the success or failure of theories.

Traditionally, historians of science were divided into two camps: internalists, who studied purely scientific factors in the development of knowledge, and externalists, who concerned themselves with the practical applications of that knowledge. The sociology of science makes this distinction meaningless because it forces us to accept that, even in the most detailed scientific research, external factors may influence the way scientists behave. Sociologists find it relatively easy to demonstrate the existence of external factors when they study the debates which occur in the course of current research. Here, no one knows which theory will eventually be accepted, and we can see that scientists' choices are determined by a range of factors, including professional loyalties, assessment of career prospects, and awareness of theories' broader implications (Latour and Woolgar 1979). The historian faces a more difficult task, because the scientists themselves are inclined to say that, in any earlier debate, one side was proved to be right and one wrong. This promotes the assumption that, in the end, success was determined by objective factors. Scientists argue that if external factors did play a role, it can only have been in promoting misguided support for the wrong theory. Those who chose the right one did so for the right (i.e., purely objective) reasons. The sociologist of science argues that "right" and "wrong" are to some extent decided by social activity within the scientific community. The historian must be willing to consider the possibility of external forces determining support even for the ultimately successful theory.

The history of evolution theory offers clear illustrations of the need to take such factors into account. We are no longer quite so sure that unsuccessful theories were defended only by scientists whose judgment was warped by their religious beliefs and moral values. We sometimes can see that, in their heyday, theories later judged to be false served a valuable role in generating knowledge incorporated into a current paradigm. Scientists such as Lyell and Darwin, who are supposed to have hit on the right approach, were themselves influenced by their wider beliefs. A few years after publication of Darwin's theory, Marx and Engels had pointed out the analogy between natural selection and the competitive ethos of nineteenth-century capitalism. The modern sociology of science has passed on from the cruder forms of Marxist determinism, in which a certain kind of society was thought to generate a corresponding worldview in science. But the questions raised by Marx and his followers form the basis for much of the later analysis of the social values underlying evolution theories. Whatever our feelings about the ultimate success of Darwinism as an explanation of evolutionary change, as historians we cannot afford to ignore the possibility that the formulation and popularization of the theory were to some extent reflections of changing values in Western society. Some would argue that the same ideological factors underlie modern debates on related issues.

THE HISTORIAN'S PROBLEMS

The sociology of science has underpinned a revolution in the way we think about history. We are now far more concerned about the social environment within which scientific knowledge was generated, and far more willing to admit that the development of science is not the inevitable triumph of a series of factually true assertions about the natural world. Especially in areas such as evolution theory, where scientists are trying to reconstruct the past from fragmentary evidence that has survived to the present, debate about fundamental issues persists. The religious and ideological issues discussed above still affect people's attitudes to rival theories—and rival theories still abound despite the improvements in observational techniques. Professional constraints also influence the positions scientists take up when they begin their research, and we may presume that similar factors were at work in the past, at least from the time that science became a professional activity in the mid–nineteenth century. Thus, the historian must balance the need to take full account of the technicalities of scientific debates with an awareness of the ways in which scientists' approach to technical issues may be swayed by other concerns.

One beneficial product of the new historiography has been the demise, at least among professional historians, of the old triumphalist approach, in which history was presented as a sequence of almost inevitable progressive steps toward our present state of understanding. Scientists in particular used to present the history of their field in this way, singling out "heroes of discovery" whose work is supposed to have laid the foundations of modern knowledge. Historians now refer to this approach as a form of Whig history, borrowing a term once used to describe that version of British history which presented everything as a step toward modern liberalism. The sociological approach to the history of science makes it less easy to identify key discoveries, because it turns out that the scientists who first advanced important new ideas often did so for reasons that we no longer accept today. The heroes of discovery seem a little less heroic when studied in detail, and we are forced to look more closely at the ways in which they interacted with the scientific community of their time in order to gain acceptance for their ideas.

To some extent, the choice of major theories as topics for historical investigation has also become a casualty of the growing awareness of social factors. The first historians of science saw their discipline as part of the history of ideas and thus were encouraged to treat the Copernican and Darwinian revolutions as important steps forward both in science and in the development of Western culture. Now attention has switched to the social relations of the scientific community, focusing more on the emergence of research traditions and professional disciplines (L. Laudan 1977). Some historians are inclined to ask if any theory can be all that important if it is not associated with the creation of a particular professional grouping (e.g., Maienschein 1991). From this perspective, genetics makes a suitable topic: in America at least, the new theory was created by a nucleus of biologists who defined themselves as practitioners of a new discipline. But evolution theory does not map onto such neat professional boundaries, because its effects were felt in an indirect way across a variety of existing disciplines. Only in the mid–twentieth century did a science defining itself by the label "evolutionary biology" emerge (Smocovitis 1996). By the standards of this approach to the history of science, it is simply misleading to concentrate on the emergence of evolutionism in the nineteenth century: we should be looking instead at paleontology, biogeography, and similar themes which defined scientists' professional identity. The fact that this history of evolution theory is being written at all is a protest against those who carry the sociological definition of science this far. We should not throw the baby out with the bathwater: theories are important even when they are not articulated by a coherent professional group—although we should be more aware of the professional interests of those who get involved. Evolutionism had a major impact on nineteenth-century science and thought precisely because it transformed so many areas of knowledge, scientific and nonscientific alike.

For historians who accept the emergence of evolutionism as a legitimate topic, the following themes stand out as those where significant developments have been made over the last few decades.

The Precursors of Darwin

When the old-fashioned "great man" style of history was applied to science, it generated a curious by-product in the form of a search for the precursors or forerunners of each major discoverer. This was certainly the case with Darwin (see, for example, Glass, Temkin, and Straus 1959): naturalists as far back as Aristotle have been hailed as the real founders of evolutionism. More sensibly, eighteenth-century naturalists such as Georges Buffon and Lamarck have been depicted as prototype evolutionists who *nearly* put together an outline of the modern theory. This approach seems to arise out of a conflict between the individualistic notion of the hero of discovery and the image of science as steadily accumulating knowledge. It assumes that if one person was able to think up a major new perspective, then surely a few earlier figures must have been able to see at least partway to the truth. If Darwinian evolutionism is the correct answer to so many biological problems, some earlier naturalists must have been able to see where the evidence was pointing. According to this model of history, there can be no genuine scientific revolutions, no period at which the modern way of conceptualizing an issue was literally unthinkable. So the writings of pre-Darwinian naturalists are searched for occasional glimpses of the truth emerging from the fog of traditional misconception. Precursor-hunting occasionally has less flattering motives. Samuel Butler's attack on Darwin (1879) exaggerated the achievements of the forerunners of evolutionism in order to denigrate the official "hero" by showing that he was not as original as his followers believed.

Some mid-twentieth-century histories of evolutionism indulged in precursor hunting (e.g., Eiseley 1958). Isolated passages from the writings of Buffon, for instance, are hailed as evidence that he was almost in a position to fit everything together to give the theory of natural selection. The fact that the relevant passages had to be strung together out of their original context was ignored. Modern historians insist that a scientist's writings must be read as a whole in order to reconstruct the context within which he or she studied nature. Instead of assuming that previous generations were interested in modern problems (an essential prelude to the claim that they were able to formulate modern answers), we accept that they may have been thinking along lines different from ours. In some cases, pre-Darwinian naturalists did begin to explore the possibility that nature might not be entirely static. But instead of trying to force their work into a Darwinian mold, we should appreciate the very different ideas they came up with as they grappled with this new perspective. Eighteenth-century natural history was shaped by the cultural values of the period we know as the Enlightenment, and these values may have made the Darwinian view of evolution literally unthinkable (Foucault 1970). Some historians would go as far as to argue that it is misleading to use the modern term *evolutionism* to denote ideas of natural change that may differ fundamentally from the theory we now call by that name.

These earlier ideas may have played a role in shaping the climate of opinion in which Darwinism emerged. Darwin did not present his theory to a public that had never considered the possibility of evolution. On the contrary, the issues had been openly debated in response to radical anatomists who presented Lamarck's theory (among others) as a foundation for a new, more materialistic worldview (Desmond 1989). In the 1840s Robert Chambers popularized the idea of progressive evolution as the foundation for a new middle-class system of values (Secord 2000). These early evolutionists made little impression on the scientific community, and this has usually been interpreted as a consequence of the "immaturity" of their ideas. But historians now realize that scientists' attitudes toward radically new ideas are affected by the wider opinions expressed by the general public, as well as by the emergence of new evidence. Darwin certainly proposed a more sophisticated explanation of evolution, but even he had to project his theory in a way that would take account of the concerns that already had been raised in the earlier debates.

The Geological Background

It has always been recognized that earlier developments in geology and paleontology laid the foundations for the emergence of modern evolutionism. Because Darwin acknowledged Lyell as an important influence, it has been assumed that uniformitarian geology was a step toward modern evolutionism. From this, it was a short step to the claim that the opposing catastrophism served only to obstruct the development of science. Lyell himself argued that the catastrophists were led astray by their desire to preserve some aspects of the biblical creation story, especially a role for supernatural events in the earth's history. The need to assess Lyell's role more carefully was pointed out by R. Hooykaas (1959) and reinforced by Martin Rudwick (1971, 1972) and Gould (1987). What Darwin got from Lyell was the principle of continuity, that is, the idea that all changes are both natural and gradual—but Lyell himself associated this principle with a steady state viewpoint which denied the possibility of progressive evolution. The catastrophists upheld the notion of a developmental trend in earth history and used their model of discontinuous change in their reconstruction of the sequence of geological periods. The fact that they did work of enduring value, along with the reemergence of catastrophist views in modern science, warns us not to view too negatively their theoretical perspective.

The recognition that Lyell's uniformitarianism was not the sole positive force in nineteenth-century geology illustrates a number of points that have become commonplace in the history of science. The legends that built up around major figures have had to be reexamined. The temptation to present one theory as right, and its rival as wrong, has had to be reassessed. The past cannot be evaluated by modern standards—even if scientists could agree what those standards are. We cannot present one theory as the product of objective study and disparage its rival as having been upheld by nonscientific prejudice. The catastrophists did some good science despite their religious beliefs, while Lyell himself was by no means free from the desire to link his scientific position to a wider philosophy. No theory escaped the influence of the cultural values debated at the time.

The Origins of Darwinism

Following the consolidation of the modern Darwinian synthesis in biology, a great deal of historical work focused on how Darwin was able to formulate so successful a theory more than a century earlier. The centenary of the *Origin of Species* in 1959 prompted a flood of books (e.g., Eiseley 1958; Greene 1959a). The following decades saw the emergence of what became known as the Darwin industry, which focused on analyzing his published works, as well as his abundant notebooks and private papers. This industry was still active in the 1980s (see Oldroyd 1984), although the flow of publications now seems to have slackened, apart from the project to publish all of Darwin's correspondence.

The work of the Darwin industry reveals how difficult it is for historians to agree on how a great discovery was achieved, even when plenty of evidence is available. In part, the disagreements reflect the fundamentally different views on how science works, outlined above. The internalist approach, often favored by the scientist-turned-historian, stresses the objective nature of Darwin's method, picturing his theory as the most plausible solution he could find for a series of technical problems (de Beer 1963; Ghiselin 1969). Continued study of Darwin's views on technical matters has, however, suggested that his work was often shaped by factors that once would have been dismissed as the residue of outdated theories. Even if we still marvel at the modernity of his insights in areas such as biogeography, we have been forced to recognize that his thinking about the reproductive process by which new variations are generated remained firmly rooted in an older tradition (Hodge 1985). This challenges the image of the "great discovery" by revealing a complex of seemingly modern ideas interacting with a decidedly nonmodern interpretation of heredity.

At the opposite extreme of the historiographical spectrum are the modern exponents of the view that Darwinism must represent an extension of the social philosophy prevalent at the time (e.g., Young 1985). The impact of Malthus's population principle—acknowledged by Darwin himself—remains critical because it provides a direct link with social attitudes. Some historians think they can trace elements of the old natural theology in Darwin's early formulations of his theory; others stress his awareness of the materialistic implications of what he was doing. Michael Ruse (1996) links the rise of evolutionism to the prevailing faith in the idea of progress. The best chance of a synthesis will come from the effort to show how Darwin tried to articulate his response to technical issues under the influence of more general attitudes derived from his social and cultural background.

The Reception of Darwinism

Traditional accounts of the response to the *Origin of Species* contrasted the backward-looking and ultimately unsuccessful reaction of the natural theologians with the progressive attitude of converts such as T. H. Huxley. On this model, one was encouraged to believe once again that religious prejudice held some scientists back from accepting evolutionism, while those who welcomed a materialistic approach exploited the advantages of the new theory. Later work, some of it my own, has shown that the conversion of both the scientific community and nineteenth-century culture to acceptance of evolutionism was a far more complex process. Many "Darwinians"— Huxley included—could not accept what modern biologists regard as Darwin's greatest insight, the theory of natural selection. Much late-nineteenth-century evolutionism was non-Darwinian in character. Thus, acceptance of the general idea of evolution in the 1860s must be dissociated from the process by which the selection theory came to dominate biology nearly a century later (Bowler 1983, 1988).

Recognition of this fact forces us to reassess the impact of Darwin's ideas on his own time. We can no longer assume that evolutionism triumphed because Darwin presented the "correct" explanation of how the process works. Scientists cannot have been converted merely by the superiority of Darwin's arguments and evidence. To explain the conversion, we must take account of social changes both within the increasingly professionalized scientific community and within Western society as a whole. Perhaps evolutionism triumphed at least in part because it was adapted to the increasingly popular idea of progress. But if this was the case, the prevalence of non-Darwinian evolutionary ideas symbolizes a general failure to come to grips with what we now regard as the more radical (and hence the more materialistic) aspects of Darwin's thought. Evolutionary biology in Darwin's time would have had research priorities significantly different from those of modern Darwinism. Social Darwinism may not have reflected the aspects of natural selection we focus on today. From the historian's perspective, it is illegitimate to evaluate Darwin's impact in his own time by standards derived from current concerns about the social applications of biology.

If there was an initial phase of non-Darwinian evolutionism before the emergence of modern Darwinism, we must also identify the later steps by which the theory of natural selection was at last recognized as the most plausible explanation of evolutionism. These steps in turn will have to be correlated with developments in early-twentieth-century society and culture. It has long been recognized that Darwin failed to anticipate modern ideas on heredity, but the traditional historiography of evolutionism presented the synthesis with Mendelian genetics as merely the filling-in of a missing piece in the jigsaw puzzle that Darwin almost completed. I have suggested that the emergence of genetics represents the collapse of a pre-Darwinian "developmental" view of nature with consequences that were at least as profound as those associated with the initial conversion to evolutionism (Bowler 1989b). The social consequences of biological determinism, although often portrayed as a form of social Darwinism, are a product of this later era and may only partially reflect the concerns of Darwin's contemporaries.

Was There a Darwinian Revolution?

The phrase *Darwinian revolution* has been used to denote a transformation in science widely seen as a turning point in Western culture (e.g., Ruse 1979a). Kuhn himself mentioned this episode as an example of a scientific revolution (1962: 171–72). But Kuhn based his assessment on the belief that Darwin destroyed the teleological view of nature, and we now know that the breakdown of that older worldview was a much more gradual process. Greene (1971) suggests that the history of evolutionism should be depicted not as the replacement of one paradigm by another but as an ongoing clash between two rival worldviews. The continued opposition to Darwinism in the modern world supports this interpretation, although most scientists would argue that the revolution is essentially complete. The revolution took place, however, in a series of reasonably well-defined stages spread over a century or more. Several major turning points can be identified, of which Darwin's conversion of the scientific community to evolutionism is one of the most significant. Other stages may have had almost the same level of influence, including the Mendelian revolution. Looking further back in time we can see developments in other areas such as geology which paved the way for the advent of evolutionism. Pre-Darwinian evolutionists such as Robert Chambers shaped both the scientific and the public reception of the *Origin of Species* (Secord 2000). If a Kuhnian revolution is a paradigm shift within an established science, the Darwinian revolution does not fit the bill. Any attempt to define the pre- and post-Darwinian paradigms founders on the fact that major developments took place on either side of the conventional revolution.

If there was no Darwinian *revolution*, should we also abandon the idea that there was a transformation (however protracted) that should be called *Darwinian*? Our perceptions have been distorted by the fact that Darwin was both the discoverer of natural selection and the figure who precipitated the conversion of the scientific community to evolutionism. Historians have found it difficult to shake off the belief that the discovery of what turned out to be the "right" mechanism was responsible for the more general conversion. We now know that this was not the case: evolutionism became popular despite the fact that natural selection was widely rejected until the synthesis with genetics in the early twentieth century. We are impressed with Darwin's creative genius—but if his major insight was not appreciated at first, how much attention should be paid to it in the story of nineteenth-century evolutionism? Should we not give more attention to those anti-Darwinian ideas that did so much to shape nineteenth-century values, even though they were rejected in the twentieth?

A modern Darwinian may still wish to tell the story from a Darwin-centered perspective (e.g., Mayr 1982). On this model, the scientific work that led toward currently accepted theories is important; anything else is a blind alley that may be noted as a curiosity but is not worth serious study. The introduction of Darwin's theory and the steps by which it was turned into modern Darwinism define the main line of scientific development. Paradoxically, those who distrust modern Darwinism also tend to see the introduction of the theory as a major turning point (e.g., Løvtrup 1987). The Darwinians and their critics thus unite to preserve the myth that the introduction of the selection theory was the defining moment in the history of modern evolutionism.

There is certainly a role for this kind of history driven by hindsight, especially for those whose main concern is the creation of the modern theory. Scientists favor this approach, and no one can deny that the emergence and clarification of the Darwinian theory are major aspects of the story of evolutionism. But as a historian whose main concern is an appreciation of the past, I remain suspicious of this Whiggish approach. I prefer to study the nineteenth (or any other century) primarily to understand what actually happened then, not to pick out those events that hindsight tells us were important. If an earlier generation worked with ideas that we find unacceptable today, we should study those ideas because they help define the climate of opinion at the time—and show us the changes that had to occur subsequently to bring about the modern world. If Darwin's immediate followers sidelined the theory of natural selection, we must learn why they were attracted to the more general idea of evolution. If we do not make an effort to understand why they preferred alternative mechanisms, we are left with an impoverished view of how the more general idea gained so important a foothold in science. This approach is particularly important when we seek to discover resonances between science and culture at large, including social

Darwinism. All too often, cultural historians have meekly accepted the Darwin-centered view of history arising from the older historiography of science. Instead of projecting modern ideologies onto the past, we should recognize that the context of debates over the social implications of biology has changed.

This book strikes a balance between the two approaches to history. By including chapters covering the Darwin industry's reassessments of the familiar story of his discovery and its subsequent development, I have respected the desire of the modern Darwinians to seek the origins of their unique perspective on nature. But I have also made a serious effort to include those areas of evolutionary science and thought which are traditionally dismissed as side branches, especially where it can be shown that those areas helped define worldviews widely accepted by earlier generations. Darwin's theory had a major influence on modern science and thought, but we must take account of historical revisions suggesting that this influence was far more subtle than the traditional picture allows.

2   The Pre-evolutionary Worldview

Modern creationists accept a literal interpretation of the Genesis creation story in which the earth was formed only a few thousand years ago. All animal and plant species were miraculously created by God, and all the fossil-bearing rocks were deposited during Noah's flood. Many assume that the European cultural tradition was founded on the same picture of creation. But there has never been an unbroken period of consensus on this question. Medieval Christians were aware of complex debates among the ancient Greek philosophers on such issues, and of divergent opinions among the early church fathers. As Catholic evolutionists in the post-Darwinian era showed, no authority could be found among the earliest Christian writers requiring acceptance of a simple creationist perspective (Dorlodot 1925). The model of creation accepted by modern fundamentalists was first articulated in the seventeenth century by biblical scholars inspired by the Protestant Reformation. Almost immediately, this model was challenged by insights derived from the newly emerging scientific tradition.

Our story begins with the attempt to demonstrate that the new science inspired by Galileo and Newton could be made to harmonize with the contemporary understanding of the Bible. The attempt was doomed to failure, and we can identify some of the sources of the tension that would fragment the synthesis and pave the way for the Darwinian revolution. Copernican cosmology had already shattered the comfortable image in which the human race inhabited a world lying at the center of a small and neatly ordered universe. Now the stage was set for further conceptual revolutions that would dismantle the earliest efforts to show that the study of nature endorsed a simple model of divine creation. These developments in the natural sciences fitted into a wider intellectual revolution that established the foundations of modern thought (Hazard 1953; Wade 1971).

Some of the earliest theories simply assumed that the creation of Adam coincided, more or less, with the creation of the world. This assumption now seemed increasingly unlikely. It was threatened by the cosmologists, anxious to extend the new worldview in ways that would explain the origin of the earth in terms of natural processes operating in the larger universe. Some of the earliest theories were deliberately formulated in ways that allowed parallels to be drawn between steps in the natural process of formation and events recorded in Genesis. But there was always a temptation to follow the logic of a naturalistic train of speculation when it led away from the Bible.

The new science wanted to explain everything in mechanical terms: not only the earth itself, but also the structure of living things. At first the attempt to incorporate the study of living nature into the new worldview seemed to enhance the creationist perspective. Only God Himself could have created the complex structures of living organisms and given them the power to perpetuate their species. Attempts to create a system for classifying species were based at first on the assumption that there was a rational plan of creation that the human mind might hope to understand. But the very fact that living things might now be seen as mechanical systems only fueled the expectations of the more radical thinkers, who began to wonder if the laws of nature might actually create, as well as maintain, organic structures. Some eventually would reject the belief that there is an orderly plan underlying the bewildering variety of natural forms.

The increased reliance on the observation of nature also revealed facts that seemed difficult to reconcile with the biblical creation story. The discovery of fossils entombed in rocks that seemed to have been laid down in water suggested that the earth and its inhabitants had not always been the same. Again, reconciliation was attempted: perhaps the fossil-bearing rocks had been laid down under the waters of Noah's flood. But the more these naturalists looked at the earth's structure, the more complex it seemed, and the less likely to have been the product of a single catastrophic event. Soon theories based on extensive transformations of the earth's surface began to appear, dovetailing with the more speculative accounts of the planet's origin. From this combination of influences inspired by the new science, modern geology and paleontology would eventually emerge. The earth acquired a history that seemed to extend far beyond the biblical timescale.

These new initiatives, and the tensions they generated, were a product of what has often been called the Scientific Revolution. Building on Copernicus's sun-centered astronomy, Galileo, Kepler, and Newton created both a new cosmology and a new physics. These developments have traditionally been seen as the centerpieces of the Scientific Revolution, the life and earth sciences having lagged behind. But modern historians have argued that the transformation of knowledge was much more widely based (Cohen 1994; Lindberg and Westman 1990). Enthusiasm for the new science was inspired as much by the drive toward the minute study of nature as it was by the new physics. The empiricist philosophy of Francis Bacon was the foundation for a new methodology which began to transform the life and earth sciences. And from these areas would come the most obvious challenges to the orthodox view of history. Only by seeing the Scientific Revolution as a complex process inspired by complementary (and sometimes even rival) methodologies will we gain a balanced view of the developments that would transform our culture's view of the world and its origins.

HUMAN HISTORY

New ideas about the history of the earth were inspired in part by challenges to the conventional account of human history. As suggested by Paolo Rossi (1984), the growing willingness to modify the biblical account of creation was a by-product of scholarly doubts about the conventional view of how the human race itself had originated. If the story of Adam and Eve could no longer be taken at face value, then the six days of creation themselves might have to be interpreted as a metaphorical, rather than a literal, account of the earth's origin.

The 4004 B.C. date of creation once printed in the margins of many Protestant Bibles was arrived at by eminent scholars who took the sacred record seriously as a guide to the early history of the human race. The best known—and hence the subject of most of the subsequent ridicule—is James Ussher, archbishop of Armagh (Gould 1993: chap. 12). Ussher fixed the date and time of the initial act of creation at midday on Sunday, October 23, 4004 B.C. He arrived at this date by counting back through the generations of Hebrew patriarchs to the creation of Adam, and by employing a literal interpretation of the text in which Adam was formed as the culmination of six days of creative activity. This was respectable biblical scholarship by the standards of the period, but by the time Ussher published his date in the 1650s, doubts were already being expressed as to whether Genesis offered a complete account of human origins. In the traditional interpretation, Adam and Eve were taught the arts of civilization by their Creator. The human race was civilized from the start, and the great empires of the Babylonians and Egyptians stretched back almost to the dawn of time. There was, as yet, no systematic study of prehistoric archaeology to test this assumption.

Alternatives to the traditional story came from a variety of sources. Philosophers such as Thomas Hobbes and John Locke were beginning to think about the origins of society in a nonbiblical way. In an effort to understand the relationship between the nature of the human individual and society as a whole, they assumed that people once had lived in a "state of nature" with no social organization. From this idealized beginning, philosophers could seek to understand how the laws governing society had been established. Locke and Hobbes had different pictures of what humans living in a state of nature might have been like. Locke's *Two Treatises of Government* (reprint 1960) was based on the assumption that people were naturally well-disposed toward one another. But Hobbes's *Leviathan* of 1651 (reprint 1957) postulated a state of constant struggle in which life was "nasty, brutish and short" until monarchs imposed order by armed force. Neither had any hard evidence for the primitive state of the earliest humans, but their attempts to treat civilization as something built up by human activity provided an alternative to the biblical story.

No one as yet had imagined that humans, even in a primitive state of nature, were indistinguishable from animals. Europeans were beginning to learn about the great apes, whose similarities to humans were certainly disturbing. Travelers' tales abounded of apes that showed almost human capabilities and even a preference for human sexual partners. A detailed anatomical study of a young chimpanzee by Edward Tyson in 1699 confirmed that the animal seemed to bridge the gap between humans and lower animals such as monkeys (Greene 1959a: chap. 6). But Tyson, like most of his contemporaries, was sure that the human mind existed on a higher plane than that of any other creature. For the time being, at least, the possibility of a link between apes and primitive humans was pushed aside.

The concept of uncivilized humans nevertheless threatened traditional ideas about our origins, at least indirectly. For the first time, philosophers were beginning to think about the possibility of social progress. Culture was a human artifact, something *we* had created in the course of time, not a set of rules and skills taught to our forebears by God. This threw doubt on the traditional interpretation of the Genesis account of human origins. Those doubts were reinforced by other factors deriving from Europeans' wider knowledge of the world. It was reported that the Chinese civilization claimed to have been in existence since *before* the date conventionally given for the creation of the universe itself. Questions were asked about how the inhabitants of the Americas could have arrived there after dispersing from Noah's ark. Since many American animals and plants differed from those of the Old World, was it possible that the "Indians" too were of a stock different from the human inhabitants of Europe and Asia? In 1655 Isaac La Peyrère published a controversial book supporting the existence of "preadamites," humans created before Adam (Popkin 1987).

These ideas remained at the fringes of orthodoxy, yet acceptance of a literal interpretation of Genesis was being systematically undermined. For La Peyrère and others, the Bible became merely a history of the Jewish people. It was not a universal history, and the Genesis account of the creation of the earth in six days did not have to be taken literally. It was no coincidence that, at the same time, the first attempts were made to provide a naturalistic explanation of how our planet had been formed.

THEORIES OF THE EARTH

The old earth-centered cosmology had been based on Aristotle's distinction between the earth and the heavens. Because the two areas were held to be fundamentally different, it was inconceivable that the earth had been formed along with the planets by a physical rearrangement of matter within the universe (S. Kelly 1969). The Copernican revolution established a universal physics and thereby opened up possibilities that could not have occurred to the medieval mind. The triumphs of the new physics allowed Galileo and Descartes to establish the mechanical philosophy: the whole universe was to be understood as a system of matter in motion. One could argue that God had designed and created the machine exactly as we see it today. Descartes, however, was increasingly tempted by the possibility that the laws of mechanics might have rearranged matter into its present distribution. His philosophy would explain the origin of all things in physical terms. Instead of designing the details of the universe as we see it, God merely established the laws of nature and made them responsible for all future developments. The earth, and indeed the whole solar system, could have been created by a purely physical process (Jaki 1978b).

Descartes and his followers realized that the stars were other suns like our own. They noted that stars might occasionally die, perhaps choked by a crust of dense matter formed originally as sunspots on their surfaces. Dead stars would wander through space until captured by still-burning stars, thus becoming planets. In his *Principles of Philosophy* of 1644, Descartes explicitly applied this idea to explain the formation of the earth (R. Laudan 1987:

41–43). To avoid criticism from the Church, he claimed only to have shown how the earth *might* have been formed, admitting that revelation tells us that God actually did it by a miracle. Not surprisingly, as the influence of the mechanical philosophy grew, its supporters insisted on taking such explanations seriously. By the last decades of the seventeenth century, mechanistic theories of the earth's origin had become a source of wide public debate (J. Greene 1959a: chap. 2; Haber 1959: chap. 2).

Initially, most of these theories were offered not as alternatives to Genesis but as reinterpretations of the creation story in naturalistic terms. The title of Thomas Burnet's *Sacred Theory of the Earth* (English version 1691), originally published in Latin in 1681, makes this intention clear (see fig. 1). The Cartesian philosophy would provide a complete history of the earth compatible with the events recorded in Genesis (Gould 1987: chap. 2). Burnet took up Descartes's idea that a dead star would be covered by a shell of solid matter and argued that, in its initial state, our planet had had a perfectly smooth solid surface *above* the waters surrounding its core. This surface was the paradise on which Adam and Eve and their children had lived. When they turned away from God, they were punished by the great flood— but the theory provided a natural explanation for this event in the form of a collapse of the crust into the waters beneath. Only irregular fragments of the original surface were left standing out of the water, forming the mountainous terrain of the present landmasses. Our punishment thus became permanent, because Noah's descendants were forced to live among the ugly and dangerous mountains (few as yet anticipated the romantic view that mountains are beautiful). In Burnet's view, we live on a ruined planet that matches our sinful state.

By the time Burnet wrote, Newton's physics was replacing Descartes's. In a letter to Burnet (reproduced in Brewster 1855), Newton himself speculated about the formation of the earth. Later theories of the earth were based on Newtonian physics, but they continued the search for mechanical origins that Descartes had begun (Vartanian 1953). The first Newtonian cosmogony was William Whiston's *New Theory of the Earth* (1696), in which the earth condensed from a cloud of material particles under the influence of gravity. The deluge occurred when a comet swept by the earth, depositing large quantities of water on the surface and distorting the originally circular orbit.

The cultural climate of the time made it important for Burnet and Whiston to show that the new science was compatible with religion. Despite their efforts to match the Genesis story, though, there were critics who warned that the new trend was dangerous to orthodoxy. Could the Christian believe in a God who punished sinful humans by means of a natural (and

1.   The frontispiece to Thomas Burnet's *Sacred Theory of the Earth*
(1691). Christ is depicted astride the beginning and end of the sequence
of events in the earth's history. Beginning as a dead star, the earth
acquires a smooth crust which then collapses into the waters of the
deluge, eventually revealing the modern continents. Finally, the whole
planet ignites in a conflagration and becomes a star once again.

hence inevitable) sequence of events? Burnet argued that an omniscient God would have foreseen the need for punishment and designed the natural world so that it would produce a catastrophe at the appropriate time. The critics felt that his mechanistic theory was merely the first step in a program that would lead to the elimination of all concern for divine providence. Deists would soon begin to argue for a God who was merely a designer, and who left the universe to function without involving Himself in its history (Torrey 1930).

There was also a growing feeling that natural processes could not have produced the present state of the earth within the time span allowed by Genesis. Burnet resisted this trend. Although his theory was based mostly on cosmological speculation, he was aware that empirical evidence could be brought to bear on the same questions. He noted that natural erosion was gradually wearing away at the mountains: at least some aspects of the earth's surface were a product of this decay (Davies 1969). Eventually the mountains would be leveled—and for Burnet, the fact that they still existed meant that their original formation could not have been too long ago. Others looked at the evidence from nature, however, and found it more compatible with an extended timescale.

European thinkers were already seeking to combine cosmological speculation with the fruits of observation, and some were more willing to challenge orthodoxy. In 1691 the philosopher G. W. Leibniz published his *Protagaea*, arguing that, if the earth was a dead star, it must have gradually cooled down from its initially much hotter state. Leibniz noted that the existence of fossils suggested that many rocks had been formed by natural processes in the course of the earth's history. A far more radical approach was taken by the French writer Benoît de Maillet in his *Telliamed*, written around 1700 but not published until 1748 (translation 1968; see also Carozzi 1969). This theory made no reference to the deluge and took it for granted that the earth was enormously old. De Maillet thought it prudent to pretend that he was merely recounting the views of an eastern philosopher whose name was his own, read backward. Although writing within the Cartesian tradition, he ignored the idea of a cooling earth and assumed that the planet originally had been covered by a great ocean. The sedimentary rocks were laid down when the surface was covered with water, and since had been exposed by a decline in the sea level. This retreating-ocean theory would become popular in the eighteenth century, but seldom in so explicitly an antibiblical form. De Maillet even speculated about a natural origin for life and a process by which aquatic creatures could adapt to the emerging dry land.

De Maillet's comprehensive theory reflected the ambitions of the

Cartesian philosophy to provide a mechanistic account of how the world had been formed. Yet his attempt to explain how sedimentary rocks had been laid down reflects a concern to interpret the observed state of the earth's crust. The planet's origin might be a matter for cosmological theory, but its history had to be read not from the Bible but from the rocks themselves. It was becoming increasingly obvious that the rocks were formed by natural processes such as sedimentation under water. Unless these deposits could all be attributed to the great flood, this would mean that natural processes—almost certainly acting over a long period of time—had shaped the earth's surface since its original formation.

THE MEANING OF FOSSILS

The rocks of the earth's surface were an appropriate object of investigation for the natural philosophers who followed Francis Bacon's philosophy of empiricism. Science would be built on a fund of information derived from the minute study of every aspect of nature, including the rocks and minerals of many different regions. From these observations, a new science of what later would be called geology gradually emerged (J. Greene 1959a; R. Laudan 1987; Oldroyd 1996; Porter 1977). But the implications of these observations were sometimes profound. The objects we call *fossils* (the term originally covered anything dug up from the earth) seemed to be the petrified remains of once-living things. They were found embedded in rocks laid down in layers or strata superimposed on one another. The most obvious implication of these facts was that the stratified rocks were of sedimentary origin: they were laid down as layers of mud, sand, and so forth at the bottom of a lake or ocean, and then they had hardened into stone. The fossils were merely the remains of animals and plants that had lived during the period of deposition. For this interpretation to be plausible, however, one had to accept that sedimentary rocks now exposed on dry land had been laid down underwater. There must have been significant changes on the earth's surface: either the oceans had retreated (de Maillet's theory) or earth movements had elevated areas of seabed to form dry land.

These implications became apparent only after it was accepted that fossils really were the remains of living things (Rudwick 1972: chaps. 2, 3). Some seventeenth-century collectors treated them as no different from any other mineral specimen: they were merely curiously shaped rocks, perhaps formed by some mysterious "plastick virtue" that could mimic living structures. Several naturalists of the late seventeenth century challenged this position to establish the modern view of fossils. They insisted that the resemblances to living things were too close to be coincidental and sought various explanations of how the remains of animals could become entombed in deposits now exposed on dry land. Even more problematic was the fact that some fossilized creatures seemed no longer to be found among the earth's living inhabitants. This raised the disturbing possibility that species might have become extinct (i.e., have died out altogether) or might even have changed over time.

These new studies were not intended to subvert Genesis—indeed, a recent study suggests that it was the Protestants' insistence on stripping the Bible of all the symbolism accumulated in the medieval period which encouraged naturalists to take a similarly realistic view of the earth and its inhabitants (P. Harrison 1998). The best hope of reconciling fossils with the biblical story of earth history was to assume that the sedimentary rocks had been laid down beneath the waters of Noah's flood. This solution was adopted by John Woodward in his *Essay toward a Natural History of the Earth* (1695). Woodward collected and described many fossils, but he also engaged in what we would call archaeology: he dug up human artifacts which had become buried in the earth (J. Levine 1977). He was an antiquary, someone who studied ancient objects, and within the biblical timescale there was no need to separate the early history of the earth from the early history of the human race (Rappaport 1997). Woodward thought that the flood had completely reshaped the surface of the earth, producing vast amounts of sediment that had settled down onto the bottom in layers determined by their density. This was why different layers of rock contained different fossils: some of the dead bodies had sunk faster than others.

Woodward's explanation is still adopted by modern "young earth" creationists, but even in the seventeenth century, most naturalists realized that the complex nature of the sedimentary rocks made it unlikely that they were laid down in a single event such as the deluge. This position was adopted by two of the earliest defenders of the view that fossils are the remains of living organisms, Nicholas Steno and Robert Hooke (Oldroyd 1996: chap. 3). Steno was trained as an anatomist and once had the opportunity to dissect a stranded shark. He noted the close resemblance between the shark's teeth and certain fossils and argued that the fossils must be the remains of sharks that had become buried in the sediment that formed the strata. His *Prodromus* of 1669 (translation 1916) reconstructed the geological history of Tuscany in Italy from the sequence of strata, deducing that the lowest strata must have been formed before the upper ones were laid down on top of them. Steno postulated two episodes of deposition, one following the creation, the other during the deluge, with earth movements and erosion both acting to distort the strata once they were laid down.

Robert Hooke was an influential member of the group of scientists associated with the Royal Society of London. He published a series of observations made with the newly discovered microscope, including a study of fossil wood which showed its resemblance to the wood of living trees. Hooke's "Discourse of Earthquakes" was read to the Royal Society in 1668 (published in Hooke 1705). Here he stressed that fossils were the remains of animals and plants buried in sediment, and argued that earth movements could raise what was once the seabed to form new areas of dry land. Like Steno, Hooke could not imagine a vast period of time before human history began. He thus cited legends of massive earth movements, including the sinking of Atlantis, to show that earthquakes had been much more violent in the past.

Hooke noted the puzzling fact that some fossils seem to have no living equivalents. The curled shells of ammonites, for instance, certainly resemble the modern pearly nautilus, but they are not identical (see fig. 2). He was thus prepared to admit

> that there may have been divers Species of things wholly destroyed and
> annihilated, and divers others changed and varied, for since we find that
> there are some kinds of Animals and Vegetables peculiar to certain
> places, and not to be found elsewhere; if such a place have been
> swallowed up, 'tis not improbable but that those Animal Beings may
> have been destroyed with them. (1705: 327)

The rock strata showed that the earth itself had changed, but the fossils suggested that the population of animals and plants was no longer the same as that produced by God at the creation.

To those with deep religious convictions, this conclusion was unacceptable. The English naturalist John Ray doubted that a caring God would have created species only to let them die out in a natural catastrophe. Ray was well aware of the geological evidence for change, and in his first account of this (1692) he argued that the unknown species found among the fossils must still be alive in some unexplored part of the globe. In a later book, though, he became so concerned over the problem of extinction that he abandoned the thesis that fossils are the remains of once-living things. He now accepted the view of his friend Edward Lhwyd that they originated from seeds that somehow grew within the rocks and thus mimicked living structures (1713: 203–4). Rather than face up to the prospect of extinction,

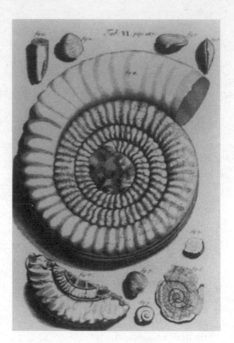

2. Ammonites, as depicted in Robert Hooke's *Posthumous Works* (1705, Plate 6). Although Hooke recognized these fossils as the petrified remains of shellfish, he conceded that they were a form unknown on earth today and, hence, must have become extinct.

Ray abandoned what most of his contemporaries took to be the only scientific explanation of fossils.

NATURAL THEOLOGY

Ray's concern over extinction reflects an attitude that became very common among the exponents of the new science. The work of Galileo, Descartes, and Newton was reducing the world to a gigantic machine, a system of matter in motion. Their followers perceived an underlying threat to traditional religious values: if the world consisted only of material particles (atoms, as the ancient philosophers had called them), on what grounds could one believe there was a God who both created and cared for the system? Was it possible that the objects we observe in the world (including ourselves) are merely random and transient products of the everlasting whirl of atoms? It was vital to defend the new natural philosophy against the charge that it would lead to atheism. Nowhere was this more urgent than in Britain, where radical new religious and political ideas had led to a vicious civil war. When the Royal Society was formed under the patronage of the restored monarch, Charles II, its members had to show that their scientific activities did not threaten the stability of traditional religion. Along with the chemist Robert Boyle, Ray became the chief exponent of natural theology, the claim that the study of nature led to better knowledge of God (Westfall 1958; on Ray, see Gillespie 1977; Raven 1942).

Boyle was an architect of the mechanical philosophy, the worldview in which nature was demystified and turned into nothing more than a complex machine. The Cartesian interpretation of this philosophy assumed that it was legitimate to seek mechanical explanations of how the structures we observe were formed. Burnet and others had followed this lead to explain the origin of the earth itself by a physical process. But it was precisely this trend that worried many Christians. If God became merely a lawgiver, in what sense was He responsible for creating the details of the world we live in? Boyle wanted God to be seen as the Creator of the actual structures we observe. His tactic was to argue that, although nature is a machine, it is obviously a complex and purposeful structure that could only have been produced by an intelligent designer, that is, God. The scientist who studied these complexities and showed how they were subordinate to a higher purpose was, in effect, illustrating the power of God.

A later exponent of this tradition, William Paley, used the analogy of a watch and a watchmaker to explain the logic of what became known as the argument from design. Anyone finding a watch would know that such a complex piece of machinery had been designed and constructed by an intelligent agent, the watchmaker. It was inconceivable that so complex a structure could have originated through the random motion of atoms. Given that nature contained many structures far more complex than watches, all apparently designed for a purpose, it was necessary to postulate a Designer of superhuman intelligence for the whole universe.

Ray's *Wisdom of God as Manifested in the Works of Creation* of 1691 was an enduring exposition of this argument, reprinted throughout the following century. Ray's targets were the atheists who thought that everything could be the product of chance, and also the Cartesians who reduced the Creator to a distant lawgiver. His preface noted, "That by the Works of the Creation in the Title, I mean the Works created by God at the first, and by him conserv'd to this Day in the same state and condition in which they were at first made." This was a static worldview: nature could preserve structures but not originate them. Ray's technique, imitated by generations of disciples, was to give innumerable examples of complex and purposeful structures in nature, defying his opponents to suggest that the structures could have been produced except by intelligent design.

Ray was a naturalist renowned for his work in classifying animal and plant species, and from these he took the majority of his examples. Species were assumed to be parts of the original creation, and only minor modifications were possible by natural processes. The perfection of the design of the human body provided many illustrations: for example, the eyes were not only designed to provide clear vision, they were placed on the highest part of the body to ensure a good field of view. But Ray was determined not to limit himself to an anthropocentric perspective. The Creator must have had good reasons for designing the vast range of animal and plant species we see around us. Some, of course, were useful to humans, and even the harmful ones might act as scourges to punish the wicked. Others seemed to have been created merely to display the versatility of their Designer. In every case, the naturalist could trace "the exact Fitness of the Parts of the Bodies of Animals to every one's Nature and Manner of Living" (Ray 1717: 124). The complexity of the animal's body proved the wisdom of its creator, but the adaptation of structure to function was an indication of His benevolence. This was what we might call a utilitarian argument from design: it stressed the *usefulness* of each structure to the organism. Natural theology retained a role for what Aristotle had called "final causes," a system in which it is legitimate to explain the existence of a structure in terms of the purpose it is intended to fulfill. For Ray and many others, adaptation was a fixed state designed by God.

Ray described in detail the wings of birds and many other examples of adaptation throughout the animal kingdom. These included the claws and teeth of predators, structures which might seem to be designed to cause suffering. But for the natural theologians, predators actually minimized suffering, since it was assumed that they always chose sick and aged animals as prey and thus gave them a quick, clean death. Suffering was admitted as a part of nature but given a subordinate role in a system that had been designed to produce overall well-being. Within such a philosophy, only parasites were a problem, and some natural theologians did agonize over the role of intestinal worms (Farley 1977: chap. 2). Ray admitted he was puzzled by God's decision to create animals whose only purpose was to cause suffering to others.

Ray's book was merely the most popular version of an argument that received wide publicity through the eighteenth century. But on the Continent, natural theology increasingly came under threat from the radical thinkers of the Enlightenment (see chap. 3). Only in Britain did it survive as an active view of nature, to be revived in Paley's *Natural Theology* of 1802—a book which impressed the young Charles Darwin.

THE NEW NATURAL HISTORY

By modern standards, the argument from design appears to be a backward-looking theory which would have prevented the life sciences from participating in the seventeenth-century scientific revolution. Biology has all too often been dismissed as incapable of responding to the new techniques based on mathematics and experiment—if it experienced a revolution, it was only at a much later date. Yet Ray was an important member of the group of natural philosophers who saw themselves as transforming our view of nature. At the time, his work in natural history was seen as an integral part of the new worldview. Physical scientists such as Boyle and Newton shared his vision of a designed universe and were prepared to admit that serious efforts were being made to extend the new program of science to include the study of living things.

What was revolutionary about Ray's approach was not the direct application of the mechanical philosophy to the structure of living organisms. A mechanistic biology was indeed proposed at the time (see next section), but with limited success. Ray and a group of like-minded naturalists set out to revolutionize our view of the relationships between the various natural kinds of animals and plants. Their belief in a divine creator assured them that nature was a rationally ordered system, and it was the task of the naturalist to discover the plan that connected all the diverse forms of life. To do this, they had to establish the basic units that had to be taken into account and then find a way of putting the commonsense view of natural relationships onto a scientific footing (Atran 1990). Modern biology explains these relationships as the consequences of evolution; but without an unambiguous way of expressing the relationships themselves, the transition from belief in design to belief in common descent would have been impossible. The emergence of a good system for classifying species (that is, a system that accurately reflects degrees of relationship) was itself a revolutionary innovation—even if our understanding of what those relationships mean is still developing today.

The first step was to determine the basic units in the system. Some naturalists argued that nature consisted only of individual organisms that did not fall into well-defined groups. But casual observation suggests that for the most part individuals do fall into distinct groups, traditionally known as *species*. We do not expect to find hybrids combining the characters of dogs and cats. If species are real, natural entities—groupings defined by something more than chance resemblances—then they will form the basic units in a natural system of classification. In practice it is not always easy to establish the boundaries (the gap separating dogs from cats is obvious enough, but what about differences between dogs, wolves, and foxes?). The members of what is widely accepted as a single species may vary distinctly between one region and another: how do we tell the difference between this phenomenon and the existence of two distinct but closely related species? Ray invoked the argument from design to solve this problem. He thought that, in principle, it should be possible to distinguish trivial local differences from the more fundamental gaps separating the real species originally created by God. It was ridiculous to add a new species to the list every time some minor variant of a known form was discovered. The Creator would have ensured that natural modifications could not go so far as to blur the distinctions between true species. Ray thus established the difference between *species* designed by God and *varieties* formed within the species as a result of changed local conditions.

It seems paradoxical, but the modern theory of evolution accepts the reality of species (although not their fixity). It is wrong to think of evolution as a process that must blur all natural distinctions. Opponents often ask why, if humans have evolved from apes, there are no ape-human intermediates still walking the earth (why is the "missing link" missing?). There have been theories which use the idea of transformation to eliminate species as real entities (see the discussion of Lamarck in the next chapter). But Darwinism was based on the assumption that animals and plants form distinct breeding communities that are reproductively, and usually physically, distinct from one another. The populations may change, and they may divide to form two subpopulations that eventually move so far apart that they can no longer interbreed. But in each case the whole population changes—it does not leave isolated individuals behind as links to its past state. We no longer accept the rigid distinction between species and varieties (because varieties may change further and eventually become reproductively isolated from the parent form), but establishing the reality of species was a step toward the modern view, not a step backward (Glass 1959a).

For the naturalists of the seventeenth century, species were both real and fixed: each was defined by its original design, just as all the cars of a certain model are defined by the manufacturer's blueprint. But species are not a series of unrelated individual designs. We can see relationships, and degrees of relationship, between them. Everyone accepts that the dog is more closely related to the wolf than either is to the cat. The lion, tiger, and leopard are related to the domestic cat, which is why we call them the "big cats." The dog and the cat are more closely related to each other (as carnivores) than either is to the horse, while the dog and the horse are more closely related (as mammals) than either is to a lizard. To seventeenth-century naturalists, these relationships could not be accidental—they suggested that God had ordered the world by designing species with built-in family resemblances. If our intuitive sense of these relationships could be put on a formal basis, naturalists would, in effect, be seeing into the heart of the Creator's great plan. The modern evolutionist reinterprets the concept of "relationship" by taking it literally in a genealogical sense. Two species are closely related because they share a recent common ancestor. But a more abstract concept of relationship had to be defined before the evolutionary interpretation of the phenomenon could be recognized.

The achievements of the early taxonomists can be appreciated by comparing their accounts of animals and plants with those offered a century earlier. Conrad Gesner's *History of Animals* of 1551–1558 (partially translated in Topsell 1608) listed a wide range of animal species, but some were purely mythological, and even the real animals were described along with their roles in mythology, heraldry, and other aspects of human culture. For Gesner's generation, the physical form of the animal was only part of its interest, and in these circumstances a purely natural classification was impossible. Only by eliminating this emphasis on the "emblematic" significance of animals and plants could a scientific approach to natural history emerge (Ashworth 1990; as noted already, Peter Harrison [1998] sees this movement as a by-product of the Protestants' biblical literalism).

Andreas Cesalpino had made a pioneering effort in 1583 to classify plants according to their natural characters. His efforts were largely ignored because the Aristotelian philosophy on which he based his system was becoming discredited. A new initiative was made a century later under the influence of the firmly established mechanical philosophy. According to this philosophy, the only real, or primary, characters of a natural object were those defined by its physical form (even color and smell were dismissed as secondary qualities produced only in our sense organs). It thus seemed self-evident that a scientific classification of species should be based only on a comparison of physical properties. John Ray and other pioneers of the new methodology openly repudiated Gesner's emblematic approach. Their work was becoming all the more urgent given the influx of specimens of animals and plants from all the territories that Europeans were exploring around the globe. These newly discovered species were not included in ancient European accounts and thus lacked emblematic significance. The need to reduce this vast flood of new data to order could only encourage the adoption of a more materialistic approach to classification.

Two rival classification policies arose within the new natural history. One problem with the sheer number of new species was that the amount of information available was very difficult to process. A workable system of classification has to allow a reasonably fast determination of where a species fits in. One way of dealing with this was to follow Cesalpino in using one or two supposedly essential characters to form the basis of the classification. This systematic approach was adopted by Joseph de Tournefort in his *Elements of Botany* of 1694. Tournefort insisted that the Creator must have given us such a guide to simplify our task. It was now recognized that flowers were the sexual or reproductive organs of plants, and Tournefort thus adopted these structures as the most essential (since they are responsible for preserving the species).

John Ray traveled widely as a young man collecting species, and even after he lost his academic post he was supported in his study of plants and animals by a wealthy patron, Francis Willoughby. Although at first tempted by the fast-track approach of the system, Ray soon began to suspect that this was not adequate. He followed the philosopher John Locke, who argued that we have no means of seeing the essence of a species: all we can observe is its outward appearance. Ray now insisted that (whatever its advantages in simplicity) a system based on a single character was bound to be misleading because there was no way of being sure that the chosen character was truly important (Sloan 1972). He advocated a method based on the comparison of as many characters as possible, applying this technique first to plants in 1682 (see also Ray 1724) and later to animals (for details, see Raven 1942).

In theory, Ray's method should yield a more natural classification than would a system based on a single character (because two species might be similar in the "essential" character and differ widely in all others). In fact, some of Ray's groupings were less natural than Tournefort's, perhaps indicating the difficulty of processing the vast amounts of evidence required by Ray's method. The rival systematic approach played an important but temporary role in the creation of modern biological nomenclature (see chap. 3). In the end, though, Ray's approach would become universal: classifying by a single character might be a useful first approximation to the real pattern of nature, but as many characters as possible would have to be processed to reveal the most natural groupings. The taxonomic debates of Ray and Tournefort played an important role in creating the modern framework for understanding organic relationships.

THE PROBLEM OF GENERATION

The new taxonomy rested on the assumptions that nature is a stable, orderly system and that species retained their characters unchanged from the creation. Such assumptions were not, however, part of the worldview favored in medieval and Renaissance times. On the contrary, nature often had been seen as an unstable and almost capricious system. It was widely believed that monstrous births were the result of unnatural unions between species. Many naturalists accepted the popular view that life could be produced by "spontaneous generation" from disorganized matter (Farley 1977). Flies were produced by rotting meat, rats and mice from dirty clothes. Such ideas would play into the hands of materialists intent on proving that natural objects were no more than random combinations of atoms. Who could be sure that the organisms produced by such a process would fit into the neat categories required by the new system of classification? In 1668 Francesco Redi performed a famous experiment in which he showed that meat produced no maggots unless flies laid their eggs on it. John Ray hailed this demonstration as a decisive blow against the old image of an essentially unstable nature and against modern efforts to revive materialism. Living organisms were produced only from parents of the same species, thus guaranteeing the stability of the forms that God had created.

Yet the process of generation (reproduction) remained a problem for the mechanical philosophy. Descartes had proclaimed that animals were nothing more than complex machines. At one level, this was a useful insight: Giovanni Borelli's *On the Movement of Animals* of 1680 (translation 1989) showed how the bones and muscles worked on mechanical principles. But when it came to explaining physiological processes within the body, Borelli was reduced to wild speculations. Generation was even more problematic. Was it really conceivable that a mechanical system, however complex, would be able to manufacture small copies of itself? More seriously for the new natural history, was it conceivable that the copies would preserve the basic character of the parents' bodies faithfully enough to ensure the stability of the species over many generations? Ray admitted that a purely mechanical theory of generation would be ludicrous, preferring to accept that in this area, at least, God had appointed nonphysical powers to shape matter into the required organic structures.

For many of Ray's contemporaries, however, the appeal of the mechanical philosophy was so strong that they attempted to extend its principles into the area of generation. Their task was to some extent made easier by the newly invented microscope and the discovery of minute organic structures (C. Wilson 1994). Regnier de Graaf had seen what was thought to be the mammalian ovum or egg (actually it was the follicle in which the egg is contained). Antonie van Leeuwenhoek observed the spermatozoa within male semen. More promising still, Marcello Malpighi and Jan Swammerdam thought they could detect faint rudiments of structures at very early stages in the development of the embryo (Adelmann 1966). Was it possible that the whole organism was somehow preformed within the egg or sperm, merely waiting to expand when provided with the right conditions? This would certainly solve the mechanists' conceptual problems: it was far easier to imagine a material process that merely had to expand an existing structure. On such a model it was difficult to see how the egg and sperm could play an equal role, however. The miniature organisms must be within one or the other. Some favored the sperm, but this would mean that millions of potential organisms were wasted in every ejaculation. The view that the new organism was preformed within the female ovum soon became dominant.

But preformation by itself was not enough. How could the mother's body manufacture her offsprings' miniatures? The full development of the preformation theory (better known as the theory of preexistence) required the assumption that offspring were already contained within the mother's ovaries when *she* was born. To preserve the logic of the argument, each generation must be contained within the other and so on back through the generations to the first created female of the species. The whole human race, for instance, must have existed, enclosed one within the other like a series of Russian dolls, in the ovaries of the first woman, Eve (Bowler 1971; Gasking 1967; Pinto-Correia 1997; Roger 1998).

Such a theory sounds ridiculous today, but it was taken seriously at the time because it fitted the combination of the mechanical philosophy and the argument from design. There was no true generation in living nature: all organic structures were literally created by God at the beginning. Mechanical processes merely brought the miniatures to life one by one to maintain the successive generations. The theory guaranteed the absolute stability of the world by ensuring that natural processes could create nothing new by themselves. It also ensured that the species would remain fixed and distinct, every member of each series having been created to the same pattern.

The preexistence theory was a radical departure from previous ideas about reproduction, but it was an essential component of the new natural philosophy's effort to retain the idea of divine creation. Nor should we dismiss the problems it was designed to solve as illusory. Modern biochemistry allows us to postulate molecules, such as DNA, which are complex enough to be encoded with the information needed to transmit characters to future generations. Some biologists talk of genetic "preformation," recognizing that there is a sense in which the characters of the offspring are stored within the parents' genetic material. The naturalists of the seventeenth century cannot altogether be blamed if their more primitive version of materialism forced them to conceive of preformation in a more literal sense. Inevitably, though, they understood the theory in the context of a worldview in which each species was fixed by a design established by its creator.

3   Evolution in the Enlightenment

The late seventeenth century saw established a worldview that compromised between the demands of the new science and the traditions of the old religion. Nature was a largely material system, but it had been designed by God in the fairly recent past. In the next century, many aspects of this compromise came under threat. Developments in the study of the earth's crust had already suggested that the planet had been significantly modified since its origin. Now it became increasingly difficult to believe that those changes could be accommodated within the old timescale. More seriously, an increasing number of radical thinkers were willing to challenge not only the biblical story of creation but also the very foundations of the Christian religion. This was the Age of Enlightenment—a period in which triumphs of the new science were seen as evidence that reason might throw off ancient superstitions and allow us to come to a new understanding of both the natural world and human nature. Christianity was pilloried as the bastion of a repressive social order (classic surveys of Enlightenment thought include Cassirer 1951; Gay 1966, 1969; Hampson 1968; Hazard 1963; and Willey 1940; for modern accounts, see Israel 2001; Outram 1995; and Porter 1990).

Small wonder that such a period saw major challenges to the worldview of natural theology. The materialist philosophers of the Enlightenment sought explanations of how life originated on the earth that did not depend on supernatural intervention. Life must have originated by a natural process (spontaneous generation), and since there was no divine guarantee of stability, species might be subject to change through time. The Enlightenment saw the emergence of the first worldviews containing an element of what we would call evolution. But this statement must be treated with caution, lest we exaggerate the degree of similarity between these early ideas and those of the Darwinian period. Eighteenth-century thinkers did try to grapple with the concept of a changing universe, but they did so within a conceptual system that was very different from that of Darwin's century. The results do not always map neatly onto our modern concepts, and some historians refuse to use the term *evolution* when speaking of this period.

There was an element of transformism (to use an older term for the mutability of species) in some eighteenth-century ideas. But we have to accept that there are versions of transformism that are fundamentally different from those we now accept. The attempt to treat Enlightenment naturalists as forerunners of Darwin (Glass, Temkin, and Strauss 1959) must be treated with caution. The historian must avoid the temptation to impose modern values onto the past by, for instance, focusing only on extracts from eighteenth-century writings which make sense in modern terms. The past must be read within its own context and allowances made for the fact that this context may differ from our own interpretation. (For a more nuanced account of these developments, see J. Greene 1959a; and Ritterbush 1964; for a detailed study of the life sciences in France, see Roger 1998; on eighteenth-century science generally, see Clark, Golinski, and Schaffer 1999; and Rousseau and Porter 1980.)

English-speaking historians have concentrated on the rise of materialism, but the French philosopher Michel Foucault (translation 1970) stresses the enormous gulf separating the classical thought of this period from modern ideas (see Gutting 1989, 1994). For Foucault, the key to understanding this earlier style of thought lies in its emphasis on the desire to classify rather than to explain. He insists that the Enlightenment was dominated by the belief that the ordering process by which the human mind seeks to classify natural objects corresponds to the real structure of the world. The system of classification defined all the possible combinations of natural characters. Unlike the Darwinian view of things, such a system is not open-ended: the range of possible forms is predictable from the system used to classify them. In effect, the mind sets up a network of pigeonholes into which all natural objects must fit. The idea of change could be accommodated by supposing that not all slots in the system were filled by real organisms at the same time. But the production of new forms could never be anything other than the unfolding of a predictable pattern, the filling in of the empty pigeonholes. Starting from such a perspective, the eighteenth century could never aspire to anything resembling the Darwinian view of evolution.

Some aspects of the eighteenth-century view of nature support Foucault's thesis. When the Swedish naturalist Karl von Linné, better known by his Latinized name Carolus Linnaeus, created the modern system of biological classification, he did so by building on the image of a divinely ordered universe. Linnaeus eventually conceded that new species might appear in the course of time, but his chief explanation of the process was hybridization—which merely recombines the characters of existing species. This period saw the last flowering of interest in the ancient concept of the chain of being, a system of classification in which the pigeonholes are arranged in the simplest possible way, a straight line. The first really comprehensive transmutationist system, that of Lamarck, still contained elements of this linear way of thinking. Only a handful of the most radical Enlightenment materialists dared to speculate that new organic forms might appear by a process involving the random combination of atoms—and they were more interested in spontaneous generation than transmutation.

Equally puzzling for the simplified view that the Enlightenment represents a step toward the modern way of thinking is an almost total lack of interest in the possibility that the human mind itself might have evolved. Materialists might insist that the mind is a mere by-product of the brain, but they failed to address the question of how an animal's brain might expand its capacities to become human. Most philosophers remained convinced of the uniformity of human nature; for them, all humans at all times have thought in basically the same way. Even the idea of social progress was slow to become fully articulated. Many social theorists during the Enlightenment adopted an essentially ahistorical search for the laws of human behavior. Only at the very end of the century did Marie-Jean-Antoine-Nicholas Condorcet and others begin to explore the idea that the earliest humans had lived in a state of savagery. The possibility that humans might have emerged from an animal ancestry was unthinkable for most eighteenth-century thinkers.

HUMAN NATURE

We have seen that, by 1700, the similarities between humans and apes were obvious enough to be a source of concern. As Europeans explored the world, they encountered races of humans living under conditions so basic that they questioned their humanity. There were tales of apes showing almost human capabilities and even a preference for human females. How, then, could naturalists distinguish between the highest animals and the lowest humans? In the end, most agreed to follow the conventional view that humans' mental and moral powers lifted them above the brutes—although the most despised races were already being treated as having degenerated into apelike beings (Corbey and Theunissen 1995; J. Greene 1959a: chaps. 6, 8; for collections of primary sources, see Augstein 1996; and Bernasconi 2001). For the time being, it was left for materialist philosophers to raise questions about the origins of the mind.

The tensions are evident in the attitude Linnaeus adopted when he had to fit the human species into his system of classification. He included humans and apes together in the order Anthropomorpha, or primates. When criticized for thus implying a link between humans and brutes, he challenged naturalists to show him any *physical* character by which humans and apes could be distinguished more clearly. Although prepared to admit that humankind was on a different moral and intellectual plane, Linnaeus insisted that naturalists must classify by physical resemblances alone, and at this level the relationship was clearly evident.

Linnaeus was a conservative thinker, and it is surprising that his views on the similarities between humans and apes were not exploited by radical thinkers to challenge the traditional view of our place at the head of creation. Linnaeus's rival Buffon was developing a more materialistic view of nature in his *Natural History*, and in his fourteenth volume, published in 1766, he too tackled the problem of the apes' status. He accepted a physical resemblance between apes and humans but insisted that this was as far as it went; stories of intelligent actions by apes were fabrications, and in fact the dog was closer to humanity in mental powers. Buffon thus backed away from seeing a genealogical relationship between humans and apes and preserved the conventional view that our mental faculties lift us above the animal kingdom. James Burnet, Lord Mondboddo, argued in the first volume of his *Of the Origin and Progress of Language*, published in 1774, that the apes might represent the earliest form of humanity. But he regarded humans (including savages and apes) as quite distinct from the rest of the animal kingdom. The first suggestion that the human species was descended via the apes from the lower animals did not come until Lamarck's *Zoological Philosophy* of 1809 (translation 1914). Lamarck is popularly supposed to have seen the orangutan as our direct ancestor, although in fact he specifies the orang of Angola (the chimpanzee), not the orang of the East Indies (171–72). Contemporary descriptions of apes had appropriated the Malay term *orang*, which means "man," as in *orang utan*, "man of the woods," for other apes. In any case, Lamarck's derivation of humans from apes came too late to have any influence on Enlightenment thought.

The problem had been effectively headed off by naturalists who sought to minimize the physical resemblances between humans and apes. Buffon drew upon Edward Tyson's 1699 description of the chimpanzee, which had already noted some important distinctions. Petrus Camper's study of the orangutan's vocal organs (1779) confirmed that it was certainly incapable of speech or of walking fully upright. This last point was seized upon by the German anthropologist J. F. Blumenbach and by Georges Cuvier to make the distinction that Linnaeus had proclaimed impossible. They divided humans from the apes by creating two separate orders: Bimana and Quadrumana. The term *quadrumana* (four-handed) was meant to indicate that the apes' lower extremities were not really feet in the human sense— they were an extra pair of hands. Only humanity was two-handed, with feet fully adapted to an upright gait.

Although there was little enthusiasm for the possibility that they themselves might be derived from apes, Europeans were increasingly willing to depict other humans as having apelike features. Camper studied the various human races and arranged them into a hierarchy with whites at the top. To measure the shape of the skull, he defined the *facial line* joining the jaw, nose, and forehead and the *facial angle* between this line and the horizontal (see fig. 3). In the most perfect human features, represented by classical Greek statues, the facial angle was almost ninety degrees. In the ape this angle was much smaller. When Camper applied this technique to the various races, he found that Europeans approximated the classical ideal most closely, while the facial angle decreased in other races, especially blacks. He insisted that since apes and humans were quite distinct, the "lower" races could not be hybrids between the two types.

How was the apelike character of these races to be explained? The answer was to assume that the human species was capable of degeneration from its original purity of form. In 1745 Pierre Louis Moreau de Maupertuis published *The Earthly Venus* (translation 1968), which suggested that the white form of humanity was the original, and that colored races had been produced by the accumulation of accidental variations in the course of time. Buffon and Blumenbach assumed a similar process of degeneration but explained it in terms of exposure to unsuitable conditions. Blumenbach used his world-famous collection of skulls to consolidate a system of racial classification (translation 1865). Linnaeus had already suggested four basic races, and now Blumenbach established five. The purest was the Caucasian, or white (called Caucasian because it was assumed that Europeans had migrated from the original center of humanity somewhere near the Caucasus Mountains), followed by the Mongolian, American, Malay, and Ethiopian. The fact that the degeneration led toward an apelike form could be explained by the chain of being—the widely popular view that there existed a single hierarchy of animal forms in which the human species stood immediately

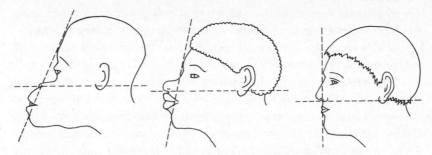

3. Petrus Camper's facial line and facial angle. The facial line joins the jaw, nose, and forehead. The angle between this line and a horizontal line joining the nose and ear is the facial angle. Note how the facial angle for the classical figure on the right is ninety degrees, whereas for the ape it is about sixty degrees. Camper and many of his later followers claimed that the facial angle for the black races was between that of the European and the ape.

above the ape (see below). The Ethiopian, or black, was frequently identified as the most degenerate race, allowing Europeans to use the naturalists' arguments as a justification for their exploitation of Africans as slaves (Banton 1987; Snyder 1962; Stanton 1960; Stepan 1982).

Despite the tendency to disparage the mental powers of other races, most European thinkers remained faithful to the traditional view that human nature was uniform around the globe. The human mind had powers not found in the animals, and the most popular philosophy of the Enlightenment sought to explain those powers. The "sensationalist" philosophy derived from John Locke's *Essay Concerning Human Understanding* of 1690 (reprint 1975) proclaimed that all knowledge was derived from the senses. The mind was a system for processing these sensations and the ideas derived from them. Locke insisted that there were no innate ideas, thus deflecting attention from the possibility that the mind might be governed by instincts derived from our bodily nature. The question of how the body generates sensations was sidestepped in order to concentrate on the laws governing the mind itself. Etienne Condillac extended this approach (translation 1756; see also Knight 1968), as did Claude Helvétius (translation 1810), who argued that, by controlling the input of sensations via education, the growing mind could be molded in any desired way. In Britain, David Hartley (1749) emphasized the principle of the "association of ideas" to explain learning: if two ideas were presented together often enough, the mind would expect an automatic link between them (Oberg 1976). Once Joseph Priestley had freed it from the religious overtones retained by Hartley, this principle became the basis for a new science of human nature (Hartley 1775). Social reform was possible through better understanding of how the mind worked: the individual's behavior could be molded to fit society by psychological manipulation. This possibility was seized upon by the "philosophical radicals" of the late eighteenth and early nineteenth centuries (Halévy 1955).

There was a more radical alternative that challenged the traditional view of the mind as something functioning on a level different from the body. Descartes had proclaimed that animals were nothing more than machines, but had insisted that in humans the material body was somehow linked to a mind. The materialist philosophers of the Enlightenment took the animal machine doctrine and boldly applied it to humanity. Descartes's view that animals had no sensations seemed ridiculous—but if the physical organization of their bodies generated some level of awareness, why could this not be true for humans as well? Julien Offray de La Mettrie's *Man a Machine* of 1748 (translation 1991) proclaimed the elimination of the concept of the soul as an entity that added life to an otherwise inanimate body. Life was a product of the organized matter of the body, a view supported by Abraham Trembley's discovery of the regenerative powers of the "polyp," or freshwater hydra (translation 1973; J. Baker 1952; V. Dawson 1988; Vartanian 1950). La Mettrie argued that the same principle applied to the human rational soul: it was merely a by-product of the activity of the brain and nervous system. Enlightenment materialism thus anticipated many of the questions later raised by the prospect of an evolutionary origin for the human mind, but did so from a purely philosophical analysis of the relation between mind and body. Given the limitations of the biological techniques available, however, eighteenth-century materialism often degenerated into a kind of animism. Humans and animals could only be seen as machines by attributing the properties of life to matter itself (T. Hall 1969; Heimann and McGuire 1971; Moravia 1978; Rosenfield 1968; Schiller 1974).

THE ORIGIN OF SOCIETY

The new models of human nature were also political in nature. The old view had been used to support not only the idea of absolute moral values but also the "divine right of kings" and the claim that the social hierarchy was established by God. The Church was seen as a bastion of the old social order, and an attack on Christianity served as ammunition in a campaign for reform. The hope that human reason could now uncover both the moral and physical foundations of the universe led to expectations that a better society could be conceived by taking account of the true nature of humanity. New policies were proposed which used the new ideas about human nature to ensure the greatest happiness of the greatest number of people. Some eighteenth-century thinkers pinned their hopes on reform by an "enlightened despot," but others were radicals belonging to a variety of subversive groups (Jacob 1981). The whole program was threatened by the marquis de Sade's rejection of all moral values: if personal happiness was the only goal, why should the individual worry about anyone else (Crocker 1963)?

The ideas about human nature developed by Condillac, Helvétius, Hartley, and Priestley were thought to offer a practical means by which social reform could be effected. Priestley joined radicals like William Godwin in arguing for the prospect of a future in which human happiness could be greatly increased. Not everyone, however, saw reform as something to be imposed by state control (even if well-intentioned). The reformers were often from the rising middle class, which derived its income from commerce. For them, the most important route to progress lay in removing the restrictions the old hierarchical society had placed on individual freedom. Economic and social progress would then be brought about by market forces, not by legislation. The laissez-faire school of economics founded by Adam Smith in his *Wealth of Nations* of 1776 (reprint 1910) held that prosperity resulted from individual initiative, not state intervention.

The expectation of social progress in the future influenced ideas about history. The Enlightenment saw a growing faith in the inevitability of progress, which looked to the past for evidence of the earliest stages in the ascent toward a better way of life (Bury 1932; R. Jones 1965; Nisbet 1980; Pollard 1968; Van Doren 1967). But Enlightenment thinkers only slowly groped their way toward this faith. Most now accepted that modern science and learning were superior to anything developed in the ancient world. But they were also aware of the collapse of ancient civilization into the Dark Ages. Giambattista Vico hoped to found a "new science of history," but for those races outside the biblical story he saw only a cycle of progress and degeneration (translation 1948; Rossi 1984). Voltaire presented the age of Louis XIV as a pinnacle of achievement, without seeing it as a step in a more general advance. The Enlightenment's most respected social thinker, the Baron de Montesquieu, wrote his *Spirit of the Laws* of 1747 (translation 1989) to explain societies as the outcomes of historical forces (Vyverberg 1989). But he did not see European society as more highly developed than others, and he saw each form of society as a blend of human and physical factors—there was no sense of inevitable progress. To develop a full theory of progress, in which the rise of civilization has been interrupted by only temporary setbacks, required a self-confidence that Western culture developed only at the end of the eighteenth century.

Seventeenth-century thinkers such as Locke and Hobbes postulated a "state of nature" in which the earliest humans lived before the development of social customs and laws. If such a state were seen as the earliest period in social history (rather than as a philosophical abstraction), then the subsequent development of successively higher levels of civilization would serve as the foundation for a theory of progress. It was by no means obvious, though, that social change was progressive. J. J. Rousseau's *Discourse on the Origin and Foundations of Inequality among Men* of 1755 postulated a peaceful state of nature in which the earliest humans had been free. Whatever the benefits of civilization, they were bought at a cost. Rousseau's name was often linked to the cult of the "noble savage," which glorified freedom from the restraints of civilization.

A theory of social progress was finally articulated in the later decades of the eighteenth century. A. R. J. Turgot made the first effort in his lectures of 1750 (translation 1973), but it was left for Turgot's friend and biographer, the marquis de Condorcet, to provide the best-known expression of the theory (translation 1955; see also K. Baker 1975). Condorcet wrote his sketch of human progress while hiding from the French revolutionary government, but he presented future social progress as the inevitable continuation of a long-standing historical trend. The state of nature was soon swept away as humans developed agriculture. Superstitions had always been used by the powerful to consolidate their grip on society, but reason gradually uncovered the true nature of humanity and showed the way to social progress. Condorcet thought that modern technology, especially the printing press, guaranteed future progress by ensuring the diffusion of knowledge.

Similar ideas had been developed by a group of "philosophical historians" in Scotland (Bowler 1989a: chap. 1). The economist Adam Smith's *Lectures on Jurisprudence* (1978), based on lectures delivered during the 1760s, postulated four stages of social development defined by technological developments: the ages of hunting, herding, agriculture, and commerce. As the apostle of free enterprise, Smith saw the growth of freedom as the driving force of progress. Adam Ferguson's *Essay on the History of Civil Society*, published in 1767, compared social progress with the individual's development toward maturity.

The naturalist Buffon presented a brief but effective picture of the earliest humans in the supplementary volume of his *Natural History* of 1778 entitled *The Epochs of Nature:*

Naked in mind as well as in body, exposed to the injuries of every
element, victims of the rapacity of ferocious animals, which they were
unable to combat, penetrated with the common sentiment of terror,
and pressed by necessity, they must have quickly associated, at first to
protect themselves by their numbers, and then to afford mutual aid to
each other in forming habitations and weapons of defence. They began
with sharpening into the figures of axes those hard flints, those *thunder-
stones*, which their descendants imagined to have fallen from the clouds,
but which, in reality, are the first monuments of human art. (1791:
9:381)

The harsh conditions of the state of nature were a spur to human ingenuity
and the source of progress.

Buffon saw the possibility of providing what we would call archaeological evidence of the primitive state of early technology. Stone Age tools had
been unearthed in the eighteenth century, but few were prepared to admit
that these gave evidence of how our ancestors lived. It was left for the nineteenth century to develop a science of prehistoric archaeology to demonstrate that humanity had progressed from a state of savagery. Before the
1850s few scholars accepted that the human species was more than a few
thousand years old, which left too little time for the kind of progress postulated by Buffon's model.

THE HISTORY OF THE EARTH

One point, at least, was established by Buffon and his contemporaries: the
appearance of humankind on the earth was merely the last act in the long
drama of the planet's development. Seventeenth-century theories of the
earth had retained a biblical timescale which drastically restricted naturalists' ability to imagine a long sequence of geological changes. Now the history of the earth began to open up in a way that laid the foundations of
modern geology (Haber 1959; Toulmin and Goodfield 1965; Rossi 1984). For
Buffon, the appearance of humankind marked the seventh and last epoch of
the earth's development: human history occupied only a small proportion of
the history of the earth. By the end of the century, geologists were beginning to work out the historical sequence in which the stratified rock formations had been laid down. The timescale now postulated for the earth's history had expanded to an extent that left even radical thinkers such as Buffon
astounded by "the dark abyss of time."

As yet there was no detailed analysis of the fossil record for a clue to the history of life on earth. After much controversy, most naturalists were now willing to accept that fossils were the remains of once-living creatures. Some were even beginning to admit that species known from the fossil record might no longer be in existence. But the debates that raged in the late seventeenth century were not settled with sufficient clarity for geologists to begin a systematic use of fossils to reconstruct the history of life on earth. As influential a figure as Voltaire could still ridicule the "systems built on shells" (Rudwick 1972: chap. 2). Construction of the modern outline of the fossil record would be left for nineteenth-century paleontologists to begin.

The complexity of the rock formations was itself an indication of the planet's long history, leading geologists away from the religious concerns of earlier writers such as Burnet and Whiston. As noted in chapter 2, one of the earliest works to abandon the biblical viewpoint completely was Benoît de Maillet's *Telliamed*, written around 1700 and published posthumously in 1748 (translation 1968; Carozzi 1969). De Maillet postulated an earth of enormous age and made no mention of the great flood. In the first volume of his *Natural History* (1749), Buffon himself estimated that the earth had taken seventy thousand years to cool from its originally molten state, and insisted that the flood had had no visible effect on its surface. Both authors were aware of the dangers of speaking out so openly against the biblical timescale: Buffon was censured by the Church and formally had to retract his opinions. Yet, thirty years later, he was able to return unmolested to the same theme in his *Epochs of Nature* (reprint 1962; see also Roger 1997), although he took the precaution of dividing the earth's history into seven epochs corresponding to the days of creation. By this time, new theories which took for granted a long timescale were becoming commonplace. Although at the end of the century there were some attempts to revive a biblical view of the earth's history, they were peripheral to the development of geological science.

Buffon's theory was traditional in one respect: he felt obliged to begin with an account of the origin of the earth. His theory supposed that a comet had collided obliquely with the sun, knocking globules of molten matter out into space, where they cooled to form the planets. A rival theory, the "nebular hypothesis," was suggested by the philosopher Immanuel Kant in 1755 (translation 1969). Exploiting Newton's theory of gravitation, Kant argued that a cloud of dust particles would condense under its own gravitational pull, with the majority of the material concentrating in the center to form the sun, while smaller bodies were left orbiting as planets. Because Kant's book was never issued, his idea received little attention in the eighteenth century. It became known only after Pierre-Simon Laplace discussed and elaborated it in 1796 (translation 1830). The theory seemed to be supported by the telescopic observations of William Herschel, which revealed nebulae or clouds apparently at various stages in the process of condensation (Hoskin 1964). In the early nineteenth century, the nebular hypothesis would provide an evolutionary model that inspired a wide range of thinkers to imagine that natural processes could create order out of a random distribution of matter (see fig. 4).

Kant had little interest in geology, and Buffon's was one of the last theories of the earth to explain the planet's origin. In the late eighteenth century, geology emerged as a distinct branch of science concentrating on developing an explanation of how the earth's crust had been brought into its present form. The empirical study of rock formations, fossils, and visible agents of geological change was its central concern. Different schools of thought emerged, in part because each region had its own geological structure and thus focused attention onto a particular set of problems (Davies 1969; J. Greene 1959a; R. Laudan 1987; Porter 1977).

Buffon's explanation of the planets' origin locked him into a particular model of earth history: there was a built-in direction of change imposed by the fact that each planet had to cool down from an initially molten state. Once the crust had solidified and cooled, massive rainstorms deposited an ocean of water onto the surface so deep that it covered the highest mountains. Following de Maillet, Buffon believed that the sedimentary rocks had all been laid down beneath this ancient ocean and had then been exposed on dry land as the sea level declined. He was aware that some of the older rocks contained fossil shells, some of them quite different from those of shellfish today, but made no systematic attempt to use fossils to chart the history of life on earth. Buffon assumed that the early land surface was much warmer than it is today, so that in the fifth epoch even the northern regions were inhabited by tropical creatures. Fossil bones of elephant-like creatures—the mammoth and the mastodon—had been found in Siberia and North America. Some anatomists were beginning to suspect that these were the remains of extinct species, but to Buffon they indicated that animals resembling the living elephants had once been able to roam the north (J. Greene 1959a: chap. 4).

Buffon's and de Maillet's postulations of what we may call the retreating-ocean theory indicated their commitment to another directional process in the earth's history. They could not accept Hooke's suggestion that earth movements were powerful enough to raise new continents and mountains; to explain how sedimentary rocks were found exposed on dry land, they were forced to imagine that the mountains once had been covered by

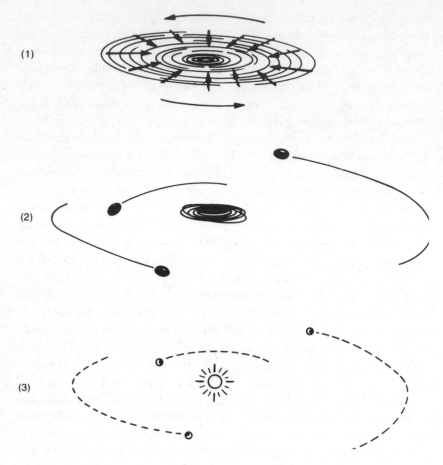

4. Nebular hypothesis of planetary origins. (1) A nebula—a great cloud of slowly rotating dust and gas—begins to collapse inward under its own gravitational attraction. (2) Most of the material has begun to condense into a central body, but lesser amounts have formed rings rotating around the center, which are also beginning to condense. (3) The central body has collapsed far enough that sufficient energy has been released to make it incandescent, forming the sun. The smaller bodies have consolidated to form the planets, which continue to orbit around the sun. Originally, the sun's high temperature was supposed to have been the result of potential energy released by its condensation. This energy immediately began to dissipate by means of radiation into space, leading to a steady cooling. In the modern version of the theory, the temperature and pressure reached in the center of the collapsing sun were eventually great enough to trigger a thermonuclear reaction capable of maintaining its high temperature for billions of years.

the ocean. The theory that the sea level had dropped over time eventually became known as Neptunism, after the Roman god of the sea. Its greatest exponent was Abraham Gottlob Werner, who taught at the mining school in Freiburg, in Saxony. Werner wrote only one short account of his theory (translation 1971) but attracted students from all over the world. Primarily a mineralogist, he first classified the rocks and then postulated a historical sequence in which they had been laid down beneath the retreating ocean. The younger rocks were deposited on top of the older ones, so the order of formation could be read from the observed sequence. Because the ocean was deeper when the first rocks were laid down, they are often found exposed on the tops of mountains, girdled by overlying younger rocks deposited when the ocean had already begun to retreat.

The retreating-ocean theory was decisively rejected early in the nineteenth century, and has since been widely ridiculed. But recent historians have shown that Werner's theory played an important role in establishing modern stratigraphy (Hallam 1983; A. Ospovat 1969). One reason for Neptunism's poor reputation is that the theory was hijacked by a small number of geologists anxious to revive a biblical view of earth history in which Noah's flood played a key role (e.g., Kirwan 1799; see also Gillispie 1951). But Werner himself did not stress this factor, and neither did the majority of his continental supporters, so the charge that Neptunism was accepted almost solely on religious grounds cannot be sustained.

The Neptunists minimized the role of earthquakes and volcanoes. The rival school of Vulcanism recognized that volcanic activity had played a far more significant role in shaping the earth's surface. P. S. Pallas developed a position resembling Hooke's. Pallas hypothesized that massive earth movements had elevated the land surface after the sedimentary rocks had been laid down. The most controversial extension of the Vulcanist position was proposed by the Scottish geologist James Hutton, who developed what later would be called the uniformitarian approach to geology, in which past changes are explained only in terms of observable processes (Dean 1992; Gould 1987). Both the uplift of new land and its destruction by erosion were immensely slow, and thus, to explain the creation of the earth's existing structure, Hutton's theory required the passing of vast amounts of time. For this reason he has been hailed by some historians as the father of modern geology. Hutton's theory certainly looks modern at first sight, but he applied his cyclic view of the earth's history so rigorously that the earth had to be seen, in effect, as a perpetual motion machine. The creation and destruction of landmasses had gone on for so long that all traces of any original state of the earth's surface had been obliterated. Hutton declared that he could find "no vestige of a beginning—no prospect of an end" (1795: 1:200). There was no "direction" in Hutton's system: the past was governed by exactly the same forces as those we now observe, however far back in time the geologist could trace the rocks.

Hutton's cyclic model of history was inspired by his belief in a God who had designed a world that was perpetually fit for habitation. His elimination of a "creation" and anything resembling a great flood led geologists such as Richard Kirwan to attack his theory as antibiblical. John Playfair's *Illustrations of the Huttonian Theory* (1802) eliminated the teleological aspect of his views and brought out those aspects which would be picked up by nineteenth-century geologists. But even so, the Wernerians in France and Germany ignored Hutton and found their own way to replace Neptunism with a new, directionalist vision of earth history which would become known as catastrophism (see chap. 4).

THE CHAIN OF BEING

Since there was no true science of paleontology in the eighteenth century, the question of the origin of life was addressed mainly by naturalists studying the vast diversity of living species. Following the work of Ray and others in the seventeenth century, new ideas were developed about how the relationships between species should be expressed. The emergence of a modern system of classification posed the question of why species seem to arrange themselves naturally into well-defined groups. For most naturalists it was obvious that the relationships must be part of a divine plan: a rational God would create the living world according to a meaningful pattern that the human mind could comprehend. The species were the original elements in that plan. They were maintained by the process of reproduction, which ensured that each generation preserved the basic character of the species. For many conservative naturalists, the theory of preexisting germs developed in the previous century provided an explanation of this stability. If each new organism came from a tiny miniature originally created by God, then His power to ensure the stability of the creation was absolute.

Eighteenth-century naturalists such as Linnaeus established the system of biological taxonomy by grouping the species according to their visible resemblances. But this period also saw the last flowering of a much older vision of how species are related: the chain of being, an idea that goes back to the ancient Greek thinkers (Bynum 1975; Lovejoy 1936). We have an intuitive sense that some living organisms are more highly developed than others. We see ourselves as the highest form of life on earth and assume that the categories "high" and "low" can be applied throughout the animal kingdom. Dogs and cats are more advanced creatures than fish, and fish are more advanced than shellfish (invertebrates). Lowest of all are single-celled creatures like the amoeba. The chain of being was an attempt to put this intuitive sense of hierarchy onto a systematic basis. It postulated a single, linear hierarchy of species stretching down from humanity to the simplest living form.

The attractions of the chain of being came not from its practical applications but from its articulation of certain key assumptions. It was conceived as a static plan of nature: all the links in the chain were created together by God. It was a highly structured system in which every species had an unambiguous position, with its closest relatives placed one above and one below it in the sequence. It was a *complete* system: every link in the chain, from amoeba to human, had to exist, and God's perfection was displayed by the fact that He had created every link in the chain: this is what Arthur Lovejoy calls the "principle of plenitude." There was perfect continuity from one end of creation to the other: each species fitted close to its neighbor, and there were no visible gaps in the pattern. In such a worldview there was no room for extinction, since this would destroy the unity of creation. The poet Alexander Pope best expressed the logic of the system in his *Essay on Man:*

> Vast chain of being! which from God began,
> Natures aethereal, human, angel, man,
> Beast, bird, fish, insect, what no eye can see,
> No glass can reach; from Infinite to thee,
> From thee to nothing.—On superior pow'rs
> Were we to press, inferior might on ours;
> Or in the full creation leave a void,
> Where, one step broken, the great scale's destroy'd;
> From Nature's chain whatever link you strike,
> Tenth or ten thousandth, breaks the chain alike.
>   (1751: vol. 3, epistle 1, lines 237–46)

Note that Pope imagines a spiritual hierarchy above the earthly one: we have a crucial position in creation because we link the spiritual and physical sections of the chain.

The most philosophical of the eighteenth-century naturalists took the chain seriously. The leading figure was Charles Bonnet, who gained fame through his discovery of the parthenogenesis of aphids, but who then turned to philosophical studies in works such as his *Contemplation of Nature* of 1764 (Anderson 1982; Bowler 1973; Glass 1959c; Rieppel 1988). The outline of Bonnet's chain (see fig. 5) shows how he used superficial re-

5. The chain of being. This is a simplified version of the chain in Bonnet's *Contemplation of Nature* of 1764.

semblances to create an impression of continuity. He considered bats to be intermediate between mammals and birds, while flying fish linked the birds to the fish. This meant that reptiles had to be placed lower in the scale than fish—allowing snakes to serve as a bridge to the slugs and other invertebrates. Even Bonnet allowed one side branch to the chain, despite the fact that this undermined the principles of continuity and plenitude. Eventually the logic of the linear pattern broke down under the evidence for a multiplicity of relationships that can be handled only by a model with many branches.

Bonnet's chain was not a model for evolution through time. For him, each species was an eternal element in the divine plan: "The centuries transport from one to another this magnificent spectacle; and they transmit exactly what they receive. Nothing changes, nothing alters; perfect identity. Victorious over the elements, time and death, the species are conserved, and the period of their duration is unknown to us" (translation from Bonnet 1779: 5:230).

It is no coincidence that Bonnet was also the leading defender of the theory of preexisting germs to explain reproduction. He held that natural forces were incapable of creating organic structures: each organism developed from a germ or miniature created by God and stored, one inside the other, within the ovaries of the first females. Bonnet's discovery of parthenogenesis in aphids seemed to confirm the theory: if several generations of females could produce offspring without access to a male, everything necessary for generation must already be contained within the female. Not only was the species designed by God, but every individual member of the species was separately created at the beginning. Bonnet conceded that the germ-sequence defined only the basic character of the species: individual differences were formed by parental or environmental influences acting on the growing embryo. Such influences may have helped explain the observed facts of heredity and local adaptation, but they were powerless to alter the basic form of the species. There were powerful philosophical arguments in favor of the theory of preexisting germs that outweighed the difficulties of providing empirical support by microscopic observation of embryos. The theory flourished into the later eighteenth century and was only gradually displaced by the rival theory of epigenesis, in which the parts of the body were thought to be produced in sequence by a genuinely developmental process (Roe 1981).

Another philosophical writer who combined the chain of being and the germ theory was Jean-Baptiste Robinet (Murphy 1976). In his *On Nature* of 1761–1766, Robinet extended the logic of the principle of continuity so that it effectively eliminated the existence of distinct species. For species to be recognizably distinct, there had to be gaps between them—but gaps are exactly what the principle of continuity could not tolerate. Robinet held that what we perceive as distinct species are actually sections of a continuous sequence of forms. In effect his was a rope, rather than a chain, of being: it was not divided into separate links. By this theory, it is only because the intermediates are very rare that we think there are distinct species. A number of eighteenth-century naturalists were attracted to this view, which some historians once saw as an anticipation of the modern idea of evolution. But although Darwinian evolution assumes that species are transformed gradually in time, it does not deny the reality or the distinct nature of species. Because evolution is now seen as a branching, not a linear, process, gaps open up between the branches as they diverge. Robinet's view was a development from an older philosophy, not an anticipation of Darwinism (Glass 1959a).

Like Bonnet, Robinet originally assumed that the chain of being was a static pattern of creation existing for all time. It is significant, though, that both writers modified this position toward the end of their careers and accepted a limited notion of development. This is what Lovejoy calls the "temporalization" of the chain—the injection of a time element into an originally static plan. Bonnet and Robinet both conceded that some species might not have been manifested in physical form on the earth during its early history. Their germs might have been created, but they had not yet developed into physical bodies. Both accepted that the simpler species would be manifested first, followed by successively higher forms. The precise means by which the new developments were supposed to take place was highly speculative. Bonnet imagined souls being reincarnated in higher bodies in different geological epochs, offering the prospect that the humans of today could expect to be reincarnated in superior bodies at some point in the future.

Some historians have maintained that Bonnet's theory was a sham, a superficial image of progress imposed on a fundamentally static worldview (e.g., Glass 1959c). His views certainly fit in with Foucault's assertion that the thinkers of the classical era could only imagine change as the filling-in of slots in a rigorously ordered system. The chain of being has existed since the creation, but its elements achieve physical incarnation only at successive points in time. This is preordained development rather than open-ended evolution. But it may not be wise to dismiss Bonnet completely. He made a genuine effort to respond to the geologists' demonstration of a lengthy history for the earth, invoking catastrophic upheavals to explain why different physical conditions triggered the manifestation of new kinds of germs. And, as we shall see, the evolutionists of later generations found it difficult to eradicate the concept of linear development. Despite the establishment of branching models of evolution, the very human tendency to see the animal kingdom as a hierarchy with ourselves at the top would continue to influence naturalists' thinking for a long time to come.

THE NEW CLASSIFICATION

The chain of being assumed that similarities between species naturally would define a linear pattern. In practice this view was already losing support as exploration revealed an ever wider range of species that had to be fitted in. There were too many different degrees of relationship involved. By the end of the seventeenth century, naturalists such as John Ray had sought to understand the relationships between species by ignoring the chain of being and creating a more flexible system of biological taxonomy (classification).

Linnaeus built on those foundations to create the basis for modern taxonomy. He ignored the debate over whether life could be explained in mechanical terms, and sought to revolutionize the life sciences by understanding the pattern of creation. Like Bonnet, Linnaeus assumed that God had created an array of distinct species which perpetuated themselves unchanged to the present. If the whole system was a divine creation, it must have a rational order, and that order ought to be discernible by the human mind. Linnaeus believed that he had been privileged to see the outline of the Creator's plan. His technique was outlined in his *System of Nature* (1735)—a work that began as a slim pamphlet but grew over the decades into a multivolume classic that gained its author worldwide fame (Blunt 1971; Broberg 1980; Frangsmyr 1984; Hagberg 1953; Koerner 1999; J. Larson 1971; Stafleu 1971).

Linnaeus realized that the world is not simply based on a formal pattern of relationships: it also had to work in practice. Natural theology held that the Creator designed each species to fit its particular way of life, but Linnaeus and his followers knew that the environment to which the species was adapted did not consist solely of the physical conditions: it also included other species. These scientists were interested in what we would call the ecological relationships between species. Each species depended on the whole for its livelihood, and God ensured the stability of the system by building in a series of checks and balances. If a temporary fluctuation in the climate caused one species to expand its numbers, it predators also would increase to check the expansion. This concept of the balance of nature allowed natural theology to explain relationships that later would be seen as part of the struggle for existence (see the dissertation by one of Linnaeus's students, Bilberg 1752; see also Egerton 1973).

Ecology was only a sideline, however, to Linnaeus's main program of discovering and representing the pattern of creation. Instead of trying to impose a fixed structure such as the chain of being, he constructed the pattern from the most obvious relationships visible between species. He grouped the most similar species into a higher category called the *genus* (pl. *genera*) and grouped the genera themselves by even more basic resemblances. John Ray tried to do this by taking into account every visible character of every species. Ray's natural method was based on the assumption that we have no way of knowing which are the most important characters. Many naturalists agreed that in principle the natural method was preferable, but some were so concerned about the sheer amount of information that would have to be processed that they opted (at least as a first resort) for an artificial system based on a single important character. This was Linnaeus's policy, although he freely admitted that it would eventually be superseded. In his own field of botany, he drew upon R. J. Camerarius's discovery of plant sexuality, arguing that the reproductive organs—the flowering parts—were the most important because they were responsible for perpetuating the species. Relationships among the flowers, then, formed the basis for his system of classification.

Linnaeus's system was successful because this policy made it easy for even an amateur to assign any species to its correct position. Any newly discovered species could soon be located alongside its closest known relatives. Linnaeus grouped the genera into higher level categories called orders, and grouped these in turn into classes. To determine the class and order of any particular plant, it was necessary only to count the stamens and pistils in its flower. It could then be assigned to a genus and species by closer inspection of the structure of these parts. Linnaeus then introduced the same taxonomic ranks into the animal kingdom, where he distinguished six classes. Modern biologists find it necessary to use a larger number of ranks, so the genera are first grouped into families before being combined into orders. Figure 6 shows examples of the modernized Linnaean system from the animal kingdom. It also illustrates another of Linnaeus's innovations, the binomial nomenclature that identifies each species by two Latin (or pseudo-Latin) names. The first name defines the genus, and the second the species, so that all species in the same genus have the same first name. By international convention, the starting points for botanical nomenclature are Linnaeus's *Species of Plants* of 1753 and the fifth edition of his *Genera of Plants* of 1754. Zoological nomenclature originates from the tenth edition of *System of Nature* of 1758–1759, in which the binomial nomenclature was first applied to all known species of animals.

Some naturalists still hoped that the natural method would reveal a chainlike arrangement (e.g., Adanson 1763), but Linnaeus adopted a more pragmatic approach. If observation revealed that the Creator's plan was not a simple linear pattern, then Linnaeus would let the arrangements build themselves, so to speak, from the visible resemblances between species. There was no need to assume that one species must always be either higher or lower than another: cats are not superior (or inferior) to dogs, they are merely different kinds of carnivores. Instead of each species having only two immediate neighbors, as occurs in the chain (one above and one below), there may be several equally close relatives, all different. Linnaeus drew an analogy between these relationships and the way countries fit together on a map: the pattern requires two dimensions to be represented, not one. The

Family	Genus	Species	Common name
Canidae (Dog family)	*Canis*	*familiaris*	Dog
	Canis	*lupus*	Wolf
	etc.		
	Vulpes	*vulpes*	European fox
	Vulpes	*fulva*	American fox
	etc.		
	Etc.		
Felidae (Cat family)	*Felis*	*catus*	Cat
	Felis	*lynx*	Lynx
	etc.		
	Panthera	*leo*	Lion
	Panthera	*tigris*	Tiger
	Panthera	*pardus*	Leopard
	etc.		
	Etc.		

6. Classification and the binomial nomenclature. Using familiar animals, this table shows how species are grouped into genera, and genera into families. The most closely similar species fit into the same genus and share the same first name. The system can be expanded to include newly discovered forms at any level. Higher up the taxonomic hierarchy, Canidae and Felidae belong to the order Carnivora (flesh eaters) of the class Mammalia.

Linnaean system is hierarchical in a very different sense: we regard the more basic taxa as higher because they *include* the lower ones.

The close resemblances between the species in a genus at first were not seen as indications of a relationship in the evolutionary sense. *Relationship* still meant something existing in the mind of the Creator, who had determined the degrees of similarity between the species. In some cases, however, the resemblances were so close that naturalists were tempted to suppose that one species was derived by natural transformation from another. Ray and others had acknowledged that the effect of local environmental conditions could modify the original form of a species to give varieties. These local varieties were seen as trivial and reversible modifications of the species' original form. But even Linnaeus was tempted by the possibility that the process of transformation might occasionally have gone much further. A local variety might become so strongly marked that it would have to be treated as a new species formed by nature rather than by divine creation. This possibility threatened the principles on which the system was built: how could one tell which was the original and which the newly formed species? Most of Linnaeus's followers continued to believe that the number of species was fixed at the creation.

Linnaeus's hint that a few new species might be produced by transmutation represented a small move toward the belief that species are related because of descent from a common ancestor. But his most ambitious attempt to allow for the production of new species postulated a very different process of change: hybridization (Glass 1959c; H. Roberts 1929). Species do not normally interbreed, although hybridization does occasionally occur. The mule is the best-known example, although mules are usually sterile and thus cannot form a new species intermediate between the horse and the ass. Linnaeus and his pupils eventually came to believe that fertile hybrids were sometimes produced in the plant kingdom. Since the hybrid bred true, it constituted a new species combining characters from the two parent species. The first example was a new form of toadflax, described by one of Linnaeus's pupils under the name *Peloria* (Rudberg 1752; for other examples, see Hartmann 1756). In his "Disquisition on the Sexes of Plants" of 1756 (reprinted in Linnaeus 1749–90: vol. 10), Linnaeus suggested that in the original creation God formed only a single species as the founder of each genus; all the other species were formed by hybridization between the original forms (see fig. 7).

Other investigators dismissed Linnaeus's hybrids as trivial modifications or as infertile forms. Only one is accepted as genuine today, although botanists recognize that the formation of new species by hybridization is not uncommon in the plant kingdom. Linnaeus made some effort to accommodate the growing awareness that the world had changed since it was first created. Yet his theory also supports Foucault's contention that much eighteenth-century thought remained trapped within an image of nature as a system based on a prearranged pattern of relationships. Unlike transmutation, where genuinely new characters are formed, the hybrid species merely recombined characters already present in the parent forms. Nothing truly new was produced by natural processes. The divine plan was there from the start, with only the details left to be filled in by hybrid combinations.

Naturalists paid little attention to Linnaeus's ideas on the natural origin of new species, but they continued to explore the idea that the human mind can recognize a set of meaningful relationships between species. Linnaeus's own artificial system was gradually replaced by a "natural system," one which claimed to take account of all the many characters by which species may be related. The botanist Anton-Laurent de Jussieu played a major role in this process (Stevens 1994). By the start of the nineteenth century, naturalists accepted that their goal must be the uncovering of this natural system of relationships, although opinions differed both on how to go about the

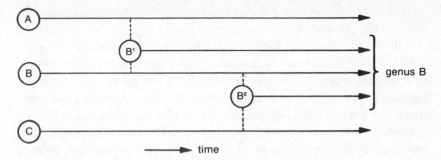

7. Linnaeus's system of hybridization. A, B, and C are originally created plant species that continue to breed true. At some point in time, a female of B is fertilized by pollen from A to give a hybrid species B1, which then continues to breed true. Since the female parent is assumed to determine the basic form of the hybrid, B1 belongs to the same genus as B. A later cross with pollen from C forms another species, B2, again in the same genus. Other hybrids belonging to the genera A and C could be formed if their females were fertilized by pollen from other species.

process and on the significance of the relationships themselves. Few as yet thought that the relationships were an indication of genealogical descent.

THE NEW THEORIES OF GENERATION

Whatever the practical success of Linnaeus's method, his efforts to deflect attention away from the mechanical philosophy had limited effect. The hope of explaining life as a function of material systems had taken too strong a hold on the more radical Enlightenment thinkers. Their hopes were raised by the fact that, in a Newtonian universe (as opposed to Descartes's clockwork model of nature), matter itself possessed mysterious powers such as gravity. If other nonmechanical forces could be discovered (electricity was soon included in the program), then it might be feasible to suppose that a complex material system such as the animal body might exhibit the activities we call life.

Descartes's original program had sought to explain the origins of all natural structures by material processes. The Enlightenment materialists hoped to extend this program to living structures. In the preformation theory still advocated by Bonnet, even the growth of a new organism did not involve the production of a genuinely new structure, only the expansion of a preexisting miniature created by God. The materialists repudiated this theory; for them it was obvious that the forces which govern the living body were capable of explaining generation. The parents' bodies actually produced the structure of their offspring by directing material processes within the reproductive organs.

This opened up the prospect of explaining how the species itself originated by natural processes. In Bonnet's theory, the series of germs created by God guaranteed the fixity of the species—but if reproduction occurred through a material process producing new structures in every generation, that guarantee was lost. If the parents' bodies changed in the course of their lives, might not such changes be passed on to their offspring? And if such "acquired characteristics" were inherited over many generations, might a sufficient number of these accumulate to produce a new species? Beyond natural transmutation lay an even more radical prospect: explaining the origin of life without supernatural creation. If a material process were responsible for normal generation, could we not assume that under certain circumstances the forces of nature might produce living bodies directly from inorganic matter? The idea of spontaneous generation had been repudiated by Ray and his generation because it undermined their explanation of the fixity of divinely created species. For the eighteenth-century materialists, this was precisely why the idea was attractive. If life was the product of natural forces, it need not exhibit the designing hand of the Creator.

De Maillet's Telliamed

An initial step along this road was taken by de Maillet, whose *Telliamed* provided one of the first theories to dispense with the biblical creation of the earth and the deluge. Although not published until 1748, the book had been written much earlier, when simple Cartesian physics made complete elimination of the germ theory implausible. Yet in keeping with his materialist outlook, de Maillet refused to invoke a miracle to explain the origin of life on the earth. Instead, he adopted a version of the theory of preexisting germs in which the miniatures existed independently of parent bodies and were found scattered throughout nature. In normal generation, a germ would find its way into the womb or egg of the female who would become its mother. But in the earth's early history, the great ancient ocean might have provided an environment in which germs could develop without parents. The first living things thus appeared by a natural rather than a supernatural process; once formed, they began to reproduce in the normal way.

De Maillet also accepted the idea that change occurred within the species thus produced. As an exponent of the retreating-ocean theory, he assumed that when life first appeared, the earth was completely covered by water. The first members of each species must have been aquatic. Each terrestrial species would have been produced by the transformation of its aquatic fore-bear as soon as dry land began to appear. De Maillet accepted sailors' stories of mermaids as evidence that the aquatic form of the human species still existed, and he thought that flying fish had been transformed into birds. Such ideas seemed ridiculous even to many of his contemporaries, but they show that de Maillet appreciated the need for life to adapt to changes in the earth's physical environment. His theory retained an element of fixity, however, in the sense that the basic form of each species is recognizable through the transformation—the mermaid is still human. The range of species we see was fixed by the kinds of germs that have existed in nature since the beginning of time.

De Maillet eliminated miracles and partly circumvented the argument from design by supposing the germs adapted to different conditions as they grew. He avoided postulating a supernatural origin for the germs by supposing that they had always existed throughout the universe. Such a theory still gave too much away to the old view of creation, however, to satisfy the more radical materialists of the Enlightenment. To maintain a completely materialist picture, it would be necessary to eliminate the germs altogether and suppose that natural processes can actually generate living bodies.

Maupertuis on Heredity

The first major challenge to the theory of preexisting germs came from a leading exponent of the Newtonian worldview, Pierre Louis Moreau de Maupertuis (Glass 1959b; Sandler 1983). In his popularly written *Earthly Venus* of 1745 (translation 1968), Maupertuis noted that many observers had confirmed that the embryo grows by epigenesis (the appearance of new parts one after another), not by the expansion of a preformed miniature. He also studied heredity, especially the inheritance of polydactyly (the appearance of a sixth finger on the hand; see Maupertuis 1768: vol. 2, letter 14). Did the exponents of the germ theory think this abnormality had been designed by the Creator when He formed the original germs, or was the extra finger an accident of growth? Assuming that God did not deliberately design malformations, how could the product of an accident be transmitted to future generations as though it were part of the germ? The only solution was to abandon the germ theory and accept that the embryo was formed by physical processes during the act of reproduction. Maupertuis adopted the ancient belief that both sexes produced a fluid semen; when male and female semen were mixed in the mother's womb, the material particles in the mixture were able to arrange themselves into the body of the new organism. Thus, both parents were able to transmit their peculiarities to the offspring.

At the end of *The Earthly Venus*, Maupertuis applied his theory of generation to the problem of the origin of the human races. He argued that the races had evolved by natural modifications accumulated in the course of normal reproduction over many generations. Because the parent's body produced the semen from which the embryo would be formed, modifications of the parent's body would be reflected in the material content of the semen and could reappear in the offspring. The modification thus could become a permanent feature of all future generations. As in the case of the sixth finger, the variation might become a permanent feature of the species. To explain the origin of the physical differences between the human races, Maupertuis postulated two processes corresponding approximately to what later would be called natural selection and the inheritance of acquired characteristics. Perhaps the colored races' dark skin appeared by some accident of the reproductive process in their distant ancestors, which was preserved by isolation as the affected individuals were driven to the less hospitable tropics. Or perhaps the climate of the tropics actually produced a dark coloration in the skin of those who came to live there, and the character had become permanently fixed by heredity. Whichever the correct explanation, new characters had appeared in the human race because the process of heredity was not locked into the perpetuation of a fixed type.

Maupertuis expanded these suggestions into a wider theory of the development of life on earth in the *System of Nature*, which he published under an assumed name in 1751 (reprinted in Maupertuis 1768: vol. 2). He now adopted a materialistic view of the origin of life based on spontaneous generation. When the earth was originally covered with water, some especially active particles of matter had been able to arrange themselves into living structures without requiring a womb. Maupertuis also noted that the original forms of life might have changed through time, so that related species today might have been derived from a common ancestor. The suggestion that racial characters could appear in the course of time thus was extended—in a highly speculative way—into what we would now call a complete theory of evolution.

Maupertuis became extremely concerned by a problem that had begun to bother him while writing *The Earthly Venus:* if there was no preexisting pattern to determine the embryo's structure, how did the particles in the semen know how to arrange themselves in the correct order? More seriously, in the act of spontaneous generation, how did material particles know how to form a living structure for the first time? Maupertuis had assumed that Newtonian forces acting between the particles could produce the necessary arrangements, but now he was forced to propose that matter had a spontaneous tendency to organize itself into complex structures. This had to involve something more than Newtonian forces, because the particles had to have a form of awareness that let them know how to behave. Here Maupertuis contributed to the strangest outcome of Enlightenment materialism: in attempting to explain life as the product of material processes, philosophers were forced to attribute the properties of life to matter itself.

Buffon: Spontaneous Generation and Degeneration

Maupertuis's investigation of heredity was an important contribution, although his later ideas were never presented as anything more than speculation. A similar theory of generation served as the basis for one of the most comprehensive pre-Darwinian accounts of organic change. We have already encountered Buffon's materialistic worldview in his efforts to explain the origin and development of the earth. But Buffon was determined to extend his Newtonianism to explain the history of life on earth. When Darwin's rival Samuel Butler (1879) tried to argue that all the significant components of the Darwinian theory had been proposed by earlier writers, Buffon was one of his major examples. According to Butler, Buffon was a complete evolutionist who had to conceal his true opinions to avoid censure by the Church, although anyone who could "read between the lines" could see what he really believed. Some later historians (e.g., Eiseley 1958) have also suggested that buried in Buffon's voluminous writings lie all the components of a complete theory of evolution. With hindsight, it is possible to identify passages in Buffon which seem to anticipate fragments of the modern theory. And he was certainly trying to set up a materialistic alternative to the biblical story of creation. But modern historians are reluctant to twist Buffon's words into an anticipation of the Darwinian theory, preferring to see his history of nature as based on quite different conceptual foundations (Bowler 1973; Eddy 1984; Farber 1972; Fellows and Milliken 1972; Lovejoy 1959a; Wilkie 1956; the most detailed studies are Roger 1997, 1998).

Buffon was hostile to the Christian view of creation, and some writers (esp. Roger 1997, 1998) treat him as little more than an atheist. He knew and interacted with the more radical materialists discussed below. Yet as R. Wohl (1960) points out, he was far from being a simpleminded materialist, and some aspects of his theory seem to retain a commitment to the belief that nature is based on an underlying, fixed pattern. The more radical thinkers postulated a universe in total flux, in which new structures are produced by the haphazard combination of atoms. But Buffon went out of his way to imply that the forms which appear in the history of life are predetermined. He rejected the simple view of design by God, yet retained a faith in the existence of an underlying pattern which constrains the activity of the material universe. Once again, Foucault's model of the Enlightenment's view of change rings true.

Buffon was appointed superintendent of the Royal Garden (the modern Jardin des Plantes) in Paris in 1739 and planned the publication of a comprehensive survey of natural history (Hanks 1966). We have seen that the first three volumes of his *Natural History*, published in 1749, included a materialistic account of the earth's origin. But this was only a prelude: the other volumes outlined a theory of generation and a general philosophy of natural history. Later volumes provided detailed descriptions of a wide range of animal species interspersed with theoretical commentaries (translation 1791; extracts in Buffon 1981). Buffon was soon established as France's leading naturalist. If Linnaeus's system was accepted because of its practical applications, Buffon provided a popular alternative for those who wanted a more materialistic account of the natural world.

The first volume of the *Natural History* launched an attack on Linnaeus and the whole attempt to reduce natural history to a search for abstract relationships. In it Buffon even seemed to deny the existence of distinct species. There are only individuals in nature, and sometimes intermediate forms linked two supposedly distinct species (the principle of continuity). He argued that the grouping of species into genera is a product of the human imagination: there was no guarantee that the system corresponds to anything in reality.

Buffon's position has puzzled many historians, because in the other volumes of 1749 he was already writing of species as distinct, fixed entities. Sloan (1976) points out, however, that his position is not really inconsistent. He certainly objected to Linnaeus's program of defining relationships that were "real" only in the sense that they existed in the mind of the Creator. In his enthusiasm he may have gone further than he really intended in dismissing the categories on which his rival's system was based. He accepted that there were relationships between individual animals that allowed them to be grouped into species, but he insisted that these were grounded in the process of reproduction, which maintained populations through time. But Buffon saw no possibility of tracing relationships *between* species. He set out to describe the known species, adopting the purely arbitrary policy of starting with the better-known ones.

The definition of the species as a population preserved by reproduction threw great emphasis on Buffon's theory of generation. Like Maupertuis, he argued that life could be explained only if its production could be accounted for in material (i.e., Newtonian) terms. The germ theory was dismissed as providing no real explanation, because it traced everything back to creation by God. Buffon shared the view that the embryo developed from a mixture of male and female semen. The semen consisted of "organic particles" derived from the food and was superfluous to the animal's nutritional requirements. As in Maupertuis's theory, though, the crucial question was: How did the particles know how to arrange themselves into the complex structure of the embryo? Buffon introduced the concept of the "internal mold," an entity supposedly capable of directing the particles into place. (The analogy is with the molds used in casting figures from metal, but the internal mold directs particles throughout the whole mass.) Buffon never explained what the internal mold is—not surprisingly, since it clearly had no physical existence and thus evaded his own requirements for a materialistic explanation. He did, however, insist that it was characteristic of the species and provided a guarantee that the next generation would take the same basic form as their parents. In the end, Buffon's internal mold was little more than a nonmaterial equivalent of Bonnet's encapsulated series of miniatures.

Buffon was thus in a position to begin his description of the animal kingdom in terms of fixed species. In his fourth volume (1753), his description of the ass deliberately raises the question of whether this species could be related to the horse in the sense of being derived from it by change through time. He answers the question in the negative, and although Butler and others have treated his rejection as insincere, the argument is developed in such detail that it is difficult not to take it seriously. Over the next decade, however, Buffon became convinced that such closely related species must indeed have descended from a common ancestor (see fig. 8). In the article "On the Degeneration of Animals" in the fourteenth volume (1766), he openly supported the position denied in 1753. The relationship between certain Linnaean species was now to be regarded as a real one, based on divergence from a common ancestor in the course of the earth's history. The anatomist who provided Buffon with information, J. L. M. Daubenton, did not share this opinion (Farber 1975). And Buffon himself accepted this apparently "evolutionist" position only in a very limited sense—so limited that he was still able to insist on the *fixity* of species.

Buffon supposed that all the forms united within a Linnaean genus (or modern family) were derived from a single original population which became divided into separate groups through migration to different parts of the world. Each separate population was influenced by the climate of its own area and gradually changed its form. Buffon argued that the effect of the external conditions was produced by the different qualities of the organic particles taken up as food. The overall effect of the process was to produce local varieties—but Buffon was now arguing that the changes were pro-

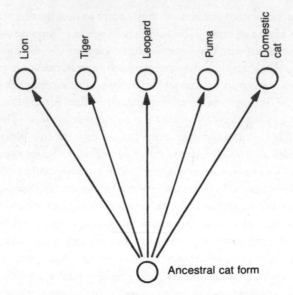

8.   Buffon's theory of degeneration. Buffon supposed
that the two hundred or so modern species of mammals had been formed from thirty-eight original forms
by a process of divergence, illustrated here with the cat
family. The ancestral cat population, formed by spontaneous generation, eventually migrated into several
different areas, where it degenerated under the influence of changed conditions to give the different modern species. Note, however, that for Buffon these were
not true species, because the divergence could never
cause enough change to render the products infertile
with one another. The amount of degeneration adds up
over a number of generations until it reaches the limits
defined by the fundamental nature of the type.

found enough to generate the forms normally described as distinct species.
Yet he continued to insist on the fixity of species, maintaining that each had
its unique internal mold. It is evident that he now wanted to treat the
Linnaean genus as the true species. Linnaean species were not true species at
all, but merely varieties so strongly marked that everyone had been misled
into treating them as species. This meant that, in theory, hybridization between the so-called species within the genus ought to be possible. Buffon argued that such crosses did occur, although they were seldom observed be-

cause accidental factors interfered with successful reproduction. He claimed to have produced hybrids between related species (although modern biologists are suspicious of these claims). The mule was not a sterile freak but a potentially viable product of parents whose reproductive incompatibility is only superficial. Hybrids between members of different genera were impossible, though, because their reproduction was based on different internal molds.

In arguing that a single ancestral population might diverge into a number of significantly different "species," Buffon seems to come close to the modern idea of evolution. His recognition that migration to different parts of the world might cause the divergence represents a pioneering insight into the role that geographical distribution plays in explanations of the history of life. There is nothing like natural selection in his theory, only an assumption that new environments act directly to change the individual and, eventually, the species. The most important difference between Buffon's theory and modern evolutionism, however, is evident in his views on the source of the ancestral population whose offspring first diverged. In the modern theory, the founders of each genus have in turn diverged from some more-remote ancestral form. This was not possible in Buffon's theory, because the mold of each species (Linnaean genus) was fixed. Yet to sustain his materialist program, Buffon must have supposed that the ancestors appeared on the earth by some natural process. Since his theory of the earth presupposed that the planet was originally too hot to support life, the earliest living forms must have been produced when the conditions eventually became suitable. To avoid the need for divine creation, Buffon turned to the concept of spontaneous generation (Bowler 1973; Roger 1997, 1998; Wilkie 1956).

Spontaneous generation had been part of Buffon's original theory of reproduction. He believed that a mass of disorganized organic particles could arrange itself into at least a simple living form without there being any parent organism. In a series of experiments done with John Turberville Needham, Buffon claimed to have found microorganisms growing in flasks of meat gravy that had been sterilized and then sealed from the air (Needham 1748). These experiments were subsequently criticized by Lazarro Spallanzani (1769), who argued—quite rightly, as we understand things today—that the flasks had not been sterilized well enough to destroy all the spores from which the microorganisms grew. The situation was complex, however, given the prevailing debate over the nature of generation, and modern historians have taken a more sympathetic view of the position adopted by Buffon and Needham (Farley 1977; Roe 1983, 1985). Many naturalists at the time believed that spontaneous generation might occur, and Buffon felt confident enough of his experiments to extrapolate a theory of how life first appeared on the earth. He supposed that at certain periods in the earth's history, spontaneous generation produced even the higher forms of life directly from free organic particles.

In 1778 Buffon refurbished his theory of the earth in a supplementary volume titled *The Epochs of Nature*. He believed that in the third epoch the earth had cooled enough to allow the formation of organic particles, and that these had spontaneously organized themselves into the first living organisms. These were not the creatures we know today, because the earth was then too hot to support modern kinds of life. The first population comprised a different set of species adapted to the hot environment. As the earth cooled, these early species eventually became extinct. In the fifth epoch, more organic matter formed, and this, in a second wave of spontaneous generation, gave rise to the ancestors of the species we now observe. Since that time, these species have changed both by diverging into geographical varieties (mistakenly called species) and by generally becoming smaller. Tropical species like elephants migrated southward as the climate continued to cool.

Thus, each Linnaean genus is derived from ancestral forms produced by spontaneous generation. Buffon still insisted that the species (Linnaean genera) were basically fixed in form, with only superficial modifications being possible. This marks off his theory from the more radical view of the origin of life proposed by the materialists discussed below. They too thought that spontaneous generation could produce complex organisms, but proposed that the organic particles come together at random, producing vast numbers of monstrous forms and only (by chance) a few viable ones that could survive and reproduce. Buffon had no time for the design concept, yet he continued to believe that species were fixed and that only certain kinds of living organisms were possible. In his second supplementary volume (1775), he argued that the same species would appear on each planet of the solar system as it cooled down to the appropriate temperature. The internal mold of the species was fixed not only on this earth but also throughout the universe. Buffon seems to have believed that only certain combinations of organic particles are stable, just as in chemistry the elements can combine to give only a certain range of stable compounds. For all his materialism, he remained locked into a worldview in which natural developments were constrained by predetermined patterns and thus endlessly reproduced the same basic forms.

THE MATERIALISTS

Among the Enlightenment materialists were thinkers more radical than Buffon who wanted to eliminate all trace of the argument from design by showing that the world we observe is the product of purely random activity within material nature. For these thinkers, everything we see arises from the ceaseless activity of nature proceeding without any direction or plan. In such a worldview, there could be no room for the concept of the fixity of species; the living forms we see are produced by trial and error, and there is nothing to guarantee that their structure will be perpetuated accurately by reproduction. The Enlightenment materialists came close to the philosophy of modern Darwinism in their refusal to accept that there are fixed forms in nature and in their assumption that species are produced by an undirected material process. They are certainly the most obvious exception to Foucault's model of eighteenth-century thought. Yet it would be wrong to treat them as forerunners of Darwin in a straightforward way. They were speculative thinkers, not scientists—although they were in contact with the best scientific minds of the day—and they presented no coherent theory of evolution. As with Buffon, their attention was deflected away from transmutation by an interest in the origin of life by spontaneous generation. Even complex living structures, they believed, could be generated directly from nonliving matter.

The aggressive side of the materialist program was pioneered by Julien Offray de La Mettrie in his *Man a Machine* of 1748 (translation 1991; Vartanian 1950). La Mettrie argued that humans themselves must be treated as purely material entities. The mind or soul was not a distinct spiritual element but merely a by-product of the physical body. Living matter was capable of maintaining the processes of life without the presence of mysterious vital forces or spirits. La Mettrie had been fascinated by Abraham Trembley's discovery of the ability of the polyp, or freshwater hydra, to regenerate two complete organisms when cut in half (J. Baker 1952; Trembley, translation 1973). If so simple an organism could regenerate itself, life must be a basic property of organized matter. The whole traditional apparatus of souls and spirits, including the human soul, could be swept away. Man becomes a machine, provided we transfer the properties of life to matter itself. Like Maupertuis, La Mettrie shows how the inability of Enlightenment radicals to provide adequate materialistic explanations of life led them toward hylozoism, the belief that matter itself is alive (Roger 1998; Yolton 1983).

When La Mettrie did turn to the origin of life, in his *System of Epicurus* (reprinted in La Mettrie 1774), he was unable to break away from the theory of preexisting germs and adopted a theory similar to that proposed by de Maillet. The two most accomplished materialists of the later Enlightenment pioneered a completely naturalistic approach to the origin of life: Denis Diderot and the Baron d'Holbach both saw nature as a totally flexible system in which there can be no absolutely permanent structures. Of the two, Diderot is perhaps the more sympathetic figure, a complex personality torn by emotional distaste for the atheism he found intellectually inescapable. As an editor of the highly subversive *Encyclopedia*, Diderot challenged the establishment by publishing critical commentaries on all topics. He ranks with Voltaire as a major figure in the awakening of conscience that paved the way for the French Revolution.

In his *Philosophical Thoughts* of 1746, Diderot still wrote as a deist, accepting the argument from design and the theory of germs. But by the time he wrote his *Letter on the Blind,* he had moved on to an extreme form of materialism (Crocker 1959; Vartanian 1953). Here, he followed La Mettrie by arguing that the mind is merely a by-product of the body. He discussed the case of a blind mathematician, Nicholas Saunderson, to argue that such a person would live in a mental world unlike that of those who have sight. In his conclusion, Diderot created a fictional impression of Saunderson on his deathbed, with the blind man rejecting the platitudes of the attending clergyman. How, he asked, could there be benevolent design in a world that could produce creatures like himself, lacking the most important organs? Diderot then had Saunderson articulate a vision of the origin of life which harks back to the materialists of antiquity. At the beginning of the earth's history, nature *experimented* with the spontaneous generation of many forms of life. Since these were produced without plan, most lacked essential organs and died out. Only by chance would nature occasionally hit on a structure that could survive and reproduce. The species we see are thus the survivors of a process of trial and error. Superficially, this seems to resemble the theory of natural selection, but Diderot was thinking of the survival of spontaneously generated forms, not of variations within existing species.

At this point Diderot saw matter as creative only in the sense that it could assemble itself at random to give potentially living structures. Once successful forms were produced by chance, they continued to breed true apart from the occasional "mistake," which we would regard as a monstrosity. After reading Buffon, Diderot realized that if living things were produced by a material process, there was no reason they should preserve their character through successive generations. In effect, he dismissed Buffon's internal mold as a fiction that bore too much resemblance to the old idea of designed species. In his *D'Alembert's Dream* (translation 1966), he again used a fictional device, presenting his speculations as the raving of his sleeping friend, the mathematician d'Alembert, recorded by his mistress. In this vision, nature was spontaneously active, and matter itself was alive. Diderot cited Needham's experiments as evidence for the spontaneous generation of life. And, like Buffon, Diderot speculated that in some circumstances even complex living forms might be produced by this means. But the forms produced in this way have no stability as they reproduce. In response to their needs, animals can develop new organs, which can be inherited by their offspring and thus become fixed in the type.

Diderot was also fascinated by the production of monstrosities (Hill 1968). Nature did not breed true but was always experimenting with the production of bizarre new forms, some of which might be able to perpetuate themselves. By accepting the chance production of new organs, Diderot broke completely with Buffon's vision of a universe constrained to produce only preordained forms. He anticipated the "trial and error" view of the origin of living structures that would become characteristic of Darwinism, although he can hardly be regarded as a precursor of Darwin at the scientific level. Diderot had no sense of adaptation as the driving force of transmutation, and his vision of hopeful monsters (to use a modern phrase) missed Darwin's emphasis on the continuous nature of the process.

Diderot was far from the kind of hardheaded materialist so despised by some later thinkers. When supporters of the Romantic movement expressed distaste for materialism, they were more likely to be thinking of that described in *The System of Nature* by the Baron d'Holbach. Published under an assumed name in 1770, this book became known as the Bible of atheism (new edition 1821, translation 1868). Diderot was friendly with d'Holbach but distrusted his militant dismissal of all religion as a fraud designed to uphold social repression. *The System of Nature* provided a new utilitarian social philosophy designed to increase human happiness, but d'Holbach insisted that this must rest on a materialistic conception of life that eliminated the traditional religious view of human nature and divine creation.

D'Holbach refused to compromise his materialism by supposing that matter itself had some primitive level of awareness. It had to be accepted as a fact that, when matter is organized into complex structures such as living bodies, mental processes begin. If matter itself was not alive, though, it was still a far more complex substance than the old mechanical philosophy had assumed. D'Holbach followed Diderot in picturing the material universe as an inherently active system that could organize itself to generate living structures wherever the conditions were suitable. All forms of matter were governed by affinities which dictate how combinations would take place. The spontaneous generation of life became little more than a chemical reaction, taking place whenever the appropriate substances were brought together. D'Holbach used Needham's experiments to support this view, although d'Holbach shared Diderot's belief that, under the right conditions, even complex forms of life could be generated in this way. He also agreed that once living things had been formed, they would be subject to constant change. Nature was never at rest; it was constantly experimenting with new forms, as in the production of monstrosities.

Of all the Enlightenment thinkers, Diderot and d'Holbach came closest to seeing the production of living forms as an open-ended process. Because they wanted to replace the concept of the Creator's design with that of a world governed by forces of nature itself, they gave unlimited power to those forces. If the universe was in a constant state of flux, there could be no fixed species and no predetermined plan of development. Yet they formulated no general theory of evolution. Their primary interest was philosophy rather than science, even though they stayed abreast of the latest scientific developments. Their ideas were speculative, intended to shock the orthodox rather than serve as the basis for research in natural history. Their primary interest was in the ultimate origin of life: spontaneous generation replaced the miraculous creative power of God. If microorganisms could be generated in the laboratory, then it was legitimate to assume that higher forms of life could be produced in the wild. Because their real concern was the relationship between matter and life, they allowed spontaneous generation to force the idea of transmutation onto the sidelines. New species might be formed from old, but, for the materialist, that was less interesting than the possibility that they might be formed directly from inanimate matter.

THE FIRST TRANSMUTATIONISTS

At the very end of the eighteenth century lived two figures whose ideas seem to come closer to the modern concept of how life on earth developed: Erasmus Darwin and Jean-Baptiste Lamarck. Both accepted the notion of spontaneous generation but realized that only the simplest forms of life might be formed in this way. They thus were forced to postulate a process by which living structures became progressively more complex. Both have been hailed as the founders of modern evolutionism, although much of the later enthusiasm for their work arose as a by-product of opposition to

Charles Darwin's theory of natural selection. As in the case of Buffon, it was Samuel Butler's efforts (1879) to discredit Darwin which led to an exaggerated view of the earlier writers' contributions. At the end of the nineteenth century, a school of neo-Lamarckism emerged in opposition to Darwinism, and some members of this school tried to present Lamarck as the founder of their alternative approach (Packard 1901). The neo-Lamarckians, however, concentrated only on those aspects of their hero's work that could be fitted into the post-Darwinian framework (see chap. 7).

Later historians have looked more closely at both Erasmus Darwin and Lamarck and have concluded that their view of organic development differed significantly from the modern one. Their ideas were developments of the eighteenth-century worldview and are only superficially similar to our own. At one point it was assumed that their ideas played no role in the later development of evolutionism, but historians have since realized that Lamarck's ideas significantly influenced the debates of the early nineteenth century (Desmond 1989). Although rejected by the established scientific community, the theory was widely promoted by more radical thinkers. The resulting debates helped shape the climate of opinion within which Charles Darwin formulated his own theory.

Erasmus Darwin

A colorful personality and perhaps the only writer we shall encounter who put forward some of his ideas in the form of poetry, Erasmus Darwin occupies a unique place in the history of evolutionism. His *Botanic Garden* (1791) and *Temple of Nature* (1803) were popular in their day, although not appealing to modern tastes. Erasmus was also the grandfather of Charles Darwin, and because his *Zoonomia* (1794–96)—a product of his medical interests, not his poetic fancy—proposed a theory of organic development, he has naturally become a focus for the kind of historian who tries to show that it was only by gleaning insights from his precursors that the younger Darwin was able to formulate his theory. Charles Darwin's endorsement of a biography of Erasmus (Krause 1879) sparked his feud with Samuel Butler (see C. Darwin 1958). Modern enthusiasts have followed Butler in finding passages in Erasmus's work which seem to anticipate the theory of natural selection (E. Darwin 1968; King-Hele 1963). But care has to be taken not to confuse the elder Darwin's recognition that predators turn the world into "one great Slaughter-house" (1803: line 66) with the younger's theory of the "struggle for existence" taking place among the individuals of each species.

Erasmus Darwin's views must be interpreted within their own context

(Browne 1989; J. Harrison 1972; McNeil 1987; Porter 1989). As a highly successful medical practitioner, he founded the family fortune. He was a deist who believed that God had designed living things to be self-improving. In their constant efforts to meet the challenges posed by the external environment, they developed new organs through the mechanism that the Lamarckians would make famous as the "inheritance of acquired characteristics." The results of each individual's efforts were inherited by its offspring, so that by accumulation over many generations a whole new organ could be formed. Erasmus Darwin developed his ideas not from a study of natural history but from his medical interests, which had led him to examine David Hartley's account (1749) of how the habits of life affected the soul. Darwin assumed that the overall result of this constant effort to adapt would be a gradual progress of life toward higher levels of organization and greater mental powers:

> ORGANIC LIFE beneath the shoreless waves
> Was born and rais'd in Ocean's pearly caves.
> First forms minute, unseen by spheric glass,
> Move on the mud, or pierce the watery mass;
> These, as successive generations bloom,
> New powers acquire, and larger limbs assume;
> Whence countless groups of vegetation spring,
> And breathing realms of fin, and feet and wing.
>   (1803: lines 295–302)

Lamarck

Apart from this poetic rhapsody, the only detailed account of Erasmus Darwin's theory is the single chapter in his *Zoonomia*. By contrast, Jean-Baptiste Pierre Antoine de Monet, Chevalier de Lamarck, was a professional naturalist who wrote extensively on his own, very similar theory. As a consequence he was vilified by his more conservative contemporaries but hailed by radical thinkers. Better knowledge of this radical support for Lamarck creates an interesting problem of interpretation for the modern historian. It has always been evident that his work represents the first major attempt to construct a comprehensive theory in which all living things have developed from primitive ancestors. The neo-Lamarckians of the late nineteenth century assumed he had proposed the modern view of evolution in which all the main groups of animals diverged from a single common ancestry. Later historians who have read Lamarck more closely have realized that his theory is based on principles significantly different from our own. It is not merely that twentieth-century genetics rejects the inheritance of acquired characters; Lamarck may not even have foreseen the modern concept of divergence from a common ancestor. Stressing these differences leads to an image of Lamarck as merely a precursor, not a true contributor to the rise of modern evolutionism (Barthélemy-Madaule 1982). Yet if his theory, far from being ignored, was widely debated in the early nineteenth century, we cannot ignore its possible influence on the debates which led to the emergence of Darwinism. Even if Lamarck is not considered a modern thinker, his ideas may have shaped the course of the Darwinian revolution in indirect but significant ways.

Charles Gillispie (1956, 1959) was one of the first to note there were deep differences between Lamarck's theory and the views accepted in post-Darwinian times. Darwinism treats evolution as a process of divergence (the tree metaphor), but Gillispie notes a strong element of linear progress in Lamarck (the ladder metaphor). He presents Lamarck as a Romantic thinker, a view not accepted by later historians. Lamarck's concept of nature as inherently active may bear some resemblance to the Romantic worldview, but his real inspiration was Enlightenment materialism, which also stressed the creative power of nature. He invoked spontaneous generation as the starting point of living development but decided that this produced only the lowest forms of life. He then postulated an inherently progressive trend which carried life up to higher levels of organization, an idea that may owe something to the temporalized chain of being. Significantly, he did not accept extinction: species could change, but they did not die out. Lamarck called in the inheritance of acquired characteristics to explain adaptation, which he at first presented as a process separate from the ascent of life. This distinction between the mechanisms of progress and adaptation is less apparent in his later writings, but most historians see it as an indication that Lamarck's theory was built on foundations very different from those we now accept (Burkhardt 1972, 1977; Corsi 1988a; Hodge 1971; Jordanova 1984; Laurent 1997; Mayr 1972; Schiller 1971; Sheets-Johnstone 1982).

Lamarck began as a botanist, and his early works show his interest in the linear arrangement of classes, although from the first he realized that animals and plants would form two parallel hierarchies, not a single chain (Daudin 1926). At the same time, he developed an unconventional view of chemistry and a system of geology based on uniformitarian principles (translation Lamarck 1964). In 1794 he was appointed to the Muséum d'Histoire Naturelle—formerly the Jardin du Roi, now reorganized by the revolutionary government—and was given the task of classifying the invertebrates (on the museum, see Spary 2000). He became one of the founders of invertebrate taxonomy, but while developing this new skill he abandoned his early commitment to the fixity of species and began to elaborate an evolutionary worldview. He first outlined his theory in 1802 and then incorporated it into his best-known work, the *Zoological Philosophy* of 1809 (translation 1914). A later version of the theory formed the introduction to his major survey of the invertebrates (1815–22).

Lamarck postulated a vast number of chemical elements with unlimited powers of combination. He thus rejected the theory of fixed, simple compounds with which Anton Lavoisier was revolutionizing the study of chemistry. At first Lamarck believed that the elements had no power to combine: the most active chemical force, fire, was merely destructive, breaking molecules down into simpler combinations. Compounds could be built up only by the nonmaterial force of life. All compounds, including those comprising the earth's crust, were built up in living bodies (this is not as silly as it sounds, since rocks such as chalk and limestone are indeed derived from the shells of minute sea creatures). At this point, Lamarck still believed that living forms constituted a hierarchy of fixed species.

Various factors have been postulated to explain Lamarck's later conversion to the belief that life itself is created by material forces, as well as his rejection of the fixity of species in favor of a developmental worldview. Gillispie states that he simply reversed the process of degradation that was central to his original chemical theory. M. J. S. Hodge (1971) argues that the starting point was the linear arrangement of the species, which Lamarck now saw as the product of a historical process. When his studies of invertebrates convinced him that the lowest forms of life had no specialized organs, he realized that they were simple enough to be produced directly by spontaneous generation. These simple forms would be the lowest rungs on the ladder of development. Lamarck believed that the key to spontaneous generation was the power of electricity (which was also responsible for nervous activity in higher animals). This could carve out channels in gelatinous matter to create simple living bodies. Lamarck was by no means the only thinker to link the newly discovered force of current electricity with life, and this link became the central theme of Mary Shelley's gothic novel *Frankenstein*, written in 1818. The horrific consequences of Frankenstein's blundering efforts to control what was now being seen as a purely natural force were presented as a warning that some things should remain outside the bounds of human understanding.

Lamarck, at least, believed that only the simplest forms of life could be formed in this way, and thus was led to see the hierarchy of species as a ladder by which living forms advanced to more complex levels of organization. In each generation, the active power of the nervous fluid tended to carve out more complex channels, so each organism ended up slightly more advanced than its parents. Schiller (1971) argues against the assumption that Lamarck derived this position from the temporalized chain of being. It is certainly true that Lamarck had gone far beyond a simple, linear chain—but even Bonnet had conceded that the chain branched, and the possibility that Lamarck was influenced by the same tradition cannot be ruled out. The early versions of his theory, at least, assumed that if the progressive force were the sole one operating on life, the result would be a perfectly regular gradation of forms with no breaks and no branches. This aspect of Lamarck's thinking sets it off from any form of evolutionism acceptable today (see fig. 9). But if there was a steady progressive force, why did one still see simple animals in the world? If all life was derived from a single, spontaneously generated ancestor, why was not all life as complex as humanity? The answer was that all life did not share a common ancestry, because the spontaneous generation of life was occurring all the time. Organisms at different levels of the hierarchy were derived from different acts of generation at different times: the higher they were, the longer they had been mounting the scale. On such a model, humans passed through an apelike stage in their development but were not related to the living apes. The apes belonged to a parallel line of evolution that stepped into our shoes as we advanced to the next stage.

Not all historians accept this interpretation of Lamarck's thinking, but it is consistent with two other aspects of his theory which also relate more to the worldview of the chain of being than to modern evolutionism. The chain was often seen as a continuous sequence of forms with no breaks to mark off distinct species, and Lamarck was of this opinion. For him, species were convenient fictions that the taxonomist could erect only because scientists were ignorant of some portions of the sequence and could use these gaps to mark off the species. Further discoveries might fill in the gaps and blur the species one into another. Lamarck also denied extinction; for him, nature was powerful enough to ensure that no forms could ever die out completely. When it was pointed out that some fossil shells differed from those of living species, Lamarck retorted that the ancient species had changed into the modern ones, not died out. These aspects of his thinking link him back to the eighteenth-century worldview, not forward to modern Darwinism.

Yet Lamarck was a practical taxonomist, and he knew that we do not observe a linear pattern in nature. The chain of being had many branches (as illustrated by fig. 10). To explain why, he invoked a second force of change: adaptation to the environment. His geology taught him that the environment is constantly, if slowly, changing, and he realized that each living form

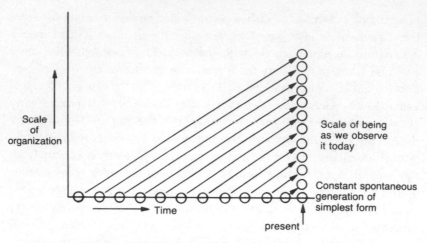

9.   Lamarck's theory of organic progress. Each point on the chain of being we observe today has been derived by progression from a separate act of spontaneous generation. The lower down the scale the modern organism is, the more recently its first simple ancestor was produced. Evolution is not a system of common descent but consists of separate lines advancing in parallel along the same hierarchy.

had to adjust its way of life, and ultimately its bodily structure, to the new demands placed on it. This secondary, adaptive force distorted the linear pattern that we would see if only the progressive trend were operating. The resulting branching arrangement looks much more modern than the linear process depicted in figure 9 (except that Lamarck inverted it and put the lowest forms at the top). And his discussions of individual relationships between forms often use the language of divergence, as when he seems to imply that humans evolved from chimpanzees (1914: 171–72). Given that Lamarck's later accounts of his theory make less distinction between progress and adaptation, it may be that his thinking was gradually moving away from the chain model. But is also possible that he imposed divergence on the parallel lines depicted in figure 9: imagine the diagram in three dimensions, with adaptation distorting the lines at right angles to the plane of the paper, and the result would look very much like the branching pattern of figure 10.

Whatever the true nature of Lamarck's theory, it was his mechanism of adaptation that caught the attention of later naturalists. The very term *Lamarckism* has come to mean the inheritance of acquired characteristics, even though he introduced this mechanism of adaptation only as a secondary, distorting factor. It had long been assumed that changes developed by

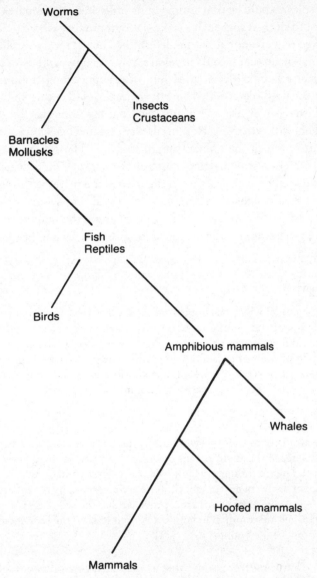

Worms

Insects
Crustaceans

Barnacles
Mollusks

Fish
Reptiles

Birds

Amphibious mammals

Whales

Hoofed mammals

Mammals

10.   Lamarck's branching chain of being. Adapted from a diagram in the *Zoological Philosophy,* this shows how Lamarck
thought the theoretically linear scale of organization had
been forced into various branches by the pressures of
adaptation to changing environments. He would not have
seen this as a tree of genealogical relationships but as a more
realistic representation of the chain produced as in fig. 9.

an adult animal could be transmitted to its offspring, but Lamarck was one of the first to suggest that such a process might produce whole new organs. He believed that an animal's needs determined how its body would change. This did not mean that it could develop a new organ by willpower alone, although many later critics accused him of arguing along these lines. For Lamarck, the needs determined how the animal would use its body, and the effect of exercise, of use and disuse, would cause some parts to develop while others withered away. New needs were created by changes in the environment to which the animal sought to adapt. Those body parts most strongly exercised would attract more of the nervous fluid, which would carve out more complex passages in the tissues and increase the size of the organ. Disused organs would receive less fluid and would degenerate. Lamarck postulated two laws, the first proclaiming the individual's ability to acquire modifications, the second stating that such modifications could be inherited:

FIRST LAW

In every animal which has not passed the limit of its development, a more frequent and continuous use of any organ gradually strengthens, develops and enlarges that organ, and gives it a power proportional to the length of time it has been so used; while the permanent disuse of any organ imperceptibly weakens and deteriorates it, and progressively diminishes its functional capacity, until it finally disappears.

SECOND LAW

All the acquisitions or losses wrought by nature on individuals, through the influence of the environment in which their race has long been placed, and hence through the influence of the predominant use or permanent disuse of any organ; all these are preserved by reproduction to the new individuals which arise, provided that the acquired modifications are common to both sexes, or at least to the individuals which produce the young. (1914: 113)

The first law is merely a somewhat exaggerated generalization of the belief that exercise develops an organ; people who pump iron to cultivate bulging muscles demonstrate its validity every day. The second law was widely accepted at the time. Although Lamarck gave no detailed theory of heredity, mechanisms by which acquired characters could be inherited continued to be proposed during the next hundred years. Charles Darwin's "pangenesis" was one such theory (see chaps. 6 and 7). The process has, however, been decisively repudiated by modern genetics. There is no way that changes to the adult body can feed back into the structure of the DNA that determines inheritance.

Lamarck turned the assumption—that acquired characters *are* inherited—into an evolutionary mechanism that adapted species to changes in their environment. In this respect it paralleled the function that Darwin would later attribute to natural selection, although Lamarckism was widely believed to place more emphasis on the initial behavioral modification. The classic example is that of the giraffe, whose long neck and forelegs Lamarck explained as a consequence of its habit of browsing on the leaves of trees (1914: 122). Adaptation to other modes of life had produced all the other specialized forms of animals. This aspect of his theory was borrowed by a later generation of naturalists, who assumed that Lamarck himself visualized all evolution as a process of divergence from a common ancestor. This relatively modern aspect of Lamarck's thinking must, however, be viewed in conjunction with the much older ideas noted above.

The most active neo-Lamarckian movement emerged much later in the nineteenth century, after Darwin's *Origin of Species* had popularized the general idea of evolution. Until comparatively recently, most historians believed that Lamarck's ideas were decisively rejected in his own time. There was thus no direct continuity between this first hesitant flowering of evolutionism and the events of the Darwinian revolution. Lamarck's theory could be dismissed as a somewhat distorted expression of the values characteristic of Foucault's classical age. Modern evolutionism emerged from a very different set of assumptions about nature established in the early nineteenth century, which paved the way for a theory of branching, divergent evolution. Recent developments in the history of science have undermined the foundations of this interpretation by showing that Lamarck's theory was not ignored by his contemporaries. It found a home within systems of radical philosophy that were either vilified or pointedly ignored by the scientific establishment. Recognition of this fact forces us to reinterpret Foucault's thesis that there was a sharp break between eighteenth- and nineteenth-century evolutionary thought.

Lamarck was certainly denied influence in the French scientific community and died in relative obscurity. His eclipse was engineered by his great rival, Georges Cuvier, whose used his new biology to gain a position of eminence within French society (Burkhardt 1970; Coleman 1964). Cuvier developed techniques of comparative anatomy which allowed him both to reform the system of classification and to reconstruct extinct animals from their fossil bones (see chap. 4). He decisively rejected the chain of being by dividing the animal kingdom into four fundamentally different types. Each type was based on a basic plan which could be indefinitely modified to give the specialized species we observe. For Cuvier, the relationships between species were based on similarities in their internal structure, not on an ordered ranking of their external characters. Foucault argues that this approach made possible the Darwinian view of evolution in which a single original form diversifies through adaptive radiation. Eighteenth-century models of development—including, to a significant extent, Lamarck's—still relied on the notion of the unfolding of a preexisting pattern such as the chain of being. Cuvier opposed transmutation because he thought each species was a uniquely coherent modification of the ground plan for its type. Yet his system of classification paved the way for the Darwinians' recognition that evolution is not predetermined but is an open-ended process of adaptation and divergence.

Cuvier dismissed Lamarck's theory as a product of outdated materialism. He held a more prestigious post at the Paris museum dealing with the vertebrates and gained political power by helping reform the educational system of Napoleonic France. He was thus in a strong position to marginalize Lamarck within the French scientific community. Materialism was again suspect in the more conservative atmosphere of Napoleonic France. In Britain, materialism was even more strongly repudiated by the political and scientific establishment horrified at the consequences of the French revolution. A new enthusiasm for natural theology swept aside the arguments for transmutation, paralleling the rejection of Hutton's geology on the grounds that it contradicted the Bible. Charles Darwin's theory of evolution was formulated in this much more conservative atmosphere, and it took Cuvier's new biology for granted. Foucault's claim for a decisive break between Lamarck and Darwin rests on this interpretation of the situation in the early nineteenth century.

Recent studies have thrown this older historiography into disarray. We now know that Cuvier's power was not absolute: radical thought continued to flourish on the margins of, and sometimes within, the French scientific community (Appel 1987; Corsi 1988a). Even in Britain, a vociferous underground of radical thinkers—including many in the medical profession—pushed a materialist philosophy and saw transmutation as a natural outgrowth of their worldview (Desmond 1989). The illusion that Lamarck was ignored arose within a generation of historians who concentrated solely on the writings of the orthodox figures who dominated the scientific establishment. The role played by radical science in the early nineteenth century is explored in the next chapter; for now, we must ask what these new insights tell us about the relationship between the ages of Lamarck and of Darwin.

Portraying Lamarck as the true founder of modern evolutionism will not work. There are substantial differences between his ideas and Darwin's, and there is no doubt that it was Darwin who finally persuaded the entire scientific community to take evolutionism seriously. There was indeed a hiatus in the development of evolutionism, and Darwin drew on a whole series of insights not available to Lamarck. Yet we can no longer accept a complete Foucaultian gap between the two episodes. It has become clear that even the conservatives' views were shaped by their (often unacknowledged) desire to block radical evolutionism. And the radicals played a role in paving the way for Darwin. When the *Origin of Species* was published, people had been discussing transmutation for years, and the earlier ideas had conditioned both public and scientific opinion. It may well be that for all the differences between Darwin's theory and Lamarck's, both the real and (equally important) the perceived character of Darwinism were shaped by earlier ideas. This warning is all the more important when we realize that evolutionism retained a link with the idea of a progressive scale. For all that Darwin repudiated Lamarck's idea of a built-in progressive force, his own theory was shaped in part by the vestige of progressionism and was greeted as a contribution to a philosophy of progress. It would be wrong to ignore the many new factors that Darwin built into his theory, but equally wrong to dismiss the possibility that Enlightenment thinking exerted some influence on thought in the century that followed.

4   Nature and Society, 1800–1859

Darwin's theory of evolution by natural selection was a product of the late 1830s, but he did not publish the *Origin of Species* until 1859, so for twenty years his ideas had no impact on the development of science. Thus, detailed treatment of his work can be postponed to a later chapter, leaving us free to explore developments in the wider debate which took place while Darwin remained silent. Whatever the role played by Lamarck's theory, Darwin drew upon scientific, philosophical, and ideological developments not available to the earlier evolutionist. His theory took off in a different direction, one that was more in tune with the aggressive worldview of industrial capitalism. Yet his theory was significantly different from anything else proposed at the time. Darwin created a unique blend of factors derived from natural history and social philosophy, establishing a theory whose full implications would not be explored until the twentieth century. To understand the complex relationship between Darwinism and the rest of nineteenth-century science, we must first understand the visions of nature proposed by his fellow scientists.

These alternative worldviews were no more isolated from the social and intellectual developments of the time than were Darwin's. Historians have long focused on the relationship between science and religion, noting that many aspects of pre-Darwinian science seemed to reinforce a link that had been threatened by Enlightenment materialism. There was no war between science and religion because many scientists were themselves both deeply religious and anxious to ensure that their scientific work could be reconciled with their faith (Gillispie 1951). In Britain, natural theology was revived and Genesis once again seen as a model (at least loosely) for the history of the earth and its inhabitants. Continental naturalists remained wary of such explicit links between science and Christianity but were still influenced by antimaterialist philosophies, including, in the German-speaking world, an idealism that verged on mysticism. This revival of conservative thinking did not go unchallenged—indeed we can now see that it was sustained by the fear that materialism was still influencing the more radical thinkers. Outright materialism was marginalized as dangerously subversive, but newer theories, including Auguste Comte's positivistic philosophy, offered a less radical alternative to the religious outlook.

These philosophical positions all had an explicitly political dimension in a world traumatized by the French revolution. Materialism was an integral aspect of a revolutionary ideology that wanted to sweep all traces of the old social hierarchy aside. Natural theology and idealism were invoked by conservatives who wanted to preserve their position in that hierarchy: the world was designed by a God who intended us all to accept our place in the preordained social scale. The situation was complicated, however, by a growing middle class making fortunes out of the new mechanized industries. They too wanted a social hierarchy that would include them in the ruling class. But they did not want a revolution that might sweep away the ownership of property. The middle class wanted reforms that would eliminate old restrictions on the individual's right to trade freely and that would allow them access to political power. Many of the scientific theories developed in the nineteenth century can be related to the desires of those engaged in these broad social movements to legitimize their preferred model of society by claiming that it was "only natural."

Beginning in Britain, the industrial revolution was transforming the social and economic map of Europe. America too was beginning to flex its muscles as a world power. The idea of progress, first articulated by Condorcet and other Enlightenment thinkers, now became the dominant model of social change. A parallel idea could also be applied to the natural world. Paleontologists uncovered a fossil sequence from simple to complex animals that even conservatives could not ignore. Liberals and radicals welcomed the idea because progress in the natural world seemed to hint that progress in the social world was inevitable.

No one could escape the implications of the new discoveries in the fossil record; indeed, some of the most conservative scientists made contributions to our understanding of the development of life on earth. They avoided the more radical implications of their findings by insisting that the history of the earth was punctuated by vast catastrophes followed by supernatural acts of creation. The traditional historiography of the Darwinian revolution treats this catastrophist position as an example of bad science upheld by religious conservatism. Darwin's thinking is traced back to Charles Lyell's alternative uniformitarian geology, in which all change is slow and gradual. Lyell's thinking reflects the liberals' faith in the gradual transformation of social institutions, but in one respect he repudiated a central plank of liberal thinking: the idea of progress. Lyell's antiprogressionism was never widely accepted, and its failure forces us to reconsider the traditional image of what constituted good and bad science at the time. Lyell cannot be seen as a pure scientist sweeping away the mists of biblical obscurantism; he had his own religious and philosophical agenda. Because both uniformitarians and catastrophists contributed to the new vision of the earth's history, we must be very careful in our assessment of the relationship between scientific knowledge and ideology.

Natural theology itself was transformed by conservative scientists. Continental idealism emphasized the unity of the divine creation. Some idealists saw the history of life on earth as the unfolding of a divinely preordained plan. But they were incapable of thinking in terms of a mechanism of change based on natural causes. Thus, their thinking evaded, perhaps deliberately, what Darwin saw as the most crucial question: how exactly did evolution work? This question had already been asked and answered by the radicals who accepted Lamarckism as a potential cause of progressive evolution. They promoted materialism at a variety of levels, including the new science of phrenology, in which the mind was seen as a product of the material activity of the brain. The conservatives successfully kept these radical ideas outside the realm of scientific orthodoxy, which is why they were ignored by historians until comparatively recently. We can now see the early nineteenth century as a period in which rival models of nature battled for supremacy, both inside science and in the public arena.

The situation was made more complex by the fact that science itself now became a force vying with more traditional sources of expertise for influence within society. The professional structure of science was rapidly taking on its modern form, shaped by scientific societies and educational institutions. A coherent scientific community had emerged which could lobby for public support, and within which the rival groups could compete for influence. But with the growth of an educated readership, scientists and others interested in important scientific concepts could also appeal directly to the general public. This was an age in which geology texts sold as well as the most popular novels. The significance of this wider audience for science is especially important in the area of evolutionism, because we know that one popular evolutionary text, published fifteen years before Darwin's *Origin*, did much to bring the topic to people's attention. This was Robert Chambers's *Vestiges of the Natural History of Creation* (1844), and according to James Secord's study (2000), the influence of this book had already established the framework within which most people perceived Darwin's work—although its impact on scientists was less clear-cut.

THE INVENTION OF PROGRESS

It was increasingly clear that the world was changing. Mechanized industries were transforming the economy and would soon be joined by the railways and other equally obvious agents of social change. Nineteenth-century thinkers responded to this sense of a world in motion by turning their attention to history in the hope of discovering a pattern that would make sense of their current upheavals (on nineteenth-century thought in general, see Copleston 1963, 1966; Mandelbaum 1971; and Willey 1949, 1956; on historical metaphors, see Bowler 1989a). The idea of progress emerged as a key theme in this effort to make sense of historical change, but it was an idea that could be explored in different ways. Liberals saw progress as gradual and cumulative—the sum of endless individual efforts at self-improvement. But they were indifferent to those left to starve in the slums, and only the most radical thinkers saw the plight of the poor as something to be ameliorated by the state. Even conservatives had to concede the fact of progress—but they could adapt it to their own ends by suggesting that at each level of development a stable social order would emerge. There were thus many different ideas of progress, each invented to fit a particular ideology.

Looking back at the early part of the century, the philosopher John Stuart Mill (reprint 1950) saw Jeremy Bentham and Samuel Taylor Coleridge as the figureheads of the two dominant streams of thought. To British eyes, Bentham typified the new liberalism of the middle classes: his philosophy updated the values of the Enlightenment, stressing the malleability of both the individual and the community, and the power of human reason to control social change. Coleridge was a conservative who saw society as something more than a collection of individuals and who appealed to spiritual values at the heart of both nature and culture. Both these strands interwove themselves through Western thought as a whole.

In France, the revolution ushered in a century in which radicalism and clerical conservatism battled for supremacy. Pierre Cabanis developed a materialist view of human nature derived from d'Holbach (translation 1981; Staum 1980). The marquis de Saint-Simon built on Condorcet's sketch of human progress, arguing for a scientific study of human affairs and calling for social justice based on equality of opportunity (translation 1952; Manuel 1956). By far the most influential French thinker, though, was Auguste Comte, whose philosophy of positivism saw an inevitable progress in humankind's efforts to control the world (Pickering 1994). His survey of "positive philosophy" of 1830–1842 (translation 1975) proposed that our understanding of the world passed through three phases. In the first, knowledge was given a theological basis, all natural events being attributed to supernatural powers. In the next, the metaphysical level, nature itself was endowed with built-in causative powers. Finally, science would emerge into its positive phase, in which the laws of nature are described without postulating underlying causes. Applying the positive method to human affairs would generate a true science of sociology and eliminate the need for religious sanctions in morality. Comte saw society as inherently progressive, but could not escape the old emphasis on the fixity of human nature. His system was immensely influential, but it was left for the British philosopher Herbert Spencer to integrate the biological, psychological, and sociological levels of progress (J. Greene 1959b).

Idealism and Romanticism

In Germany the reaction against Enlightenment materialism provided the most powerful contribution to nineteenth-century thought. J. W. von Goethe and his followers in the Romantic movement turned in disgust from the soulless philosophy expressed by d'Holbach (R. Richards 2002). They wanted to see spirit as an active force imposing its will on nature to create order and purpose. Romanticism stressed the freedom of the human spirit and the sublimity of nature. A parallel movement in philosophy created an idealist perspective that was to shape much of nineteenth-century thought. David Hume, the most skeptical of the Enlightenment thinkers, had pointed out the paradox of sensationalism: if all knowledge comes from the senses, we have no way of knowing what actually causes our sensations. We postulate laws of nature to cover the regularities in our experience, but we cannot know if there is anything in nature corresponding to our belief in causation. Immanuel Kant solved this problem by giving an active power to the mind itself: the mind does not passively receive sensations but actively organizes them to make experience comprehensible. In morality, too, Kant rebelled against Enlightenment utilitarianism by arguing for a moral sense or conscience that makes our duty clear to us (Cassirer 1945). Idealists such as J. G. Fichte and F. W. J. von Schelling maintained that the mind is the primary reality; it, in effect, creates the world in which it lives. For Schelling, the individual will was transformed into the universal Will, striving to manifest itself in the world of appearances. G. W. F. Hegel saw the universal Will, the Absolute, as the driving force of the universe, working itself out within human history to achieve its ultimate purpose.

Hegel's philosophy of history (translation 1953) reveals that idealism, although in many respects based upon a conservative ideology, had adapted to the inherently developmental framework of nineteenth-century thought. The idealists saw history developing through a series of distinct stages and driven by a universal spiritual power beyond the control of individual human beings. This approach had already been foreshadowed in J. G. Herder's survey of universal history (translation 1968). Herder even had postulated biological progress through a temporalized version of the chain of being (Lovejoy 1959b). He saw human history as progressive, but with each stage in the development constituting a stable social order. Hegel interpreted this discontinuous model of progress in terms of the philosophy of dialectic. Forces in nature or society always tend to generate their opposites, the tension being resolved from time to time by an abrupt transition to a new phase based on a synthesis.

For Hegel the ultimate human reality was the state, not the individual. The state expressed the spirit of a society or culture, often symbolized by a great leader. The individual should submerge himself or herself in the ambient culture, because this represents an essential stage in the Absolute's progress toward self-manifestation. Karl Popper (1962) sees this philosophy as the source of twentieth-century totalitarianism. One illustration of this warning is Hegel's influence on Karl Marx, who turned the idealist dialectic on its head by postulating economic forces as the true driving force of history. Marx would later welcome Darwin's theory for its materialist implications, but would also assert that the survival of the fittest was a reflection of capitalist individualism. His vision of history subordinated the individual to the social class, postulating a sequence of social developments that would culminate in the communist revolution. According to Popper, the totalitarianism of Soviet Russia, no less than that of Hitler's Germany, was an outgrowth of the idealist view of the state.

Several aspects of idealist thought had their analogs in biology. Just as Hegel subordinated the individual human to the state, so an idealist would see the individual organism as merely a copy of a blueprint representing the species (the typological viewpoint). To the idealist, the relationships between species were not products of natural forces but elements in a harmonious plan created by the divine Mind. This plan might well involve an element of development through time. Hegel's history was developmental, although he did not see a temporal sequence in the prehuman stages. The possibility of a development in the history of life itself became widely accepted among the generation of German biologists influenced by Schelling. Their *Naturphilosophie* saw orderly and purposeful development everywhere, from the growth of the human embryo to the history of life revealed by the fossil record. I examine the details of these theories below; for now, it is enough to note that the profoundly anti-individualist approach of idealism guaranteed that any form of evolutionism it inspired would be very different from Darwin's natural selection.

In Britain, individualism flourished as the political philosophy of the industrial revolution. But the Romantic-idealist tradition was not without influence, and Mill made an apt choice when he identified Bentham and Coleridge as figureheads for the two modes of thought. Coleridge, the Romantic poet and exponent of German idealism, best symbolized the efforts of conservative thinkers to stave off the challenge of radicalism. In his philosophy, a revitalized Church would symbolize the nation's cultural aspirations. The developmental aspects of idealist thought also interested historians anxious to identify stages in the emergence of Western culture. Liberal Anglicans such as Charles Kingsley (later an enthusiast of evolutionism) constructed a discontinuous model of cultural progress in which ancient Greece, Rome, and the Christianized Germanic tribes of northern Europe each had added an essential element to the flowering of the human spirit (Bowler 1989a). The rise and fall of empires marked the pulse of progress—a model of human history with distinct parallels to the antievolutionary interpretations of the fossil record I examine below.

Utilitarianism and Laissez-Faire

Jeremy Bentham's radical individualism pioneered the most active strand of British thought, because it expressed the aspirations of the rising middle class. These people shared the conservatives' horror of revolution, yet they demanded reforms that would throw off the shackles of feudalism and allow entrepreneurs the freedom they needed to do business. Progress was an article of faith because they saw their own economic activities as the driving force of social changes that ultimately would benefit the whole of humankind. Progress was not a sequence of mysterious spiritual uplifts; it was a slow but inevitable consequence of every individual's search for self-improvement. This mode of thought would ultimately propel Darwinism to the fore in the 1860s, but it was slow to develop for several reasons, including the fear of outright revolution.

Bentham was a leading figure in a group known as the philosophical radicals (Halévy 1955). He drew on the sensationalist view of nature promoted in the previous century by Hartley and Priestley, invoking the principle of the association of ideas to explain how the human mind built up connections between the experiences yielded by the senses. Like the Enlightenment reformers, Bentham hoped to design a system of laws that would manipulate human personalities by using the hope of pleasure and the fear of pain to create social habits. This "utilitarian" philosophy reduced human aspirations to a calculus of profit and loss. Everyone was assumed to judge their interaction with others in terms of what was useful in maximizing pleasure. The object was to produce the greatest happiness of the greatest number of people, happiness being defined as an excess of pleasure over pain in individual experience. If they could gain control of the country's government, the radicals hoped to put the Enlightenment's program of reform into effect.

There were others in this liberal or individualist tradition, however, who saw no need for state interference with human behavior. The ideology of industrial capitalism reduced all social interactions to the level of the marketplace. Economic (and by implication social) progress would occur naturally, without state interference, once all restrictions on individual freedom were lifted. This laissez-faire, or free enterprise, approach to economics was founded by Adam Smith's *Wealth of Nations* of 1776 (reprint 1910), in which Smith argued that prosperity depended solely on individual initiative. State control of the economy, however well-intentioned, would only interfere with the natural tendency for everyone to interact in a way that would maximize profit for all. This economic philosophy assumed the existence of an "invisible hand" guiding individuals so that actions taken for purely personal profit turned out, in the end, to be best for the economy as a whole. Smith also joined with a group of philosophical historians to formulate a scheme of social progress in which humanity had advanced from hunting and gathering through agriculture to industrial capitalism. This linear model of progress was to exert a strong influence on later evolutionary models of anthropology, although for the time being its influence was muted by respect for the biblical creation story.

Utilitarianism could also be used to block the radicals' idea of progress in the natural world. Partly to stem the threat of Erasmus Darwin's transformism, the Anglican clergyman William Paley restated the old argument from design in his classic *Natural Theology* of 1802 (Le Mahieu 1976). Paley stressed the usefulness of every organic structure by introducing the analogy of the watch and the watchmaker. If you found a watch on the ground, you would hardly conclude that it was a product of the blind laws of nature: "The inference we think is inevitable, that the watch must have had a maker: that there must have existed, at some time, and at some place or other, an artificer or artificers who formed it for the purpose which we find it actually to answer: who comprehended its construction, and designed its use" (Paley 1802: 3). Just as Ray had argued in the seventeenth century, Paley insisted that animal bodies were also carefully constructed machines that required us to postulate a Designer. The adaptation of structure to function indicated the Designer's benevolence as well as His wisdom: the utility of the structures was intended to ensure the greatest happiness of the greatest number of animals. In its utilitarianism, Paley's version of design was also materialistic: it treated the animal body as a carefully designed machine analogous to those already revolutionizing British industry (Gillespie 1990).

Malthus

Some exponents of free enterprise saw no room for progress even in the social realm. One such conservative thinker was to exert a profound influence on Darwin: this was another clergyman, the political economist Thomas Robert Malthus (James 1979; Petersen 1979). Malthus's *Essay on the Principle of Population* of 1797 (reprint 1959, second edition reprint 1990) was written to challenge the optimism of those who proclaimed that social reform would bring increased happiness for all. He argued that the "passion between the sexes" was so strong that the population had a natural tendency to expand. In principle the rate of expansion was geometrical; in practice, expansion was limited because there was not enough food to feed the extra mouths. Poverty and starvation were not the product of an unfair distribution of wealth—they were natural and inevitable, a prediction that earned Malthus the hatred of all subsequent social reformers. At best there could be a limited expansion of the food supply through better farming techniques, and the rich landowners bore the responsibility for this. The poor could do nothing but practice "moral restraint" to avoid bringing children into the world whom they could not feed (Malthus thought birth control was a vice).

Malthus was a clergyman who believed God had created such an apparently harsh situation for humankind for a moral purpose (Santurri 1982). The constant threat of starvation was designed to teach us the virtues of hard work and moral behavior, an implication that Paley himself approved. But where such lessons were not learned, the consequences were appalling, leading Malthus to coin the phrase "struggle for existence" when describing the constant wars among the wild tribes of Asia: "And when they fell in with any tribes like their own, the contest was a struggle for existence, and they fought with a desperate courage, inspired by the reflection that death was the punishment for defeat and life the prize of victory" (1959: 17). This phrase later caught Darwin's attention when he was thinking about the consequences of the principle of population for animal species. Here, there could be no moral restraint, no artificial sources of food, and the result would be a constant war of nature which, Darwin realized, would weed out those less well-adapted to the prevailing conditions (see chap. 5).

The fact that Darwin's formulation of natural selection was inspired in part by Malthus's principle has led many historians to argue that the theory reflects the competitive ideology of early-nineteenth-century capitalism. Malthus was certainly an exponent of laissez-faire: he believed that state relief for the poor merely encouraged them to breed more rapidly. There is no doubt that the principles of utilitarianism and individualism formed the common context within which both social and biological theories were debated (Young 1969, 1985). But we should avoid the assumption that Malthus was an early social Darwinist who left Darwin with nothing to do but apply his economic model to nature. Malthus wanted a static, designed universe, and he made no effort to see individual struggle as a constructive force in Western society (Bowler 1976b). Unlike the victors among the tribes of central Asia, the rich are not winners in a cutthroat competition, but the inheritors of accumulated wealth that imposes social responsibilities upon them. Malthus's metaphor of the struggle for existence among competing tribes drew more inspiration from European imperialism and colonial expansion than from the individual competition within capitalist society. Darwin borrowed ideas from both Paley and Malthus but put them together in a new way which threatened to undermine the ideology of divine harmony on which they had been based (W. Cannon 1961a).

To many radical thinkers, Malthus's indifference to the plight of the poor was as distasteful as the old privileges of the aristocracy. The more extreme radicals were little more than revolutionaries; they were suppressed by the state and held up to the orthodox as an example of what happened when the stabilizing influence of religion was removed. But there were many liberal-minded thinkers from the new middle classes who hoped that the benefits of reform would trickle down from their own class to the deserving poor. These were the exponents of gradualism in politics and in science. They also favored the idea of progress and wanted to see the efforts of individual human beings as the motor of social advance. We shall see how this radical progressionism slowly began to color the science of the time, encouraging the construction of models of nature that presented progress as inevitable. Social advance would be no more than a continuation of the universal law of progress. To a large extent, the emergence of the theory of the transmutation into respectability was a consequence of the gradual deradicalization of the idea of progress. What was subversive and revolutionary in the aftermath of the Napoleonic wars eventually became the ideology of the respectable middle classes. The philosopher Herbert Spencer provides the clearest illustration of this trend: by the 1850s, he was expounding a vision of universal progress that already included an element of Lamarckian transformism (J. Greene 1959b; Peel 1971). Spencer was an apostle of free enterprise and saw as the driving force of progress the innovative activities of individuals trying to cope with a competitive environment. His philosophy would become an integral part of the framework of thought within which Darwinism became popular in the 1860s (see chap. 6).

THE FRAMEWORK OF SCIENCE

As the industrial revolution progressed, scientists were at last able to fulfill the hope expressed by Francis Bacon back in the seventeenth century that better understanding would give better control of nature. By the mid–nineteenth century, science was generating knowledge of practical value that, according to Bacon, would transform existing industries and (as in the case of electricity) create entirely new ones. Social progress would result at least in part from the expansion of scientific knowledge and the translation of that knowledge into industrial power. Not everyone accepted this claim initially. The first phase of the industrial revolution had been based on the ingenuity of craftsmen, not on new scientific theories, and British industrialists were among the slowest to realize that the situation was changing as new industries emerged. But to the scientists, it was clear that their efforts, as a key component of economic progress, deserved support from industry and the state. In terms of government support, Britain lagged behind, because the ideology of laissez-faire held that those who benefited from science (the industrialists) should pay for it. But gradually the modern system of state support for science was established, and scientists themselves were organizing to create a better research environment and to lobby for a greater role in government and society. A recognizable scientific community was beginning to emerge, with the modern apparatus of societies and journals, and the fate of new ideas would increasingly be influenced by the politics of this community.

The pillars of the scientific community were the societies—where research could be debated and published—and the research facilities such as museums, surveys, and university departments. The British educational system did not yet provide clearly defined routes into the profession, and scientists lobbied for properly funded universities that would combine research and teaching. This combination had already been established in mining schools set up by several German states in the eighteenth century. By the 1830s the notion of postgraduate education as training for research was pioneered, again by German universities. In France, the revolutionary government created the Muséum d'Histoire Naturelle in Paris, with strong research and teaching components (Spary 2000). Lamarck worked here, as did his great rival, Georges Cuvier. The French also reformed their educational system to give a greater role to technical subjects, and it was as a leading architect of this reform that Cuvier gained much of his influence.

In Britain the Royal Society was now dominated by aristocratic amateurs, many of whom were little more than dabblers. Partly in frustration at this blockage, the British Association for the Advancement of Science was created in 1831 (S. Cannon 1978). With considerable reluctance, the government was persuaded to set up a Geological Survey and eventually a School of Mines. The ancient universities of Oxford and Cambridge taught little science, although increasingly their professors were becoming serious researchers (Darwin got his science training at Cambridge outside the curriculum). Only Anglicans could attend Oxford or Cambridge, but the University of London was founded by Nonconformists in the 1820s. Scientific societies and journals were already flourishing in particular disciplines. The Geological Society of London was the most active (Darwin served as its secretary and then as its vice president in the years following his return from the *Beagle* voyage). The British Museum already had significant natural history collections, although the modern Natural History Museum in London would not be built until the 1880s.

As the century progressed, the opportunities for scientific careers expanded, although never as fast as the scientific community wished. In the second half of the century, new universities were founded, and Oxford and Cambridge were forced reluctantly to develop a modern system for scientific education. These developments heralded a major change in the scientific community. The British Association had been founded by gentleman-amateurs, not by scientists who worked for their daily bread. They were serious about their science and demanded public recognition for it, but they were suspicious of pressures from the lower middle class to create a profession that opened career opportunities for those without private means of support (on the tensions between professionals and amateurs, see Desmond 2001). The young Thomas Henry Huxley scraped his way through a medical education, did research while serving as the surgeon on a naval vessel, and finally landed a job teaching paleontology at the School of Mines (Desmond 1994). By contrast, Darwin had been sent to Cambridge by his wealthy father and was the captain's gentleman-companion on board the *Beagle*. The transition from Darwin's generation to Huxley's marked the emergence of a modern paid scientific profession, although it was less abrupt than we might imagine.

Similar developments took place in the United States but equally slowly because here, too, state and federal governments were reluctant to spend public money on science. Much scientific work was still done by wealthy individuals, and paying positions were in short supply. American universities were starting to expand, although scholars emigrating from Europe still played a prominent role, as when Louis Agassiz was brought to Harvard to create the Museum of Comparative Zoology (Winsor 1991). Only later in the century did the United States begin to take a lead in the founding of research universities. Like their British counterparts, American scientists felt the need to unite for mutual support in their campaign for resources and influence, which led to the founding of the American Association for the Advancement of Science in 1848 (Oleson and Brown 1976).

The fact that both Darwin and Huxley did research while traveling round the world on a Royal Navy survey vessel indicates, finally, the link between the emergence of the modern natural sciences and the expansion of European power around the world. North America was, of course, already starting on its way toward economic dominance, but the old countries of Europe for the time being were able to parcel out a good deal of the earth's surface between them as colonies. Scientists expanded their areas of study outward from Europe as the colonies themselves founded geological surveys, museums, and universities, and the flow of information back to the metropolitan centers increased. The great natural history museums of the major European cities became the new cathedrals of science, symbolizing the dominance of Western civilization over what was perceived as a wild and barbaric world (Sheets-Pyenson 1989; on the changing perceptions of the natural world, see Bowler 1992).

GEORGES CUVIER: FOSSILS AND THE HISTORY OF LIFE

When Foucault (1970) wrote of a sharp break between the eighteenth- and nineteenth-century views of nature, he turned to Georges Cuvier for a symbol of the new science. Cuvier undermined the classical description of nature: he used the internal structures of animals to reveal the relationships between them and insisted that the groupings bore no relationship to a chain of being. Applying his new comparative anatomy to the reconstruction of fossils, he demonstrated that many ancient species had become extinct. Cuvier was far from being a transmutationist: he ridiculed Lamarck's theory as wild speculation and insisted that each species was a finely tuned structure that could not be modified without disruption. But he made it impossible to ignore the changes within the animal kingdom through geological time, and his system of relationships formed the framework within which the theory of common descent would be articulated. When a later generation of naturalists abandoned Cuvier's commitment to the fixity of species, they realized that each form could be seen as a superficially modified version of a basic animal type which Darwin would interpret as the common ancestor of its group. At first, however, Cuvier's influence was far more conservative. His image of carefully structured forms was seized upon by his British followers as a perfect expression of the augment from design.

Cuvier studied mollusks on the shores of Normandy while Paris was wracked by the reign of terror (Coleman 1964). His promise as a naturalist was soon recognized, and he was given a position dealing with the vertebrates at the newly organized Paris museum. Here, Cuvier consolidated his scientific and his political influence in the French scientific establishment, successfully adapting to the rise of Napoleon and to the restoration of the monarchy after Waterloo (Outram 1984). His new approach to comparative anatomy was developed around 1800, and he soon published a major survey of all the animal forms known to him (Cuvier 1805). In 1812 he proposed a new system of classification which undermined the logic of the old chain of being. His survey of the whole animal kingdom (1817a, translation 1863) articulated this new vision of nature in a more popular form, though its introduction serves as a valuable summary of his scientific principles (Outram 1986).

Buffon's collaborator, Daubenton, had worked on the comparative anatomy of different animal species, but Cuvier perfected the technique of using the internal structure revealed by dissection, rather than external appearances, as the basis of classification. He appreciated the complexity of the internal structure and stressed the "correlation of parts," the relationships which must exist between the organs to create a functioning whole. Similarly, the "conditions of existence" imposed by the animal's habits and environment must be reflected in the overall organization: in effect, the animal is a system designed to function in a certain way. An experienced anatomist could begin to recognize the kinds of relationships that must exist between the parts. The sharp teeth of the carnivore must be correlated with equally sharp claws that seize the prey. It was said that Cuvier's skill allowed him to reconstruct the whole animal from a single bone.

Better knowledge of the internal structures of different species allowed Cuvier to see that many superficially different species shared deep similarities. When classifying, he proposed a principle of the "subordination of characters": some characters were more fundamental and thus should be given more weight. In principle, the nervous system was the most important, although for one major group, the vertebrates, the skeletal structure offered the most convenient basis of classification. The possession of a backbone or vertebral column could be seen as a feature uniting the Linnaean classes of mammals, birds, reptiles, and fish. These classes thus could be grouped into the vertebrate *embranchement*, or type (Eigen 1997). The invertebrates had traditionally been given a subordinate position within the animal kingdom, but Cuvier insisted that they comprised three other types, each based on a ground plan as fundamental as that of the vertebrates (see fig. 11). The lack of skeleton was not necessarily a sign of inferiority, because other basic forms of animal structure functioned perfectly well without one.

Such a division of the animal kingdom broke the linear chain of being. Naturalists had always regarded some animals as more highly organized than others, but their rankings had been based on the assumption that the human species was at the top of the scale. The more closely a species approached the human form, the higher on the scale of organization it must be. Cuvier insisted that our being mammals did not justify the assertion that mammals are the highest vertebrate class, or that vertebrates are higher than the other three types. Vertebrates and mollusks were so fundamentally different that it was meaningless to rank one type above the other. Fish and mammals were merely different forms of vertebrates adapted to different ways of life. It would be many decades before the majority of naturalists threw off the last vestiges of the chain of being—indeed, the whole concept of organic progress was based on the assumption that there was a scale to be ascended. But in principle, Cuvier's system made it possible to represent the relationships between species as a branching tree rather than a ladder.

Yet Cuvier had no intention of explaining the diversity of structures within each class by a process of descent from a common ancestor. He ridiculed Lamarck's theory of transmutation and defended the fixity of species (Burkhardt 1970). Although each vertebrate species was a modification of the same underlying body plan, the specific forms were integrated wholes which would be rendered biologically unworkable if significantly disturbed. Bodily interactions were so delicately balanced that any major change would upset the system completely. The ability of the environment to pro-

Vertebrata (Vertebrates)	Creatures possessing a backbone: the four Linnaean classes of mammals, birds, reptiles, and fish. (The amphibians are now regarded as a separate class, but in the nineteenth century they were usually included with the reptiles.)
Mollusca (Mollusks)	Creatures with no backbone but sometimes an external shell: oysters, clams, etc.
Articulata (Articulates)	Creatures with articulated or segmented bodies: insects, spiders, worms, etc.
Radiata (Radiates)	Creatures with a radial or circular plan of organization: starfish, sea urchins, etc.

11. Cuvier's four types of animal organization. Each of the four types represents a basic pattern of organization, and each can be subdivided into classes, orders, families, genera, and species as in the Linnaean system. Cuvier's type is the equivalent of the modern phylum, still the most fundamental level of classification. But biologists now recognize far more than four phyla. The Vertebrate type has become the Chordate phylum and includes some animals with a spinal cord but no backbone. The Articulate type has been broken up into several distinct phyla. The Radiate type was used from the start as a dumping ground for creatures that Cuvier could not fit in elsewhere; it has also been broken up into a number of phyla (Winsor 1976).

duce varieties within each species was thus extremely limited. Cuvier presented the fixity of species as a scientific consequence of his efforts to understand the complexity of living structures. He made no reference to design by God, although his British followers found his system easy to adapt to the argument from design.

Cuvier's rejection of transmutation is all the more remarkable because of his contributions to vertebrate paleontology, which produced the first outline of the history of life on earth based on solid evidence. The techniques of his new biology were ideally suited to the reconstruction of fossil animals, where often only an incomplete skeleton was found (Theunissen 1986). From his experience with living animals, Cuvier could visualize how the fossil bones could have fitted together to make a living animal, and then could reconstruct its outward appearance. Napoleon's conquests gave Cuvier access to fossils collected all over Europe. He applied himself to reconstructing all the fossilized species available and soon became the acknowledged authority in this field. His collected papers (1812) are seen as the foundation of vertebrate paleontology (excerpts translated in Rudwick 1997; see also Buffetaut 1986; J. Greene 1959a; and Rudwick 1972).

The discovery of ancient and fossil bones had aroused great interest at the end of the eighteenth century. From Siberia came the remains of the wooly mammoth, an elephant-like creature which had lived in the recent geological past (see fig. 12). From America came an even stranger elephant,

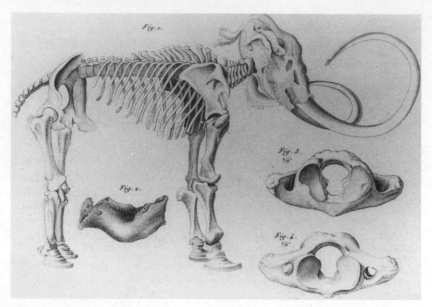

12.   Cuvier's reconstruction of the mammoth. From Georges Cuvier, *Recherches sur les ossemens fossiles* (3d ed., 1825, vol. 1, plate 11).

the mastodon, whose teeth resembled those of a hippopotamus. At first it had been assumed that these animals were only minor variants of known species, or—if they were different species—that individuals must still be alive somewhere in the world. As exploration continued, it became increasingly difficult to believe that large mammals remained unknown to science. But Cuvier confirmed that the mammoth and the mastodon must be regarded as distinct species, just as the Indian and African elephants were distinct. The mastodon had to be placed in a different genus because it was so different from the living elephants. There could now be little doubt that the mammoth and mastodon were extinct (Cuvier's reconstructions were eventually confirmed by the discovery of whole mammoth carcasses frozen in the Siberian ice). Soon, many more fossil species had been described, some quite unlike anything alive today. The fact of extinction seemed inescapable.

Cuvier realized that he had discovered a new kind of historical evidence allowing the reconstruction of the prehuman past. At first he assumed that the extinct species formed a single ancient population that had been replaced by the living species. But the Wernerian geologists had shown that the geological deposits formed a temporal sequence, with the younger strata always overlying the older. The earth was immensely ancient, and a rough outline of the sequence of geological periods could be established. The mammoth came from superficial deposits only a few thousand years old, but many fossil species came from rocks that must be much older. Cuvier and his colleague Alexandre Brongniart realized that the fossils contained within each stratum were distinctive. They provided a much better way of correlating the relative dates of the strata than the mineral differences used by the Wernerians. Different types of rock might be laid down in the same geological period, but each period had its own distinctive population of animals. In 1811 Cuvier and Brongniart published a stratigraphy of the Paris basin, with Brongniart studying the distinctive invertebrate fossils while Cuvier described the spectacular vertebrate species (new edition 1825). They established a sequence running down through the Tertiary series to the underlying chalk deposits that formed the upper limit of the Secondary series (see fig. 13). Cuvier noted that the more ancient the deposit, the more bizarre the vertebrate fossils it contained: there was a sequence in the history of life, with the species becoming successively more and more similar to those now alive.

Unlike Lamarck, Cuvier did not believe that ancient species were transformed into modern ones. Each species remained constant throughout the period in which it lived, and then it became extinct. Napoleon's expedition to Egypt had brought back mummified animals many thousands of years old, but they were identical to those now alive. There was no sign of gradual change either in the fossil record or in the modern period. For Cuvier, species did not change, and they disappeared abruptly from the record at the end of the period in which they lived. (Modern paleontologists accept that most species do, in fact, become extinct. Only a small number evolve into something else, and identifying which ancient species were the ancestors of later ones can be very difficult.)

Cuvier outlined the geological events which had influenced the history of life in a discourse on the revolutions of the earth's surface (new edition 1825, translation 1817b; and in Rudwick 1997, originally published as the introduction to Cuvier 1812). The Paris basin revealed an alternation of salt- and freshwater deposits, suggesting major changes in the relative positions of land and sea. Cuvier was influenced by Neptunism, but he realized there had been many fluctuations in the land surface in addition to any diminution of the sea. The transition from one period of rock formation to another was apparently abrupt, because the strata were sharply defined. There was evidence of massive disruption of the earth's surface in the comparatively recent past (which later geologists would interpret as evidence of an ice age). Cuvier thus was led to assume that the geological changes were catastrophic, and that the upheavals were the cause of extinction: "Numberless

Recent Deposits:	Woolly mammoth (*Elephas primigenius*) from Siberia. Mastodon (*Mastodon americanus*) from America, and also some European species of the same genus.
Tertiary Formations:	*Palaeotherium*, several species of the same genus, a mammal unlike any known today but with vague affinities to the tapir, the rhinoceros, and the pig. (Actually from the oldest Tertiary deposits, later named "Eocene" by Lyell.)
Secondary Formations:	*Mososaurus*, a giant marine lizard from Maestricht in Germany. (From the chalk or Cretaceous deposits of the upper Secondary.)

13.   Examples of fossils described by Cuvier and their geologic relations. Examples are shown with the most recent at the top. The most recently extinct forms, such as the mammoth, are more closely related to living species than to older ones, such as *Paleotherium*. The vertebrate fossils of the secondary formations turned out to be mostly fish and (more spectacularly) reptiles, such as *Mosasaurus*, and the dinosaurs, hence this era came to be known as the Age of Reptiles.

living beings have been the victims of these catastrophes; some have been destroyed by sudden inundations, others have been laid dry in consequence of the bottom of the sea being instantaneously elevated. Their races even have become extinct, and have left no memorial of them except some small fragments which the naturalist can scarcely recognise" (Cuvier 1817b: 17).

Cuvier might have accepted miraculous creation as the only explanation of how new species were introduced to replace those so suddenly wiped out. But instead he argued that, because the catastrophes were localized in particular continents, animals survived elsewhere in the world and could then migrate in to claim the vacant territory. The "new" populations were not really new, only newly appeared in the rocks of Europe. This theory implied that, in the earliest geological period, all the species both living and now extinct had coexisted, and that elements of this vast, ancient population had been wiped out one by one in successive catastrophes. Exploration outside Europe gradually rendered this theory untenable, because there were no ancient rocks containing the fossils of living European species. Cuvier's successors were forced to confront the fact that new species appeared in the course of time to replace those that became extinct.

Other French scientists built on Cuvier's work to consolidate the catastrophist interpretation of earth history. Cuvier offered no explanation of the catastrophic earth movements he postulated. By the 1820s it was apparent that the theory of a declining sea level was inadequate; uplift by earth movement provided a far better explanation of how fossil-bearing rocks were elevated to form dry land. Most geologists agreed that these movements had been more violent in the past than anything observed today. The most obvious explanation for the higher level of activity in the past was a decline in the earth's internal temperature. It was known that the planet's interior was very hot, perhaps molten, and it was natural to assume that, as it cooled down, the solid crust would become thicker and the level of volcanic and earthquake activity diminish. This interpretation of catastrophism, advanced by Léonce Elie de Beaumont, provided a theoretically plausible mechanism of geological activity quite independent of the religious support for catastrophism common in Britain. Alexandre Brongniart's son Adolphe used evidence from paleobotany to argue that the earth was cooling down (1828). He showed that fossil plants from the Carboniferous-period rocks of northern Europe resembled those now found in the tropics, suggesting that the whole earth had enjoyed a warmer climate in the earlier geological periods. In another paper (translation 1829), he suggested that the level of carbon dioxide in the atmosphere had declined as it was fixed in the rocks to form coal. This would explain the fact noted by Cuvier that mammals appeared only in the later geological formations; only then did the air become pure enough to support their respiration.

CATASTROPHISM AND NATURAL THEOLOGY IN BRITAIN

In Britain, Neptunism had been opposed by the Vulcanism of James Hutton (1795) and John Playfair (1802). But to conservative geologists, Hutton's denial of anything resembling the catastrophic effects of Noah's flood made his gradualist system unacceptable. The catastrophism pioneered by Cuvier was welcome because it allowed the retreating-ocean theory to be replaced by an equally "directionalist" model of earth history. The claim that the earth began as a molten globe of rock provided a starting point that could be equated with the biblical creation, while the assumption that change had been more violent in the past vindicated the story of the deluge. The erratic boulders and gravel left by the Ice Age were interpreted as the relics of a great tidal wave in the recent geological past. This "diluvialist" theory could be defended on scientific grounds (Page 1969) but all too easily became associated with "scriptural geology" (Gillispie 1951; Millhauser 1954). Some clergymen tried to claim that geology supported the exact story of Genesis. But the links between catastrophism and scriptural geology can be exaggerated. As the Continental geologists showed, catastrophism was a perfectly respectable scientific theory, and even the more conservative British catastrophists postulated episodes not described in Genesis (Gould 1987;

Hooykaas 1959, 1970; R. Laudan 1987; Oldroyd 1996; Rudwick 1971). Catastrophism also provided the theoretical framework for major developments in stratigraphy.

The academic environment within which some geologists worked required them to minimize the apparent challenge their science offered to religion. In many cases, though, their scientific work made only partial concessions to the Genesis story. We can see this balance in the work of the most eminent scientific advocate of flood geology, the Reader in Geology at Oxford, William Buckland (Rupke 1983). When inaugurated to his position at this highly conservative institution, Buckland delivered an address (1820) defending geology against the charge that it undermined religion. He subsequently described in his book *Reliquiae Diluvianae* (1823) what he took to be firm evidence of a geologically recent deluge. In a cave at Kirkdale in Yorkshire, he excavated the bones of hyenas and their prey buried in hardened mud. He argued that the period in which the hyenas had lived in England had been terminated by a global deluge coinciding with a change in the climate. Such deposits are now attributed to the Ice Age, but this theory was not even proposed until the late 1830s, and in the meantime Buckland's hypothetical deluge was not unreasonable. He went too far, though, in supposing that it was universal: even Cuvier had argued for only local catastrophes, and eventually Buckland admitted that he had been wrong on this point.

For Buckland, the deluge was merely the last in a whole sequence of catastrophes that had punctuated the history of the earth, but which did not appear in the sacred record. Nor did he believe that the cause of these upheavals was supernatural. Buckland eventually accepted the cooling-earth theory as an explanation of why the violence of geological activity was diminishing. He also played a major role in expanding Cuvier's efforts to describe the extinct animals of the earlier periods. The fossils allowed a powerful extension of the most positive aspect of the new science's link with religion: natural theology.

Paley's *Natural Theology* had restated the utilitarian version of the argument from design: each part of an animal's body was useful in its mode of life, and this adaptation of structure to function illustrated the wisdom and benevolence of God. The argument gained strength from the sheer number of cases of adaptation that could be cited, and reached its climax in the eight volumes of the *Bridgewater Treatises* of the 1830s, commissioned in the will of the earl of Bridgewater to illustrate "the Power, Wisdom and Goodness of God in the Works of Creation." The treatises were meant to present the orthodox view of science, although in fact they could be read in many differ-

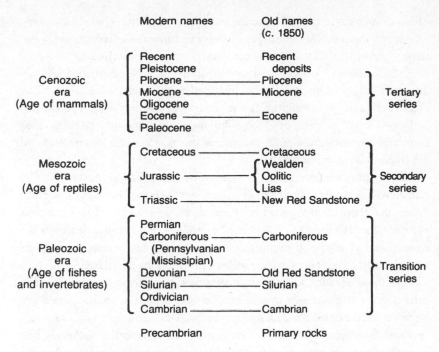

14. Sequence of geological formations. This diagram shows the sequence of geological formations established in the mid–nineteenth century and the equivalent modern sequence (adapted from Bowler 1976a). John Phillips named the three great eras in the history of life—Paleozoic, Mesozoic, and Cenozoic—in 1841.

ent ways (Topham 1998). Buckland wrote the volume on geology and paleontology (1836), applying Cuvier's techniques to show how extinct animals too had been adapted to their environment. Catastrophic extinctions coincided with abrupt changes in the environment, allowing the Creator to introduce new species adapted to the new conditions, presumably by miracles.

Although committed to creation rather than evolution, the new paleontology helped to extend knowledge of the earth's past. The use of fossils in stratigraphy greatly extended geologists' ability to identify the sequence of formations (see fig. 14). The Wernerians had identified rocks by their relative age, but it now became clear that fossils in the rocks offered the best way of fixing the sequence. Each period had been inhabited by its own unique population of living things. This point was demonstrated by Cuvier and Brongniart, and also by William Smith, the "father of English geology," whose map of the country (1815) was based on the new principles (Eyles 1969; Winchester 2001). Historians disagree over the contributions of theoreticians such as Cuvier and practical people such as Smith (who was an engineer and canal builder), but the resulting technique allowed the older and hence more distorted strata to be mapped and ordered. By the 1840s, an outline of the history of the earth as we understand it today had already emerged. Many of the geologists involved were catastrophists, including Adam Sedgwick, who taught Darwin geology at Cambridge. They expected sudden "jumps" in the fossil record which would allow it to be subdivided into distinct periods, although in practice this was never a straightforward job (Rudwick 1985; Secord 1986).

In 1824 Buckland described the first known dinosaur, a giant carnivore he named *Megalosaurus*. The term *Dinosauria* was created in 1841 by Richard Owen, and these exotic beasts have symbolized the bizarre nature of ancient life ever since (Colbert 1971; Desmond 1976, 1982). It became clear that the most recent geological formations constituted an age dominated by mammals, beneath which lay a whole series of formations populated by reptiles such as the dinosaurs, among which very few mammals were present. The term *Age of Reptiles* was coined by Gideon Mantell, who discovered the herbivorous dinosaur *Iguanodon* (see fig. 15; Dean 1999). In still older formations, there was no sign of terrestrial life at all; apparently in its early history, the earth had been inhabited only by fish and invertebrates—many of the fish being strange, armored types (described by Hugh Miller in 1841). Even the fish were absent from the oldest fossil-bearing rocks (the Cambrian) studied by Sedgwick. To most naturalists this seemed clear evidence of a direction built into the history of life on earth. Hutton's steady state worldview was refuted by the obvious progress of life from primitive invertebrates through fish, reptiles, and mammals. No human fossils had yet been found, so the progress seemed to culminate with the appearance of humankind in the modern period (Bowler 1976a, 1989a).

To those who rejected Lamarck's transmutationism, these successively appearing, ever higher types of life must have had a supernatural origin. But why did the Almighty create such a progressive sequence? Some saw a pattern in the plan of creation, with the human form as the most perfect manifestation of vertebrate life and, hence, the goal of creation (see next section). But this "transcendental, man-centered progressionism" (Eiseley 1958: 108) was not the only version. Most British naturalists felt more comfortable with Paley's utilitarian version of design. They explained the successive appearance of higher types in terms of adaptation to a changing environment. As the earth cooled and the carbon dioxide content of the atmosphere diminished, God had been able to create more advanced forms of life that could not have tolerated the earlier conditions. Buckland's contribution to

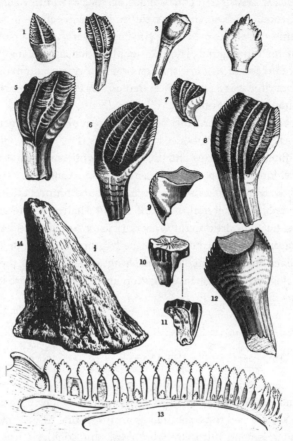

15.   Teeth of the *Iguanodon*. These fossil teeth were
discovered by Gideon Mantell, who described them in
1825 as belonging to a giant reptile similar to the
modern iguana. Richard Owen subsequently included
Iguanodon within his order of Dinosauria. The horn,
originally supposed to have been on the animal's nose,
was later found to have been situated on the forelimb.
From William Buckland's *Geology and Mineralogy
Considered with Reference to Natural Theology* (1836,
vol. 2, plate 24).

the *Bridgewater Treatises* (1836) suggested that the armor plating of the earliest fish was insulation against the high temperature of the ancient ocean. On this model, the present environment was the most perfect—a fitting habitat for humankind. The catastrophes that punctuated the earth's history were necessary to allow the creation of forms adapted to the changing conditions: this was a progression that advanced by discontinuous steps, not by gradual modification.

This explanation of the historical development of life left open the question of how the Creator introduced new species. The most obvious answer was by a sequence of miracles, and the consensus of conservative opinion favored some kind of supernatural activity (W. Cannon 1960b, 1961a; Gillespie 1979; Ruse 1975c). Yet the notion of supernatural creation did not necessarily imply the biblical image of a white-haired figure creating life from the dust of the earth. Creation was a systematic process, even if it occurred by discrete events, and it might be legitimate to talk of laws of creation. But since these laws involved the Designer's intentions, they were not a fitting subject for scientific investigation and the details were left deliberately vague.

THE PHILOSOPHICAL NATURALISTS

Some naturalists found the piecemeal approach of Paley's argument unsatisfying because it did not encourage the search for underlying laws of creation. Each species had to be counted as an individual case of adaptation to the local environment. But there were rival philosophies encouraging the view that, despite its apparent variety, nature was built according to a rational and harmonious pattern. Romanticism highlighted the sublimity of nature and its awesome powers and had a major impact on science (Cunningham and Jardine 1990). In geology and natural history, the most important manifestation came in the work of Alexander von Humboldt, who explored South America and returned to captivate Europe with his enthusiasm for nature (Botting 1973; Kellner 1963; Nicolson 1990). His personal narrative of his travels (Humboldt 1814–29) was a great influence on the young Darwin. Humboldt combined the Romantic ideal of a world based on interlocking natural powers with a determination that the interactions should be understood through measurement and scientific observation. He studied geology and what would later become known as ecology, seeking to understand how the physical conditions shaped each area's inhabitants. He encouraged the setting up of surveys and scientific voyages of exploration and was particularly influential in Britain, where this network of activities has been dubbed Humboldtian science (S. Cannon 1978: chap. 3; Nicolson 1987).

German idealists viewed the universe as a manifestation of the divine Mind and encouraged the search for an underlying unity in nature. The Romantic philosopher J. W. von Goethe sought the archetypical form of plants and speculated about a process of historical development in the vegetable kingdom (G. Wells 1967; more generally on Goethe's science, see Amrine, Zucker, and Wheeler 1987). In its most extreme form, this approach gave rise to the mystical speculations of *Naturphilosophie*, in which nature was seen as striving to perfect the human form (Lenoir 1978). Lorenz Oken (translation 1847) was the most influential spokesman of this philosophy. But even in Germany, there were more moderate philosophies which synthesized teleology and mechanism, providing foundations for ambitious research projects in biology (Lenoir 1982; Nyhart 1995). In France, Etienne Geoffroy Saint-Hilaire employed a more materialistic approach to seek unity in diversity to challenge Cuvier's authority. In Britain, a generation of "philosophical naturalists" also sought evidence of unity and harmony (Rehbock 1983). Although suspicious of Paley's natural theology, many were conservative thinkers who wanted to use the latest science to create a new version of the argument from design (Bowler 1977; D. Ospovat 1978, 1981).

The Law of Parallelism

The idealistic philosophy encouraged a developmental worldview, even though most of its supporters rejected transmutation. It was natural for idealists to see the progressive development of life toward the human form as a central theme of the divine plan. This model exploited contemporary biologists' interest in embryology, because the development of the individual organism toward maturity could be seen as another manifestation of the same progressive plan. The law of parallelism was based on the assumption that the human embryo ascended through the series of animal forms as it developed, starting as an invertebrate and then becoming successively a fish, a reptile, and then a mammal. The lower animals were, in effect, immature humans (see fig. 16). The fossil record showed that the same plan had governed the history of life on earth (Gould 1977b; Meyer 1956; Oppenheimer 1967; R. Richards 1991; E. Russell 1916; Temkin 1950). Eventually the law of parallelism became known as the recapitulation theory, in which the evolution of the species is repeated in the development of each individual organism.

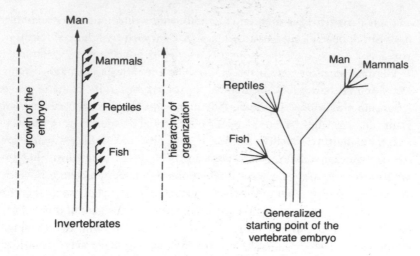

16.   The law of parallelism and von Baer's law. The law of parallelism (left) treats the stages of embryological development as a linear sequence of stages defined by the hierarchy of animal classes leading up to humanity. All embryos develop along the same path, but those corresponding to the lower animals terminate at an earlier stage in the process. Von Baer's law (right) implies no linear scale. All vertebrate embryos begin from the same point, but as they differentiate into classes, they branch off in separate directions. Further branching produces the subdivisions corresponding to orders, families, genera, and species. The human form cannot be the goal of all animal development, nor can the lower animals be treated as immature versions of the human form. The only way of defining one class as higher than another would be to show that its adult form was further removed from the initial starting point, as indicated here by the longer line drawn for the mammals.

This law was hinted at by C. F. Kielmeyer (Coleman 1973) and reached its apogee in the work of Johann Friedrich Meckel in 1821. The most active exponent of the link with the fossil record was the Swiss naturalist Louis Agassiz (Lurie 1960). Agassiz absorbed the idealist philosophy from Oken, but a spell with Cuvier in Paris taught him the necessity of tempering broad speculation with careful observation and led to his major study of fossil fish (1833–43). In 1846 Agassiz visited America and was persuaded to stay on as professor of zoology at Harvard (Winsor 1991). Having studied embryology under Ignatius Döllinger, he was convinced that the true significance of the fossil progression could be seen only when it was compared with the development of the human embryo. The teleological nature of both processes indicated their position in the divine plan. According to Agassiz, "The history of the earth proclaims its creator. It tells us that the object and term of creation is man. He is announced in nature from the first appearance of organized beings; and each important modification in the whole series of these beings is a step toward the definitive term of the development of organic life" (1842: 399). Progress was not a response to changing conditions, nor was adaptation the best indication of the Creator's wisdom. Progress was a transcendental symbol of humanity's unique position at the head of creation.

In later writings, such as his *Essay on Classification* of 1857 (reprint 1962), Agassiz qualified his acceptance of the law of parallelism, but he never wavered from his belief that the history of vertebrate life represented the unfolding of a divine plan for humanity. The chain of being was no longer accepted, yet there was a central thread running through the history of life toward the human form. The side branches of living development were merely variations on the steps in the main theme, ornamenting the plan without concealing its goal. God was a rational, almost an artistic, designer concerned more with the underlying symmetry of nature than with the details of adaptation.

This model of progress was popularized in Hugh Miller's account of the fossil fish in the Old Red Sandstone (1841). Agassiz and Miller were convinced that, in the history of life on earth, life progressed through a series of discontinuous steps of supernatural origin. The population of each period was eventually wiped out by a geological catastrophe; indeed, in 1842 Agassiz postulated another type of catastrophe—the Ice Age (translation 1967). The progressive development of life had occurred through a series of discrete steps, with no sign of a gradual or evolutionary change. Agassiz and Miller insisted that the first members of each class to appear in the fossil record were not the lowest (as an evolutionist would expect) but among the highest. When he introduced the term *Dinosauria* in 1841, Richard Owen used his reconstruction of these extinct reptiles to make the same point (Desmond 1982). The dinosaurs were, at the time, the earliest known reptiles—yet they were more highly developed than any living reptile. Transmutation was out of the question: that a step would lead directly to the highest member of a class seemed to cry out for some kind of supernatural intervention.

This support for miraculous creation owed little to biblical literalism. Ernst Mayr (1959a) links Agassiz's commitment to the fixity of species to the typological thinking that was characteristic of idealism. Individual organisms were merely copies of an underlying type, so variation within the species could never modify this fixed essence. Yet later idealists were able to accept a form of transmutation by sudden jumps, or saltations, rather than by gradual transformations. Mary Winsor (1979) argues that Agassiz's opposition to evolution was based on the apparent fixity of modern forms, coupled with the belief that each level of classification (species, genus, family, etc.) was a necessary category of analysis forced on the human mind by nature. Since those categories originated in the divine Mind, they were fixed and eternal.

New Patterns of Development

Idealism repudiated transmutation yet encouraged the expectation that regular patterns should be visible in the history of life. But the belief that the pattern must be based on a linear hierarchy was already out of date. A more sophisticated view of embryological development had emerged, one that would reinforce Cuvier's attack on the old chain of being. In the fifth scholion of his classic work on embryology, Karl Ernst von Baer attacked the law of parallelism (1828; translation in Henfrey and Huxley 1853). The lower animals were not immature forms of humanity, he asserted, because the human embryo did not develop through stages corresponding to the adult forms of the lower animals. The human embryo was never an adult fish, although there was an early stage at which a human and a fish embryo were difficult to distinguish. This was because the process of embryological development was one of *specialization:* early embryos were very generalized in structure and only later acquired the more specialized structures that identified them as a mammal or fish. Only later still did the human embryo acquire its specifically human characters. There was no single ladder of development, and to compare the development of the different classes one would have to draw a branching tree. The human form, therefore, could not be the goal of a universal trend. Higher animals might be further removed from the most general structure which appeared at the very start of embryological development, but there were many different ways of becoming more complex. There was no unambiguous hierarchy of animal forms by which progress could be measured; in some cases, animals were merely different from one another, as Cuvier had insisted.

Von Baer had no time for evolutionism, but his embryology offered a model of development that would play a major role in the emergence of Darwinism. His concept of branching development offered a new way of understanding the patterns revealed by the fossil record. An early suggestion along these lines was made by William Benjamin Carpenter (1851), and the idea was taken up with enthusiasm by Britain's most influential anatomist, Richard Owen (Desmond 1982; D. Ospovat 1976, 1981; Rupke 1994). Owen rose to prominence because he appealed to the conservative scientific establishment as someone who could modernize the traditional concepts of natural theology. In the 1830s he campaigned vigorously against radical naturalists trying to develop Lamarckian ideas of evolution. We have already noted his claim that the dinosaurs showed the ascent of life to be a completely discontinuous process. But in the 1840s, he adopted a more flexible approach which allowed him to explore the possibility that developmental trends—modeled on von Baer's concept of embryological specialization—might be visible in the fossil record.

Owen argued that the best way of illustrating the Creator's wisdom was not to stress individual cases of adaptation but to seek out the underlying unity among animal forms. In his *On the Archetype and Homologies of the Vertebrate Skeleton* (1848), he described an idealized vision of the simplest, least differentiated vertebrate form. This was an imaginary creature that exhibited the essence of the vertebrate type without any of the specializations possessed by individual species. The archetype was the foundation of the Creator's plan, and this idealist vision helped Owen to formulate the important concept of homology. He distinguished between analogies (unrelated organs that have a superficial adaptive resemblance) and homologies (different adaptive modifications of the same organ). Thus, almost every bone in the human arm could be identified with an equivalent (but differently modified) bone in the wing of a bat or the paddle of a whale. Such homologies are of great significance in classification and, in Owen's eyes, illustrated that the world was the product of rational design rather than chance. He expounded this new version of natural theology in his *On the Nature of Limbs* (1849), which he concluded with a brief indication of his belief that there might be divine laws controlling the development of life on earth.

Owen's viewpoint was consistent with von Baer's embryology: the vertebrate archetype corresponded to that point in the development of any vertebrate embryo in which the basic character of the type had only just become apparent. The different forms of specialized development produce the homologous relationships within the type. Owen soon realized that this was exactly the pattern revealed by the geological history of groups with a good fossil record. The earliest mammals, for instance, were very generalized forms, while the group's later history revealed divergent lines leading toward the highly specialized modern species (Owen 1851, 1860). Although Owen still believed that the human body was the most perfect expression of the vertebrate form, he no longer saw it as the goal toward which the whole history of life was aimed. The pattern of divergence and specialization was exactly what Darwin's theory predicted, and indeed Darwin cited Owen's work for support. Yet Owen attacked the *Origin of Species* and has subsequently been pilloried as an opponent of evolution. In fact, he had moved a long way toward an evolutionary position by the 1860s, but his evolutionism was a non-Darwinian form that saw the history of life as the gradual unfolding of a divine plan (MacLeod 1965; E. Richards 1987; for a contrary view, see Brooke 1977).

Owen was not the only philosophical naturalist thinking about the significance of the relationships between species (Rehbock 1983). The insect taxonomist William Sharpe MacLeay introduced a circular, or quinary, system of classification in which species were grouped in circles of five. This was a highly artificial arrangement that could be explained only as the product of a rational divine plan. Edward Forbes saw mysterious patterns of polarity governing the rise and fall of groups in the fossil record. More fruitfully, he looked for geographical explanations of how species had migrated to their present locations, anticipating Darwin in the explanation of the impact of the Ice Age on biogeography (Browne 1983). Taxonomists who looked for a natural system of arrangements between species began to see nature represented as a branching tree, thus paralleling Owen's ideas on the divergence seen in the fossil record (Di Gregorio 1982).

RADICAL SCIENCE

Owen's early opposition to transmutation had a specific target. Older histories (e.g., Gillispie 1951) presented the opinions of the scientific elite as though they were the only theories available at the time. There were disputes about how best to illustrate the Creator's wisdom, but no serious challenge to the view that the universe was a divine artifact. Yet throughout this period, those in high positions within the scientific community were looking over their shoulders at a threat they dared not name. We have already noted the extent to which the conservative establishment was under threat from radicals promoting revolutionary political—and scientific—ideas. Working-class revolutionaries and middle-class activists were looking for a degree of influence that would match their contributions to the newly industrialized economy. These debates frequently employed scientific ideas, and many of the ideas described above were developed quite explicitly to resist the threat of materialistic theories proposed by the radicals.

Evolutionism became part of the radical campaign to discredit the old worldview which propped up aristocratic privilege. The claim that God had designed a hierarchical universe in which everyone should keep to their allotted place was used to bolster the position of the upper classes. Both the radicals and the less strident middle-class activists saw a universe which changed through time as evidence that human conventions such as the class system could be changed. The opponents of natural theology seized upon the views of Lamarck and used them to promote the idea that progress was inevitable. Efforts were made to show that the human mind was merely a by-product of the material processes taking place within the brain. Even astronomers were proposing theories that the solar system was formed by a natural process. Materialism thus presented a concerted challenge to the conventional worldview, although opponents resisted the challenge, often successfully, during the early nineteenth century.

Historians used to assume that Cuvier's opposition had blocked acceptance of Lamarck's theory. We now know that radical French naturalists continued to develop transformist ideas along Lamarckian lines in the 1820s and 1830s (Appel 1987). A new approach to anatomical relationships proposed by Etienne Geoffroy Saint-Hilaire also challenged Cuvier's theory. In his *Philosophical Anatomy* (1818–22), Geoffroy opposed Cuvier's system of classification and declared that underlying unities between divergent forms should be sought—including between Cuvier's four basic types. Although similar in its results to the idealistic perspective, Geoffroy's was a materialistic philosophy in which resemblances were the product of natural forces. He studied the production of monstrosities to show how disturbances of the developmental process could produce new characters. In the next decade he applied his techniques to paleontology, arguing that the appearance of new species could be explained in terms of changing environments triggering modifications of development (Appel 1987; Bourdier 1969). Unlike Lamarck's theory, this was evolution by saltation, or sudden leaps: Geoffroy thought environmental pressures would produce dramatic transformations that would establish new species instantaneously.

In Britain, both Lamarckism and the new philosophical anatomy were taken up by radical thinkers trying to undermine Paley's argument from design. Even among working-class agitators, transformist ideas were rife (Desmond 1987). But the fiercest battles were fought within the medical profession (Desmond 1989). The old leadership, allying itself with the conservative naturalists at Oxford and Cambridge, kept control in the hands of the upper classes. Middle- and even working-class students, however, were demanding changes that would allow them to study medicine. They found Lamarckian transmutation and the new philosophical anatomy to be the ideal tools by which to undermine the credibility of the establishment's theories and hence, by implication, its authority over the profession.

The flamboyant anatomist Robert Knox promoted transcendental anatomy and radical political views until discredited by the revelation that he had been buying cadavers to dissect from the murderers Burke and Hare (E. Richards 1989a). An anonymous 1826 article linking Lamarckism to the progression observed in the fossil record may have been written by the Wernerian geologist Robert Jameson (Secord 1991). This article was once attributed to the radical anatomist Robert Edmond Grant, who exposed the young Darwin to his Lamarckian ideas at Edinburgh University. Grant's transmutationist views on invertebrate zoology almost certainly influenced Darwin's early work (Sloan 1985). Grant later moved to the newly founded University College in London, where he came into increasing conflict with Richard Owen, who was based at the Royal College of Surgeons. Owen intended his antievolutionist arguments of this period to discredit Grant, just as his new version of the argument from design based on unity was meant to show how the old worldview could be modernized to accommodate philosophical anatomy. In the short term, the conservatives won the battle and marginalized Grant within the scientific community.

Another materialist threat came from a new approach to psychology called phrenology (Cooter 1985; Shapin 1979; Young 1970a). A discipline founded by Franz Joseph Gall, phrenology was taken up in Britain by the Edinburgh writer George Combe. Phrenologists argued that the brain was the organ of the mind: there was no purely spiritual agent, or soul, responsible for mental functions. In this respect they carried on in the spirit of eighteenth-century materialism; but phrenology claimed scientific credentials because anatomists such as Gall thought they could identify those parts of the brain responsible for particular mental functions. Since it was supposed that the shape of the brain followed that of the skull, phrenologists offered to reveal any person's character simply by feeling the bumps on his or her skull. For Combe and many middle-class writers, this was a reforming science offering a path to social progress by allowing everyone to identify their skills. But it was also materialistic and was feared as such by the conservative establishment. Once again, the ranks of the scientific profession closed against the radicals, and phrenology was marginalized as a pseudo-science. The phrenologists' claims were indeed bogus in the sense that the skull does not reveal the shape of the brain beneath, yet their basic assertion, that mental functions could be localized in specific areas of the brain, would be confirmed by later discoveries. The marginalization of a field which had genuine potential to guide research has been seen by historians as a classic case of social values influencing the development of science.

Phrenology pioneered a materialistic view of human nature without the suggestion of an evolutionary link between humans and animals, although that link was soon proposed. Other areas of science also witnessed the emergence of materialistic theories, including cosmology. In the eighteenth century, the "nebular hypothesis" of Immanuel Kant (translation 1969) and Pierre-Simon Laplace (translation 1830) challenged Buffon's theory of the origin of planets. On this model, the sun and planets condensed out of a rotating cloud of dust particles under the pull of gravity (see fig. 4). It was precisely the theory the radicals needed to argue that natural laws could produce progress rather than stability: if the physical universe itself progressed toward higher states of complexity, why not human society? Another Edinburgh writer, John Pringle Nichol, popularized the nebular hypothesis in this context during the 1830s (Numbers 1977; Schaffer 1989). He was opposed by conservative astronomers who dismissed the theory as pure speculation. Telescopic observations of nebulae in the heavens were invoked by Nichol as evidence that other planetary systems would be seen in the process of condensation. The conservatives used information gathered by means of Lord Rosse's great new telescope at Birr, in central Ireland, to argue that the clouds were actually collections of stars (galaxies, in modern terminology).

It was no accident that Edinburgh was the center for much of this radical intellectual activity. The academic community there was not directly under the control of the churches, and middle-class political activity was rife. Many of the new ideas and theories promoted in Scotland were resisted actively when transferred south of the border, as Grant's marginalization shows. Yet the theories promoted by conservative English naturalists such as Buckland and Owen can be properly understood only if they are seen as reactions against the constant threat of political and intellectual innovation by the middle and even lower classes.

THE PRINCIPLE OF UNIFORMITY

There was a less radical attack on conservative science by a geologist whom Darwin always acknowledged as his mentor, Charles Lyell. All the theories discussed above were based on the assumption that the earth and its inhabitants had undergone a process of cumulative change through time. There was an arrow or direction to history, symbolized by the successive appearance of the vertebrate classes in the course of geological time. Lyell's uniformitarianism revived the methodology used by James Hutton to undermine the directional model of history. Theirs was essentially a steady state worldview in which one geological period was much the same as another, with the earth maintained forever by a perfect balance of creative and destructive forces. Because Lyell insisted that scientific geology should be based solely on observable causes acting at modern-day intensities, with no catastrophic interruptions, he has been hailed as a founder of modern geology. One advocate of this position has suggested that historians' efforts to paint a less negative image of the rival catastrophists has gone too far (L. Wilson 1967, 1969, 1972, 1980). As Darwin himself recognized, Lyell's methodology was a step forward, because the catastrophists had tended to invoke unknown (although not miraculous) interruptions in the natural order of things. Yet complete support for the steady state worldview was impossible to sustain then and remains so today: in its early history our planet was quite different from what it is today. The element of directionalism inherent in the catastrophists' position was valid, even if they exaggerated the catastrophes. The principle of uniformity was a major development in nineteenth-century science, but its influence must be evaluated with care (W. Cannon 1960a, 1961b; Fox 1976; Hooykaas 1957, 1959, 1966; Rudwick 1970, 1971).

Lyell's Geology

Lyell came from a wealthy Scottish family and stood firmly in the liberal camp. He wanted a society free of aristocratic privilege in which the middle classes could make their way forward, and his attack on scriptural geology was motivated by this ideology. Trained as a lawyer, he became interested in geology in the 1820s but soon turned against Buckland's catastrophism and led a successful campaign against the identification of the last catastrophe with Noah's flood. An important influence on Lyell was George Poulett Scrope's work on the extinct volcanoes of central France (1827; see also Rudwick 1974a). Scrope accepted the cooling-earth theory but applied the method of "actualism," which attempted to show that most geological formations could be explained as the product of forces still in operation. Volcanic activity in the distant past might have been more intense than today, but the French volcanoes showed evidence of having been built by occasional lava flows interspersed with periods of calm in which the exposed surface had eroded to form valleys. The whole process must have required vast amounts of time, and this was the lesson that Lyell learned: normal causes could produce great effects if only they had enough time. But Lyell went far beyond Scrope in one important respect. Actualism relies solely on observable causes to explain the phenomena but allows for fluctuations in the intensity of those causes over time. Lyell proposed the method of uniformitarianism: the rate of all natural changes is presumed to be absolutely uniform through time. Only observable causes acting at observable intensities can be used to explain past events.

Lyell tested this hypothesis by investigating the greatest active volcano in Europe: Mount Etna in Sicily. He showed that this had been built by a series of eruptions no larger than those in recorded history. To raise so enormous a cone must have taken vast amounts of time, yet the volcano rested on sedimentary rocks that Lyell was eventually able to show were very young. If Etna stood on recent sedimentary rocks, then it was geologically very young—yet it was evidently of vast antiquity by the standards of human history. The geologically young sedimentary rocks beneath must be even more ancient in human terms. How much more extensive, then, must be the geological history of the earth as a whole? If the imagination could take this leap, it would be possible to explain away all the supposedly catastrophic events that had been evoked to explain large geological changes. Just as the immense structure of Etna had been built up slowly and gradually, so might other geological formations be explained as the results of the accumulated effect of limited causes acting over vast periods of time. Convinced of the validity of this view, Lyell returned to England and began writing his classic *Principles of Geology* (1830–33).

Lyell set out to reform geology's scientific method (R. Laudan 1982). His three volumes were full of detailed information about the geological changes observed during human history, as well as ingenious arguments to show how, given sufficient time, such forces could produce major geological formations. In the first volume, Lyell argued that all advances in geological science have been made by those theorists who cited only known causes. The chief obstacles to scientific progress have been the willingness to speculate about unknown causes in the past, and the refusal to contemplate a massive extension of the history of the earth. Lyell thus tried to label catastrophism as unscientific, and his interpretation has influenced the history of science ever since. But his position has to be treated with caution. He implied that the catastrophists invoked supernatural causes, although most were now firmly committed to the cooling-earth theory. Some catastrophists admired Lyell's skill in showing how earthquakes and erosion could account for features of the earth's surface, but they could see no point in arbitrarily limiting natural forces to the intensity observed in the short span of human history. Everyone admitted that natural laws operated uniformly, but the same principle could not be extended to the complex effects shaping the earth's surface. If the interior of the earth changed by cooling, for example, then geological forces might decline—with the whole process being governed by uniform physical laws.

For Lyell's methodology to be applicable to all geological structures, he had to postulate an earth maintained indefinitely in a steady state. He had to reject not only catastrophes but also the element of directionalism built into the catastrophist synthesis. No part of the earth's crust, however ancient, could have been built under conditions differing from those of today. And to speculate about the origin of the planet itself was to go completely beyond the bounds of science. The earth had to be a perfectly self-regulating system that had maintained itself throughout the period into which inquiry is meaningful. Lyell thus revived Hutton's system, in which the slow elevation of mountains by earth movement exactly balanced the destructive effects of erosion. For Lyell, who was a Unitarian rather than an orthodox Christian, this ahistorical view of the world gave a better picture of the Creator's relationship to the universe. God was the perfect workman, and His creation was held in an eternal balance.

Historians have debated the relative significance of the two aspects of Lyell's position: his uniformitarian methodology and his steady state worldview. His opponents were aware of his opposition to directionalism. They asked how the earth, as a hot body, could be expected to maintain its internal temperature indefinitely, and Lyell had no satisfactory answer. Such criticisms increased as the science of thermodynamics became better established, and by the 1860s one of the leading physicists, William Thomson (later Lord Kelvin), was ready to launch a major attack on Lyell's steady state worldview (see chap. 7; Burchfield 1975; Smith and Wise 1989). Modern scientists know that radioactivity supplies a heat source capable of maintaining the earth's internal temperature for billions of years, thus vindicating some aspects of Lyell's methodology. But we now seek to investigate the earth's origin and early history, a project that Lyell would have rejected as unscientific. His uniformitarianism is a useful guide to the earth's later history, but he extended it far beyond the limits accepted by modern science.

The steady state theory did not imply that the earth had to be exactly the same in every period of its history, only that changes had to be small and noncumulative. Lyell was aware of evidence suggesting that some earlier periods had enjoyed climates warmer than the present, but argued that this did not support the theory of a declining internal temperature. He suggested that earth movements brought about gradual changes in geography which might affect the climate (D. Ospovat 1977). Since the ocean reflects much of the sun's heat, a period when most of the oceans were located in the tropics would be relatively cool. Conversely, a period when most of the land was in the tropics would experience a warmer climate. The warm, humid conditions thought to prevail during the Age of Reptiles were a temporary product of the endless cycle of uplift and erosion. On Lyell's theory, the Age of Reptiles might return in the future, a possibility that one of his critics ridiculed by drawing a cartoon in which a reptilian professor lectures on a fossil human skull (Rudwick 1975).

Lyell's Nonprogressionism

Lyell was also determined to show that the evidence for a progressive trend in the history of life was unsatisfactory. He maintained that the highest class, the mammals, might have existed throughout the whole period accessible to us in the fossil record. They might have been rare in periods such as the Age of Reptiles, but the fact that no one had discovered mammalian fossils in rocks dating to this period did not prove that no mammals existed. Here, Lyell was making an important point upon which Darwin would build: the fossil record is incomplete and imperfect—only a small proportion of the species that have ever lived will have left their remains in the rocks. In fact, a few mammalian fossils were soon discovered in rocks belonging to the Age of Reptiles, apparently vindicating Lyell's point. He continued to predict that they might also be found in even the oldest fossil-bearing rocks, and although he gained some comfort from later discoveries (Lyell 1851), these turned out to be misidentifications. Following the publication of Darwin's theory, even Lyell conceded that the evidence for the progressive development of life was unassailable.

For Lyell, then, the history of life was also in a steady state, with one exception: the appearance of the human species. His religious beliefs convinced him that human spiritual faculties were distinct from the mental powers of animals, so the origin of humanity had to be a unique, and presumably recent, event (Bartholomew 1973). Given Lyell's commitment to slow, gradual change, we can see why he was so anxious to oppose the prevailing belief in the progressive development of life: the combination of continuity and progression would make the appearance of humanity merely the last step in the gradual ascent of life. Lyell's desire to rule out this possibility was apparent to the conservatives, and the uniformitarian-catastrophist debate thus avoided the acrimony surrounding early suggestions of an evolutionary link between humans and animals.

The catastrophists assumed that whole populations of animals and plants were wiped out suddenly by great upheavals, after which a supernatural agency created their replacements. Lyell brought forward evidence against sudden extinctions: a significant proportion of species almost always reappeared in the rocks of the following period. He believed that, as conditions gradually changed, species migrated to follow the climate they preferred, and where that was not possible they declined to extinction. In his gradualist system, extinctions must happen all the time. But where did the new species come from to replace them? Must not the creation of species also be gradual and commonplace? Darwin applied Lyell's gradual-istic methodology to the organic world, postulating a natural process that would modify species to adapt them to changing conditions. For reasons noted above, Lyell could not go down this route. He rejected Lamarck's the-ory of transmutation, arguing that, since there had been no observed mod-ification of species in the course of human history, the basic form of each species must be immutable (see Lyell's notebooks, published in 1970; see also Coleman 1962).

Lyell's opinions on the origin of species exploited a deliberate ambiguity favored by many of his contemporaries. He accepted that new species had to appear from time to time in the natural course of events, but he retained the traditional view that the adaptation of these species to their new environ-ment was a sign of the Creator's benevolence. In an exchange of letters with the astronomer Sir J. F. W. Herschel in 1836, he declared his support for in-termediate causes in the origin of species (in K. M. Lyell 1881: 476). This would imply a lawlike process rather than miracles—yet Lyell did not accept the transmutation of species by natural laws. By modern standards, his po-sition seems to evade the real issue, although the notion of vaguely defined laws of creation would become commonplace among scientists trying to dis-tance themselves from the biblical notion of miracles without subscribing to transmutationism.

THE VESTIGES OF CREATION

Lyell's uncomfortable position on this issue, along with Darwin's reluctance to publish his own theory, must be understood in the context of the reputa-tion acquired by transmutationism during the era of radical scientists such as R. E. Grant. Endorsement by radical scientists allowed the theory to be branded as dangerous materialism, subversive of the moral as well as the in-tellectual order. The first major effort to emancipate the theory from this image was made by another Scottish writer, Robert Chambers, in 1844. Chambers came from a middle-class background, and his publishing house was active in promoting the political interests of this group. He welcomed the idea of progress and saw development in nature as the most convincing evidence that political progress was also inevitable. The trick was to show that recognition of progress up to and including the appearance of the human species did not undermine the religious beliefs still held by most respectable Victorians. Chambers sanitized the radical science tradition by using the notion of laws of creation to argue that evolution toward humanity was merely the unfolding of a divine plan. He challenged the authority of the professional scientists by going over their heads in a direct appeal to the reading public. The result was a debate which rocked the foundations of Victorian opinion and paved the way for the reception of Darwin's theory (Lovejoy 1959c; Millhauser 1959; Secord 1989; Yeo 1984; for a collection of primary sources, see Lynch 2000). A recent study by James Secord (2000) presents the debate not as a mere prelude to the Darwinian episode but as the defining factor in shaping the Victorian public's attitude toward evolution.

Chambers brought his scientific material together in an anonymously published book entitled *Vestiges of the Natural History of Creation*, published in 1844 (reprinted with additions as Chambers 1994). A response to the resulting outcry was published under the title *Explanations* (also in Chambers 1994). Although the book promoted the general idea of what we would call evolution, Chamber's concept of how the process worked did not anticipate the theory that Darwin was working on (Hodge 1972). Darwin provided a natural mechanism that adapted species to a changing environment. Chambers stressed progress rather than adaptation and invoked only the vague notion of "creation by law."

Vestiges began with the nebular hypothesis, which offered a natural (if controversial) explanation of how physical laws could generate the complex structure of the solar system (Ogilvie 1975). To explain the origin of life on earth, he invoked spontaneous generation, citing as evidence some soon-to-be-discredited experiments in which electricity was supposed to have produced small insects. Here, Chambers tapped into a popular enthusiasm for the power of electricity, including its role in affecting living things (Morus 1998). Up to this point, the theory followed an essentially materialist program, but the progress of life from simple to complex proved more difficult to explain. Chambers devoted a third of his book to a survey of the fossil record intended to show that, despite missing evidence, it lent overwhelming support for a gradual ascent to the human species. Each class began with its lowest form and then advanced. The first fish were primitive cartilaginous types, whose lack of a bony skeleton proclaimed their origin in the invertebrates. The reptiles (as Owen had argued) were difficult to fit in, but the mammals showed the same progression from the primitive creatures of the Mesozoic, which so interested Lyell, to modern types. Although Chambers was aware of von Baer's branching model of development, he thought that the branches split off from a main line that progressed to humanity. The belief in a rational pattern of creation was reinforced with a discussion of MacLeay's quinary system of classification.

Chambers paid little attention to adaptation but linked his emphasis on progress to the embryological analogy and the law of parallelism. Because there was a main line of development, the human embryo did indeed pass through phases in which it was successively a fish, a reptile, and then a mammal (see fig. 17). But unlike Agassiz, Chambers saw a real relationship between the embryological and the fossil series. A species could advance a step up the scale by simply extending its growth period a little, maturing at a point corresponding to an increased level of complexity. But the embryological development of the new species would still pass through the old form on the way to the new: thus individual development would recapitulate the evolutionary history of the species. Later editions of *Vestiges* postulated multiple lines of evolution all following roughly the same developmental pattern. Some strange genealogies were suggested on the assumption that each line began with an aquatic form. Dogs, for instance, were supposed to have evolved from seals.

Chambers did not believe that the progression was an absolutely continuous process: species were real, and the extensions of development came as small saltations, or jumps, from one to another. The jumps were possibly triggered by environmental stress, although the details were never worked out. Nor did he offer what the materialists would have regarded as an adequate explanation of how the additions to growth were directed. They were not adaptive, and his assumption seems to have been that the overall pattern of development was somehow preordained. Extending the period of development merely progressed the species one step farther up a hierarchy which was somehow built into the very constitution of nature. In this sense, his theory could be taken as a contribution to the argument from design: evolution was the gradual unfolding of the divine plan of creation. The crucial difference between his theory and the one proposed by Agassiz was that the advances did not require miraculous intervention, because God had woven the necessary laws of creation into the fabric of the universe.

Chambers sometimes implied that these laws operated within the normal course of events. He cited a popular superstition that oats, if plowed into a field and left to overwinter, would grow the following year as rye. But other discussions imply that the laws of creation were not normally observable. During the acts of transmutation, a "higher" law stepped in and overrode the normal forces of nature. Such rare interruptions could be seen as little more than predesigned miracles, but Chambers had a good authority to quote in favor of the idea that God had built such higher laws into the

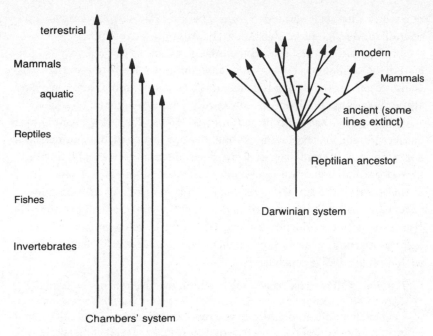

terrestrial

Mammals

aquatic

Reptiles

Fishes

Invertebrates

Chambers' system

modern

Mammals

ancient (some lines extinct)

Reptilian ancestor

Darwinian system

17. Chambers's system of linear development. In the modern Darwinian system (right), all mammals, living and extinct, are derived from one reptilian ancestor which first developed primitive mammalian characters. In Chambers's system (left), the various mammals are not directly related to each other. They consist of a series of separate lines of development which have independently reached approximately the same point on a linear scale. This is similar to Lamarck's scheme of progress, although Chambers was not clear on why some lines have lagged behind others. He largely ignored adaptation, although he assumed that aquatic forms precede terrestrial ones in the sequence. How the earliest aquatic mammals were related to reptiles was never explained.

hidden structure of nature. In his unofficial *Ninth Bridgewater Treatise* (second edition 1838), Charles Babbage had argued that apparent miracles could be the product of such higher level laws. He had invented a "calculating engine"—a mechanical forerunner of the computer—and noted that he could program his machine to change its operations according to a predetermined plan. To the casual observer, the changes would suggest that someone had interfered with the machine's operations, but in fact the event had been built in from the beginning. Chambers's series of transmutations was ideally suited to this interpretation: it was unpredictable to us, yet guided by an unseen plan which could override the normal laws of reproduction.

Had *Vestiges* merely postulated such a predesigned law of creation, it might have been tolerated even by conservative thinkers. But one implication which Chambers refused to conceal damned his book in the eyes of the scientific and religious establishment. Because the law of creation was progressive, it naturally led toward the human species as its goal: just as Lyell feared, the combination of progress and continuity led to the human race being treated as merely the last step in the ascent of animal life. Chambers argued openly that humanity's supposedly unique mental and moral characters did not mark us off from the animals—in fact they were faculties that had been gradually expanding throughout the ascent of life. Humans did not stand above natural law: we were the products of a law-bound universe. Our great intelligence was a function of increased brain size, a key feature of the ascent of the animal hierarchy. Chambers seized on the teachings of phrenology to justify the claim that the brain was the organ of the mind. He also cited statistical evidence to suggest that human behavior in a social context was essentially lawlike and predictable, however much we might think we have free will. He concluded:

> The sum of all we have seen of the psychical constitution of man is, that
> its Almighty Author has destined it, like everything else, to be developed
> from inherent qualities, and to have a mode of action depending solely
> on its own organization. Thus the whole is complete on one principle.
> The masses of space are formed by law; law makes them in due time the-
> atres of existence for plants and animals; sensation, disposition, intellect,
> are all in like manner developed and sustained in action by law. . . . The
> inorganic has one final comprehensive law, GRAVITATION. The organic,
> the other great department of mundane things, rests in like manner on
> one law, and that is,—DEVELOPMENT. (1844: 359–60)

Vestiges sold well, and the conservatives were incensed that its materialistic heresies were being communicated to a wide audience. The geologist Adam Sedgwick wrote a rambling critique for the *Edinburgh Review* (1845) proclaiming the need to protect "our glorious maidens and matrons" from such poisonous ideas. Hugh Miller, who had already challenged the development hypothesis in his *Old Red Sandstone*, wrote another book, *Footprints of the Creator* (third edition 1850), to refute this latest version of the heresy. The status of the human soul was the real stumbling block: Miller would have been willing to accept the idea of creation by law if it did not link humans and animals—but if that was the theory's implication, then it had to be rejected. If the human species was a miraculous creation, then so were all the others.

The first edition of *Vestiges* was easy to attack because of its scientific blunders. Chambers modified the later editions in response and wrote his *Explanations* to defend his position. The image of continuous progress in the fossil record presented by *Vestiges* was not consistent with the evidence available at the time. As Miller and Sedgwick insisted, the record showed sudden leaps to entirely new levels of organization. Some of these leaps remain unaccounted for today (to the delight of the creationists), but in the 1840s the evidence was less extensive, and the gaps appeared more substantial. Darwin took careful note of Sedgwick's review because it pinpointed the weak spots in the argument for transmutation (Egerton 1970b). T. H. Huxley, later one of Darwin's chief supporters, wrote a critical review of a later edition of *Vestiges* (Huxley 1854). He objected to the vagueness of the law of creation which still allowed for an element of design, but also stressed the theory's incompatibility with the fossil record.

The situation improved in the 1850s with the discovery of new fossils bridging some of the gaps in the record (Bowler 1976a). Scientists with more liberal opinions began to voice support for the idea of creation by law. Richard Owen wrote of a law of development in 1849, although he did not follow up the suggestion, because his conservative backers objected. Baden Powell, professor of geometry at Oxford and a noted commentator on science and religion, ridiculed the idea of miraculous creation and argued that development by law was a better illustration of the Creator's power (1855; see also Corsi 1988b). Some of the more philosophical naturalists began to suspect that the natural relations between species might be a product of the laws governing their production. The botanist Hewett C. Watson recognized the effect of changed geographical conditions on species and commented favorably on Chambers's effort to explain the origin of species by law (Egerton 1979). In Germany, too, the idea of continuous natural development was raised (Lovejoy 1959d; Temkin 1959). The paleontologist H. G. Bronn proposed a law of divergence similar to Owen's (translation 1859).

At one level, these developments paved the way for Darwinism. Naturalists had begun to think about the possibility that species were not individual creations but might constitute a system whose structure was determined by lawlike generative processes. Even the concept of transmutation, long damned as dangerous materialism, had been introduced into polite society by Chambers and his supporters. Yet there is little evidence that any concerted effort was under way to produce a theory of natural evolution—except by Darwin himself. The philosopher Herbert Spencer was one of the few active proponents of Lamarckism in the 1850s, and he published the essay, "The Development Hypothesis," in 1853 (reprinted in H. Spencer 1883: 1:381–87). But few naturalists were prepared to reconsider this alternative. For the most part, speculations about laws of creation were confined, as in *Vestiges*, to processes that still depended on the unfolding of a divine plan. The scientific world was trapped by an inability to explore certain avenues, because earlier speculations had rendered them unpalatable, even though the evidence for some form of natural development was becoming stronger. Darwin's theory came as a shock because it proposed an entirely new mechanism of natural transmutation, a theory whose implications were even more radical than those of Lamarckism. In the end, these more radical implications were evaded by those who converted to Darwinism. Yet Darwin's initiative broke the logjam that had built up within science as the pressure to explore ideas of natural development ran up against the blockage caused by the lack of any plausible mechanism.

Once Darwin broke this logjam by forcing scientists to reconsider, the tide then swept through to create a new worldview based on the idea of progressive evolution. The historian thus is forced to analyze these debates at various levels. Considered purely as a contribution to science, Chambers's theory was of little value—and it was widely criticized even by radicals like Huxley. But as Secord (2000) has shown, considered as a contribution to the public debate, *Vestiges* may have been even more important than the *Origin of Species*. It aroused interest in evolutionism, making it much easier for Darwin to get a hearing. But the progressionism of *Vestiges* also shaped the climate of opinion into which Darwinism was incorporated. Once the middle classes were forced to face up to the idea of natural evolution, they accepted it—but only so long as it seemed to uphold their faith in the inevitability of progress. The more dangerous aspects of Darwin's theory were temporarily shunted aside (Bowler 1988). Our modern concerns make us naturally anxious to understand the creation and reception of Darwin's theory, but the historian interested in the impact of evolutionism on nineteenth-century thought might do better to write a continuous narrative sweeping from *Vestiges* to the *Origin* debate.

5   The Development
of Darwin's Theory

While the debates of the 1830s and 1840s raged, Charles Darwin was developing a theory of evolution that was to revolutionize the whole situation when it was published in 1859. Within a few years of the publication of the *Origin of Species*, the idea of evolution was widely accepted. Darwin became one of the most famous scientists alive: when he died in 1882, he was accorded the unusual honor of being buried in Westminster Abbey (Moore 1982). Since then, his status as a hero of scientific discovery has continued to grow, especially following the emergence of modern Darwinism in the mid–twentieth century, which linked natural selection with Mendelian genetics. There are some who would now hail natural selection as the most important idea ever to be proposed, one that utterly transformed our worldview. It is no surprise that Darwin is the focus of intense public interest and the subject of much historical research. Yet to those who feel—for religious, philosophical, or moral reasons—that the theory has had harmful rather than beneficial effects on our culture, Darwin himself is more a villain than a hero. The originator of so controversial a theory must necessarily be a subject of debate rather than consensus among historians.

The highly polarized nature of the modern debates encourages both sides to focus on Darwin as a key figure in the development of modern science and culture. Not surprisingly, those who are more sympathetic to his views have played the greatest role in uncovering the details of Darwin's life and work. So great has been the level of historical interest that Michael Ruse coined the term *Darwin industry* to denote the group of historians who devoted themselves to the study of his work (Ruse 1974; other surveys include J. Greene 1975; Kohn 1985; and Loewenberg 1965). The amount of information now available is truly intimidating to the novice. In the late nineteenth century, the traditional Victorian "life and letters" was published by

Darwin's son (F. Darwin 1887) followed by a supplement (1903). Charles Darwin's autobiography, included by his son in a much edited form, was eventually published in its entirety (1958). There are now editions of Darwin's diaries (1988), notebooks (1987), and the marginalia he scribbled in the books he read (1990a), as well as an ongoing project to publish his whole correspondence (1984–). The *Origin of Species* is available as a variorum text listing all the changes to later editions (1959), while both the *Origin* and the notebooks have been provided with a concordance (1981, 1990b).

When Ruse first wrote of the Darwin industry, many detailed analyses of how Darwin developed his theory were being published. The flow of technical literature has now lessened as historians of science have turned to other topics. Theoretical revolutions are no longer fashionable: it is the emergence of new disciplines and research programs which now attracts attention, and here it is much less easy to pin down Darwin's influence. Even so, the flood of biographical studies has continued unabated, including one offering from this author (Bowler 1990).

There are also wider studies reflecting the fact that the Darwinian revolution took place both within science and within Western culture as a whole. Some historians are concerned both with the detailed evidence and arguments Darwin offered in biology and with his influence on the modern life sciences. Others are more concerned with the effects of his theory on religion, philosophy, morality, social thought, and even literature and the arts. Those who look to the sciences may have little interest in the broader issues, while those who come from the history of ideas or social history may have little patience with the complex details of evolutionary biology. Yet Darwin not only was a scientist but also took a lively interest in the social world in which he lived, so to understand his work we must be willing to take both areas into account.

The greatest division within the various interpretations of Darwin's life and work is between those which portray his theory as a major step toward a more rational view of the world and those which depict it as a force that has undermined our religious and moral values. But neither of these two camps is homogeneous; the opponents in particular fall into two mutually hostile factions. Some historians of ideas take a negative view of the revolution and tend to judge Darwin and his work harshly. To writers such as Jacques Barzun (1958) and Gertrude Himmelfarb (1959), for example, Darwin's theory was ill conceived and unconvincing, and its success must be explained as the influence of a pernicious materialist philosophy on science. The survey by John Greene (1959a) offers a more subtle account, yet still is driven by a Christian perspective which finds the materialism of Darwin's theory distasteful. But there has been a strong tradition of opposition to Darwinism based more on ideological than on religious values, which includes some whose left-wing political sympathies are most unlikely to coincide with Christian belief. The Marxist historian Robert M. Young (1985), building on the long-standing suspicion that the selection theory reflects the competitive ethos of Victorian capitalism, has undertaken a sustained critique of Darwinism intended to show that scientific knowledge reflects the values of those who generate it. This sociological perspective has been refined in the controversial biography of Darwin by Adrian Desmond and James Moore (1991).

The opposite extreme is represented by scientists with an interest in history, most of whom portray Darwin as the classic hero of discovery who established important new truths on firm, objective foundations (de Beer 1963; Ghiselin 1969; Mayr 1991). Their wider views often express an enthusiasm for the selection theory which typifies what religious and ideological opponents of the theory find so objectionable. Greene and Ernst Mayr have clashed over the philosophical and religious implications of their positions (in J. Greene 1999), while Michael T. Ghiselin has been singled out for attack as a social Darwinist. To be fair, there are some scientist-historians who take a more critical view of Darwin. Some have complained about his reluctance to acknowledge earlier writers, which in some cases amounts to a charge of plagiarism. The small anti-Darwinian wing which still flourishes in modern biology can also erupt into negative assessments of Darwin's work (e.g., Løvtrup 1987). More seriously, the philosopher Michael Ruse— a leading advocate of modern Darwinism—argues that the history of the movement can be understood only as an expression of the idea of social progress. He too believes that the scientific credentials of the theory were never strong enough to convince anyone not predisposed to accept it on ideological grounds (1996).

Both extremes encourage a blinkered view of history. The historian of ideas can misunderstand the way technical problems confronting Darwin shaped his thought, whatever inspiration he may have derived from other areas. The scientist may unconsciously read our modern understanding of the theory back into Darwin's less sophisticated pronouncements, and may be reluctant to admit that he derived even limited inspiration from nonscientific sources. The externalist becomes so involved with the cultural impact that the scientific issues are swept aside, while the internalist, who focuses only on the science, loses sight of the wider context within which the theory was formulated. The one reduces Darwin to a puppet controlled by ideological forces, mechanically translating his social prejudices into an equivalent view of nature. The other views him as the discoverer of a self-evident model of nature who merely saw what was there. Both extremes welcome the fact that another naturalist, Alfred Russel Wallace, independently developed the same theory as Darwin. To the externalist, this coincidence reflects the fact that the idea of natural selection was in the air. To the internalist, it is hardly surprising that two people should independently recognize the same truth about nature as soon as the requisite observations had been made.

Some of the detailed biographies offer a more balanced approach (e.g., Browne 1995, 2002). Darwin worked within a certain social environment and may well have been inspired in the construction of his theory by insights derived from its ideology. But he was also a scientist grappling with a set of empirical observations. He had to apply his theoretical insights to the technical problems he faced, and the way he formulated his conclusions was shaped both by those problems and by the external sources of inspiration. Such a balanced model reintroduces a role for human creativity in science without making theories into mere social constructs. Darwin was working along lines that were significantly different from those pursued by the majority of his contemporaries. Even Wallace did not discover exactly the same principle of natural selection, partly because his politics and social background were different, and partly because he was working on a different subset of scientific problems. There was no independent discovery of natural selection, which confirms that the new theory was not an inevitable product of either ideological or scientific forces simply bursting to reveal themselves.

Darwin's theory was one possible way of articulating a view of nature coherent with an individualist social philosophy, but it was not the only one, nor indeed the most obvious one, as we shall see when we turn to the reception of the theory. Darwin's theory often was linked to Herbert Spencer's philosophy of "progress through struggle." Darwin was able to catch the imagination of his age because his theory did include metaphors of struggle and competition that reflected the ideology of the rising middle classes. And he is remembered today, while Spencer is forgotten, because he framed his vision within a particular scientific context which led him in a direction Spencer could not have followed. Darwinism *is* social, as Young (1985) insists; but it is no mere copy of Spencer's individualist ideology, because natural selection is one of two very different models of evolution which can be mapped onto that view of society. Why Darwin was led to propose a different model can best be understood in terms of his scientific interests, which were by no means the same as those of most of his contemporaries. This model also helps us understand why natural selection remained problematic among scientists for so long despite the widespread assumption that social Darwinism became popular in late-nineteenth-century thought. We should never forget that our modern fascination with how Darwin developed the selection theory would not have been shared by many scientists around 1900. To them, the theory had played only an ephemeral role in the development of biology. Only since the rise of the modern synthesis of Darwinism and genetics has it seemed so important to understand how the selection theory was conceived.

Any attempt to build up a balanced picture of Darwin's work must confront the following issues (see Oldroyd 1984 for a useful overview).

1. *The role of ideology.* Most controversial is the claim that the selection theory reflects the ideology of free-enterprise capitalism—in effect that Darwin projected the ideology of his own social group onto nature itself. The possibility that the use of the metaphor "struggle for existence" represents such a social influence was noted already by Marx and Engels in the nineteenth century and is current in modern studies such as those of Young (1985). But this is not a simple issue. Darwin's reading of Malthus on population is usually presented as the source of the struggle metaphor, although this connection is beset with complications (see below). Darwin's theory is utilitarian, in that it makes adaptation the driving force of evolution: species get new structures because these structures are useful, and this focus on utility was also a component of liberal social thought. Sylvan S. Schweber (1977) argued that the individualism of Adam Smith's economics had influenced Darwin, while noting that scientific factors had already begun to make Darwin think in terms of individual variation within a population. A key factor in Darwin's innovation was what Mayr (1964, 1982) calls "population thinking"—the transition from seeing a species as based on an ideal type to seeing it as a population of distinct individuals. The liberals' rejection of the idealist model of the state in favor of individualism thus provides a more general ideological foundation for Darwinism. Population thinking also reflects a transition in nineteenth-century scientific thought noted by John Theodore Merz (1896–1903): the emergence of the statistical mode of explanation designed to cover changes so complex that the behavior of their individual components cannot be predicted. It is possible to see the rise of Darwinism as the creation of a statistical mode of explanation as opposed to the old Newtonian view of causation based on law (Depew and Weber 1995). We must balance the claim for a direct input from Darwin's social environ-

ment against the evidence for a growing awareness that, for science to tackle certain kinds of questions, a new type of explanation based on statistically modeled changes was needed.

2. *The idea of progress.* A related issue concerns Darwin's commitment to the idea of progress. It is normally assumed that liberal ideology saw individual effort (stimulated by competition) as the driving force of progress. But as Desmond and Moore (1991) argue, there was a period in the early nineteenth century when a less optimistic form of liberalism reigned, and Malthus's principle was part of this nonprogressive viewpoint. Darwinism became part of the ideology of progress only in a later, more confident era. Many modern scientists are reluctant to concede that Darwin was a progressionist, because they themselves reject the concept of progress as being too value laden, and there is a temptation to assume that the founder of the movement shared our own perception of the theory. Robert J. Richards (1991) accuses many historians of falling into this trap and thereby ignoring much evidence for Darwin's commitment to the idea of progress. His interpretation fits in with Ruse's (1996) thesis on the more general link between biological evolutionism and progressionism. Most historians now accept that Darwin cannot be seen as a nonprogressionist in the modern sense. But few would go as far as Richards, who sees Darwin as an exponent of an almost idealist vision of progress based on the model of embryological development.

3. *The argument from design.* Disagreement also focuses on Darwin's religious views and the rapidity with which he threw off the belief that nature is designed by a benevolent God (Brooke 1985). Many historians studying Darwin's early papers are convinced that by 1838 he had already recognized the materialistic implications of his theory (e.g., Schweber 1979). From this point on, he was an agnostic, and any references in his writings to the "purpose" of nature reflect a desire to conceal his true feelings. A few scholars disagree, arguing that Darwin at first tried to reconcile natural selection with design and only slowly lost his faith in a creator (F. Brown 1986; Gillespie 1979; D. Ospovat 1981). Such a model of Darwin's intellectual development presents him as someone who gradually realized the implications of his ideas. It is consistent with the claim that his theory represents a transformation of the older tradition rather than a head-on challenge to its foundations. All agree that by the time Darwin wrote the *Origin,* he had more or less abandoned any religious faith, although he was never an outright atheist. The story that he underwent a deathbed conversion, popular among creationists, has no foundation (Moore 1994).

4. *The scientific method.* A more specialized debate centers on Darwin's

scientific method. Darwin sometimes projected an image of himself as a patient observer, and his opponents have always accused him of being incapable of deep thought. But he also said the *Origin* was "one long argument," and most historians now accept that he was anything but a simple fact-gatherer (Mayr 1991). According to Ghiselin (1969), he was a follower of the hypothetico-deductive method, using his research to test the theory. Other historians focus on Darwin's desire to appear to be a good scientist as defined by then contemporary discussions of the scientific method (Ruse 1975b, 1979a). He developed natural selection by a creative process of synthesis and testing (H. Gruber 1974). Those who see the selection theory as the product of a new statistical mode of explanation also stress his innovative work in the area of methodology (Depew and Weber 1995). Darwin wanted to create a theory based on natural law in the Newtonian tradition, but by the very nature of the problems he addressed he was forced to transform this program by introducing statistical and historical elements into his explanations. Some see the historical element as derived from German idealist philosophy (e.g., R. Richards 1991), but others relate it to Lyell's uniformitarian methodology (Hodge 1982).

5. *Scientific problems.* Additional debates center on the scientific factors which shaped Darwin's thinking. Attention has always focused on three main areas: the biogeographical insights of the *Beagle* voyage and the evidence of speciation in the Galàpagos finches, his acceptance of Lyell's uniformitarian geology, and the analogy provided by the selective method employed by animal breeders. Modern research has modified our understanding of these influences, challenging some facets of the story Darwin told in his autobiography. Frank J. Sulloway (1982a) explodes the myth surrounding Darwin's finches, showing how Darwin nearly missed this vital piece of evidence (see below). Several accounts of the discovery of the selection mechanism have argued that he could not have been inspired by the model of artificial selection (Herbert 1971; Limoges 1970). He became so used to citing the work of animal breeders as a model in his later accounts of the theory that he actually came to believe that this was how he was led to the idea of selection. Historians have also focused on the strong thread that runs through Darwin's work based on a traditional view of heredity and reproduction (Hodge 1985). It has long been noted that Darwin failed to anticipate Mendelian genetics, as though this were a missing piece in his jigsaw that he really ought to have recognized. Understanding his commitment to a pre-Mendelian view of heredity forces us to accept that Darwin was not a completely modern thinker. One fact has not changed, however: it still seems vital to recognize the role of the geographical dimension in Darwin's work

and thought. The sense of the geographical diversity of life forced him to think in terms that were quite different from those employed by armchair evolutionary philosophers such as Chambers and Spencer. In a world of remote islands and barriers to migration, evolution becomes a less predictable process than anything implied by a law of development.

DARWIN'S EARLY CAREER

Charles Robert Darwin was born in 1809. His father was a successful physician, and his mother came from the Wedgwood family of pottery fame. His grandfather was Erasmus Darwin, author of *Zoonomia*, with its speculative evolution theory. The family was extremely well-off and thus had a position to maintain in society; yet they were professional people linked to the new manufacturing elite and were liberal in their politics. The men were inclined to be freethinkers, the women rather more orthodox in their beliefs. Darwin was the second son and the fifth of six children, leading Sulloway (1996) to treat him as a classic case of the more innovative later-born child. Darwin's mother died when he was only eight, and he was raised by his sisters, a situation explored in a biography by the psychoanalyst John Bowlby (1990).

Education

Darwin was not a good scholar, although he had an early interest in natural history. He was sent to Edinburgh University to follow the family tradition of medicine, but he was nauseated by the operating theater and soon gave up this career. He did, however, study natural history at Edinburgh, and this has led several historians to challenge the impression given in his autobiography that these were wasted years. In particular, Darwin claimed to have been unimpressed by the Lamarckian anatomist Robert Grant, although we now know that the two worked closely together, and that Darwin was impressed by Grant's claim that the "zoophytes" (Hydrozoa and corals) served as a bridge between the plant and animal kingdoms (Sloan 1985). More generally, Darwin's early reading focused his attention on issues that would shape his later thinking, especially his ideas about generation, or sexual reproduction (Hodge 1985; Sloan 1986). He became committed to the view that reproduction is a creative activity of the vital forces in the body, a position quite unlike our modern focus on the rigid transmission of genetic units. Here, his thinking reflects the fascination with the creativity of sex that ran through his grandfather's theories.

Having abandoned medicine, Darwin went to Cambridge to gain a bachelor's degree as a prelude to becoming an Anglican clergyman. He entered Christ's College in 1827. His decision to aim for the ministry was certainly not based on wild enthusiasm (although many a clergyman devoted himself to natural history), but neither was it hypocritical. At this point he still accepted that species were divinely created, and Paley's *Natural Theology* was an important influence on him. Paley was not central to the Cambridge curriculum, and *Natural Theology* was not required reading (Fyfe 1997)—but Darwin *did* read it and accepted its list of adaptations indicating the wisdom and benevolence of the Creator. Paley focused his attention on adaptation and utility—structures are there because they are useful to the organisms that possess them (W. Cannon 1961a). Within a few years Darwin reversed the logic of Paley's argument, turning adaptation from a fixed state into a process by which species adjust to changes in their environment by natural means.

Darwin pursued his scientific studies outside the curriculum, but he built up a close relationship with the professor of botany John Stevens Henslow and, later, with the geologist Adam Sedgwick. Henslow was liberal in politics but conservative in his religious views, and from him Darwin absorbed not only much scientific knowledge but also a sense of the role that natural theology played in the ideology of the gentlemanly elite that ran the country's science. Sedgwick took him on a geological tour of Wales in 1831, where he learned the craft of stratigraphy but also absorbed his mentor's catastrophist theoretical stance (Barrett 1974). He read Alexander von Humboldt's narrative (1814–29) of his explorations in South America and wanted to make his own voyage to the tropics. Darwin's project would become perhaps the broadest application of Humboldtian science, which encouraged the search for complex interactions underlying natural phenomena (S. Cannon 1978; Nicolson 1987).

The Voyage of the Beagle

Darwin's opportunity came late in 1831, when he was invited to travel with the survey vessel H.M.S. *Beagle*, which was being sent to chart the waters of South America. The ship's captain, Robert Fitzroy, wanted a gentleman companion to relieve his social isolation on the voyage, and this became the excuse for inviting a naturalist who would describe the areas visited (Burstyn 1975; H. Gruber 1968). Henslow proposed Darwin for this position, and after overcoming the misgivings of both his father and Fitzroy, he set out with the *Beagle* on a five-year voyage of discovery. His *Journal of Researches into the Natural History and Geology of the Various Countries*

Visited by H.M.S. Beagle (1845) is his much reprinted popular account of the voyage (see also Barlow 1946; C. Darwin 1988; Keynes 1979; Moorehead 1969).

While the ship's company surveyed the South American coast, Darwin explored the interior. The result was a body of observations which transformed his worldview. Initially, the most important results were in geology. Darwin had been given the first volume of Charles Lyell's *Principles of Geology* before setting out, and the second volume reached him on the voyage. His observations of South American geology meshed so well with Lyell's theory that he soon became a uniformitarian. What most impressed Darwin was evidence that earthquakes can permanently alter the elevation of the land surface. He saw the devastating effects of the 1835 earthquake at Concepción in Chile and noted that the coastline had risen ten feet from its original level. Even more striking was a series of raised beaches, still lined with seashells, found well above sea level. Lyell was right: the elevation of the Andes mountain chain was the product not of a single catastrophe but of a long sequence of earthquakes similar to those still occurring. Darwin proposed a new theory to explain coral reefs (1842) based on the gradual subsidence of the floor of the Pacific Ocean. He was less successful in a later attempt to explain a curious formation in Scotland, the parallel "roads" of Glen Roy. Darwin thought these geographic features were beaches formed when the land was submerged beneath the sea, although it was later shown that they were formed by a glacial lake (Rudwick 1974b).

Darwin also discovered fossils, including giant relatives of the modern armadillo, sloth, and llama. The resemblance of the extinct to the modern forms showed that there had been a continuity in the development of South American life: life there had adhered to the law of succession of types. Such a law was inconsistent with any theory in which life mounted a predetermined sequence of developmental stages toward the human form. When Darwin came to formulate his own theory, he was forced to think of each group as a distinct branch of the tree of life evolving in its own way within its own geographical region. Even within each branch there must be subdivisions, because the giant South American forms could not have evolved directly into their modern cousins: they must have gone extinct, while smaller relatives evolved into the modern forms. Thus, Darwin's theory would be conceived within a model of branching rather than linear development.

Equally important was his study of the geographical distribution of species (Sulloway 1982b). Darwin had been alerted to this topic by his reading of Humboldt and Lyell, but his observations on the *Beagle* voyage were to prove crucial. Darwin began to appreciate that the simple model of divine creation encountered difficulties when it was used to explain the facts of distribution. Eventually he would realize that those facts could be better explained by combining a theory of adaptive evolution with a study of how species are able to migrate around the world. Biogeography and ecology thus shaped the framework within which he would conceive his theory.

Natural theology implied that each species should inhabit the area to which it was best adapted. But this approach could not deal with the complex problems that Darwin now began to recognize. Why did South America have a unique collection of animal species, whereas Africa—which had essentially the same range of environments—had a different set of species? Darwin also discovered problems on a smaller scale which forced him to rethink the assumption that each species was perfectly adapted to its environment. He discovered a new species of the South American flightless bird, the rhea, on the open pampas of Patagonia, although he did not at first realize that his party had entered the territory of the new species. Why, at some undefined point on the plains, did the common rhea begin to give way to the new species? Apparently, the two species each had its own territory, which overlapped in an intermediate region. Even if each was perfectly adapted to its own main territory, neither could be perfectly adapted to the region in between. Darwin began to suspect that the two were competing to occupy as much territory as possible.

The idea of a perfectly balanced ecology thus came under question. Humboldt had already shown the necessity of studying the factors which restricted a species to a limited area (Egerton 1970a; Vorzimmer 1965). Lyell made an even greater contribution toward establishing a more dynamic view of what later would be called ecological relationships (Egerton 1968). He realized that the environment could not be absolutely uniform over any large area, and even in one spot the conditions would change slowly through time. A species could not be perfectly adapted to a single environment, nor could the species inhabiting a region interact in a perfectly harmonious way. As Darwin himself saw, the conditions at one point might favor one species, while a short distance away, a rival with a different lifestyle might have the advantage. In between, the two species would be struggling to occupy territory that was open to either of them. Lyell quoted the botanist Alphonse de Candolle: "All the plants of a given country are at war with one another" (Lyell 1830–33: 2:131). Long-term climatic modifications might alter the balance of power and lead to the extinction of the disadvantaged species (Kinch 1980).

It was not always easy to see why one species began to be replaced by another at a certain point, as Darwin discovered in the case of the rhea. But his growing realization that species compete to occupy territory was fundamental to his later ideas. This insight may well have been encouraged when he got caught up in one of the wars in which the land-hungry Europeans were exterminating the natives across South America. This extermination pinpointed the most important aspect of this new concept of struggle: the most bitter competition in nature was between the closest rivals. A species' greatest threat came from relatives in neighboring territories, which had similar lifestyles and thus threatened to exclude it from its livelihood, especially if a change in the conditions gave them some advantage.

It is a common assumption that nineteenth-century Europeans were fascinated by the image of struggle and competition (Gale 1972). But there are different levels at which struggle can occur. The often-quoted lines from *In Memoriam*, written by Alfred, Lord Tennyson, between 1833 and 1850, encapsulate a growing sense that nature was a scene of death and suffering. Nature gives no comfort to the gullible human

> [w]ho trusted God was love indeed
> and love creation's final law—
> tho' Nature, red in tooth and claw
> with ravine shriek'd against his creed.
> (1973: stanza 56)

Here, natural theology's efforts to maintain belief in a benevolent God are undermined by the assertion that suffering is integral to nature. This perspective reflects a change in the cultural climate which would encourage naturalists to challenge the idea of a harmonious balance of nature. But its focus on the predator "red in tooth and claw" is peripheral to Darwin's thinking. Tennyson's metaphor also ignores the sense of ecological competition that Darwin absorbed from Lyell and from his South American experiences. Competition between rivals is truly relentless, with extinction the penalty for failure, yet it may take place without actual bloodshed—as de Candolle pointed out in the case of plants. To treat natural selection as merely the elaboration of Tennyson's image within biology is to miss the crucial developments that took place as Darwin grappled with the implications of what he had observed.

An ecology based on struggle became integral to the Darwinian worldview, but Darwin's observations did not provide evidence of transmutation, nor did they supply the exact concept of struggle, which he would pinpoint as the driving force of natural selection. Lyell still believed that species were fixed; moreover, he did not see that de Candolle's "war of nature" might also operate *within* each species, allowing the struggle between individuals to select out those best adapted to the changes. Something else was needed—first to convince Darwin that transmutation actually had occurred and then to suggest the application of the struggle concept within a single population.

Darwin eventually was convinced of evolution by his study of how geographical barriers seemed to promote the multiplication of species. The law of succession of types implied that the oceans divide the earth into geographical provinces, each with its own inhabitants. Lyell explained this fact by postulating centers of creation, each responsible for a particular set of types. But Darwin was increasingly unwilling to accept that the Creator would make such an arbitrary division in His activities. Eventually he realized that the ocean barriers created zoological provinces because they prevented migration, which in turn ensured that on each continent all the later species would be related to those that came before them. The supposition that this "relationship" implied actual descent with modification was soon thrust upon him after he observed the same phenomenon on a much smaller scale.

The study of isolated oceanic islands brought the point home. In most cases such islands have unique species of their own, although they are usually related to those found on the nearest major landmass. While on the *Beagle,* Darwin observed how birds and insects could be blown across wide stretches of oceans, whereas seeds could be carried by ocean currents. Even land animals were sometimes carried out to sea on natural rafts of vegetation. Oceanic islands could gain their first inhabitants by such accidental mechanisms of transportation; Darwin subsequently carried out a major study of this process. But why, then, were the species on the islands often different from, although related to, those of the nearest continent? In the end Darwin was forced to admit that the most logical explanation was that the small populations established on the islands had changed through time until eventually they had become new species.

The Galápagos

This point became clear to Darwin when he thought about his experiences on the Galápagos archipelago. This group of volcanic islands lies several hundred miles off the Pacific coast of South America, straddling the equator (on their discovery and exploration, see E. Larson 2001). The *Beagle* spent some time there, allowing Darwin to collect specimens from several islands. But it was only as the ship was about to depart that he was given the vital clue: he was told that it was possible to determine which island a given specimen of the giant tortoise came from by the shape of its shell. Darwin immediately began to wonder if similar differences could be seen in other

18.   The Galápagos finches. This illustration from
Darwin's *Journal of Researches* (1845, chapter 17)
shows four of the Galápagos ground finches and the
wide differences in their beak structures produced by
adaptation to different modes of feeding.

species, and soon began to focus on the islands' birds. Ever since David
Lack's study, *Darwin's Finches* (1947), it has been assumed that these birds
provided the crucial evidence of what we now call speciation, the division of
a single parent species into several descendant species (see fig. 18). In his
published account of the voyage, Darwin himself drew attention to the wide
range of beak shapes in the Galápagos finches and commented, "Seeing this
gradation and diversity of structure in one small, intimately related group
of birds, one might really fancy that from an original paucity of birds in this
archipelago, one species had been taken and modified for different ends"
(Darwin 1845: 276). Here, "we seem to be brought somewhat nearer to that
great fact—that mystery of mysteries—the first appearance of new beings
on this earth." According to the legend, which Darwin himself thus helped
construct, seeing the variety of different finches on the various islands convinced him that isolated populations derived from a single parent species
had diverged on the different islands to produce a group of separate species,
each adapted to a different way of life.

Working on Darwin's original notes, Sulloway (1982a) undermined this
myth by showing that the Galápagos finches could not have played such a
crucial role. Although they provide a good example of speciation, the
finches' distribution is complex because each species is no longer confined to
a single island. Darwin's own specimens were not even labeled to show
which island they were collected on, and he had to rely on collections made

by others on the ship to reconstruct the distribution. Only after his return to England, when the ornithologist John Gould informed him that the finches really did constitute a group of distinct but closely related species, did Darwin realize their full significance.

The mockingbirds of the Galápagos provided Darwin with the insight that converted him to transmutationism. Here, Darwin himself identified several species with obvious resemblances to American mockingbirds. Once he was convinced that the populations on the different islands really were distinct species, not merely varieties of a single species, he was forced to confront a serious difficulty underlying the assumption that every true species derives from an act of miraculous creation. It seemed unreasonable to suppose that every one of these tiny islands lost in the middle of the ocean should have received its own visit from the Creator. To Darwin, it was more plausible to suppose that a few members of the ancestral species had been accidentally transported to each of the islands, where they founded breeding populations that remained isolated from one another by the barrier of the sea. Although the conditions on the various islands were very similar, each founding population had discovered a different way of coping with the environment, and thus each had evolved in a different direction. In the absence of normal competition, a wide range of ecological possibilities was open on each island. As each population specialized for its particular way of life, it changed further from the original form, and eventually each island had its own distinct variety. This much could have been explained by creationist beliefs, but Darwin now saw that the process had generated not merely varieties but distinct species. Once he was forced to confront this breakdown of the old paradigm, he almost immediately extended the evolutionary model to the whole history of life on earth.

THE CRUCIAL YEARS: 1836–1839

Darwin probably arrived at these insights after the *Beagle* had returned to England in October 1836 (Sulloway 1982b), although a few historians speculate that he may have accepted transmutation during the later part of the voyage. For several years Darwin lived in London, playing an active role in the scientific community and publishing his geological findings (Rudwick 1982). But behind the scenes he had begun to think hard about the question of species. Mayr (1977) calls this the first Darwinian revolution: the Galápagos results convinced Darwin that new species are formed by the natural transformation of old ones, although as yet he had no idea what the process of change might be. Over the next few years he searched for a plausible mechanism, and his development of the theory of natural selection constitutes a second revolution. By the early 1840s Darwin had formulated a primitive version of the theory, and over the next twenty years he worked to refine his idea and supply new lines of evidence.

The discovery of the selection theory has attracted a vast amount of attention from the Darwin industry. There are several ways of approaching the issue, depending in part on the purpose of the study. Some modern biologists, Mayr included, prefer a logical reconstruction of the steps by which the components of the theory were put together. They use history as a means of helping the modern reader understand the structure of the theory. The selection theory can even be expressed as a logical deduction from observed facts about nature (Mayr 1991; Ruse 1971). These include the existence of variation within each population, the tendency for such variant characters to be transmitted by heredity, and the pressure of population leading to the struggle for existence. Mayr emphasizes Darwin's move toward what he calls "population thinking" as opposed to the old typological view of species. To admit that the variability within a population is significant, one has to abandon the old idea of an ideal type for the species on which all the individuals are modeled. On this older view, individual variations are trivial, like minor imperfections in toy soldiers cast in plastic from a mold. In modern Darwinism there can be no ideal type or mold, because the species is merely the population of interbreeding individuals—and if selection changes the character of the population, then by definition the species has changed.

To the historian, though, it is more important to reconstruct how Darwin groped his way toward the new idea. His actual course of discovery may not correspond to the most logical conception. The scientists' reconstruction tidies up the process, leaving out the clues which may have led Darwin to try out new approaches. It also ignores those areas where he did not make a decisive break with the past. The modern Darwinian theory has advanced beyond the one that Darwin himself conceived, and it is easy to read into his writings a greater degree of sophistication than was really there. If we reconstruct the events by hindsight, we distort the context within which he was working. Darwin was a radical thinker, but he could not leap at one bound into the modern world, and to impose modern ideas on him for the sake of apparent clarity obscures his true creativity. This is most obvious in the case of Darwin's ideas about heredity, which did not anticipate Mendelian genetics, yet were an integral part of his thinking. Some historians doubt that Darwin fully appreciated the extent to which his theory implied a populational rather than a typological view of species. Scientist-historians such as Gavin de Beer and Mayr have also played down the extent to which the idea of the struggle for existence may have its roots in the ideological debates of the time.

One source of information that has long been available is the account of the discovery that Darwin wrote in his autobiography, published by his son Francis in *The Life and Letters* (F. Darwin 1887). But this may not be reliable, since Francis Darwin edited the text to conceal the more controversial and problematic aspects of his father's thinking, including his religious views and his increasingly outdated ideas on heredity. Even in its unexpurgated form (Darwin 1958), the autobiography is unreliable, because Darwin's memory of the discovery became distorted by his subsequent efforts to explain the theory by analogy with artificial selection.

The unreliability of Darwin's memory is revealed by the notebooks that he kept during the crucial period (Darwin 1987). The preferred method of the historian is to reconstruct past events from documents written at the time and not intended for publication (only the intense interest in Darwin has ensured that the notebooks eventually *were* edited and published). These records allow us to see every twist and turn in his thinking as he began to develop the new idea, and they also reveal his sources of information (on his reading notebooks, see Vorzimmer 1977). The "Red Notebook," begun after he returned from the *Beagle* voyage, records his first speculation about transmutation in June 1837. Subsequently, he began a series of notebooks: the A notebook is devoted mainly to geology, but B and C relate to transmutation. The D and E notebooks, begun after July 1838, continue the work on biology, while the M and N notebooks focus on the human implications of the theory.

The Creative Thinker

Darwin's notebooks confirm that the discovery of the selection theory stands as a major example of creative thinking (H. Gruber 1974). He drew on an enormous range of influences, scientific and nonscientific, constituting a unique conceptual framework that steered him toward a radical break with the past. Lyell's uniformitarian methodology was important to the process, along with the paleontological and biogeographical insights of the *Beagle* voyage. The decision to investigate animal breeding as an example of how changes might appear in a species was decisive, even though the model of artificial selection was not easy to apply. Ideas about the creative power of generation, or sexual reproduction, formed an important foundation which linked his thoughts back to Erasmus Darwin rather than forward toward

Mendelian genetics. He read widely in philosophy, psychology, and social theory in an effort to understand the wider implications of where he was going (Manier 1978).

Darwin hoped to present his theory as a product of the best scientific method. Because he was moving into unknown territory, he had to show that he was not merely speculating wildly. The methodological debates of the time acknowledged a role for theorizing beyond the known facts so long as the hypotheses could be tested by the evidence. Ruse (1975b) and Schweber (1989) emphasize the role of Sir J. F. W. Herschel (1830), who stressed the need to balance theory and observation, and of William Whewell, who pointed out that important new theories drew their power from their consilience, or ability to link diverse areas of study (Whewell 1847a,b, 1989; see also Fisch and Schaffer 1990; Yeo 1993). These methodological influences encouraged him to diversify his interests and broaden the scope of his explanatory system.

Through a combination of bold theorizing and comprehensive evaluation, Darwin came up with a concept of evolution which was unique for the time. There have been many efforts to undermine his originality by claiming that the selection theory had been developed by earlier writers, including Edward Blyth, Patrick Matthew, and William Charles Wells (on Blyth, see Beddall 1972, 1973; Eiseley 1959; and Schwartz 1974; on Matthew, see K. Wells 1973a; and on Wells, see K. Wells 1973b; for a collection of all the relevant texts, see McKinney 1971). One writer has even gone so far as to hail Matthew as the originator of the modern evolution theory (Dempster 1996). Such efforts to denigrate Darwin misunderstand the whole point of the history of science: Matthew did suggest a basic idea of selection, but he did nothing to develop it; and he published it in the appendix to a book on the raising of trees for shipbuilding. No one took him seriously, and he played no role in the emergence of Darwinism. Simple priority is not enough to earn a thinker a place in the history of science: one has to develop the idea and convince others of its value to make a real contribution. Darwin's notebooks confirm that he drew no inspiration from Matthew or any of the other alleged precursors.

Environment and Generation

Because Darwin started from his knowledge of biogeography, he necessarily approached the topic from a new direction (Hodge 1982; Richardson 1981). He began from a conviction that evolution must be a branching process in which one species is divided by geographical barriers, and the separate populations then become transformed in different directions (see the diagram in Darwin 1987, B notebook: 36). For a time he wondered if species might be "born" with a built-in lifespan, after which they become extinct. He soon realized, however, that some other, more active mechanism of change was needed, and by July 1837 he was convinced that transmutation must come about by the accumulation of individual variations over many generations. He began to explore the direction already taken by Lamarck and Erasmus Darwin: might a change in the environment produce modifications either by affecting the reproductive process or by changing the organisms' habits? The crucial break came when he decided that the Lamarckian approach was inadequate. For a variety of reasons, including his knowledge of animals bred in captivity, he decided that, although the environment might well be the stimulus, the majority of the changes it produced were not purposeful—in relation to the possibility of adaptive evolution, they were essentially random (Hodge and Kohn 1985).

Modern scholarship has revealed the extent to which Darwin was influenced by his views on generation, or reproduction (Hodge 1985; Kohn 1980; Sloan 1985). He headed his B notebook "Zoonomia," recalling Erasmus Darwin's quest for the laws of organic life. The key to transmutation was the creative power of sexual reproduction. The theory of heredity that he later called pangenesis (published in Darwin 1868) was formulated at this early stage and remained intact throughout his career. Historians used to lament Darwin's "failure" to anticipate genetics, on the assumption that, if he had developed the concept of the unit gene transmitting a character unchanged from one generation to the next, he would have forestalled the problems that beset the selection theory in the late nineteenth century (see chap. 7). Modern scholars emphasize that to understand Darwin at all, we must accept that the foundations of his thought lie in a premodern concept of heredity. Our job as historians is to understand how he made the theory work at the time, without looking ahead to see how the model of heredity subsequently unraveled. Phillip R. Sloan (1986) shows how this traditional viewpoint may have paved the way for his materialist philosophy.

For Darwin, evolution mediated between the pressure of the environment and the power of reproduction: "Why is life short. Why such high object generation—We know world subject to cycle of change, temperature & all circumstances which influence living beings" (B notebook: 2–3). He made no clear distinction between what we call heredity (the transmission of characters from one generation to the next) and development (the process by which those characters are produced in the developing embryo). As he saw it, the reproductive elements used to create the next generation were formed by "budding off" from the various tissues of the parent's body.

There were no genetic units transmitted unchanged through many generations. However, Darwin believed that, in an absolutely stable environment, the buds (called gemmules in his later theory) would be exact copies of the parental tissue. But nature ensured that the copying process was never a mechanical one. Sexual reproduction mixed the parents' gemmules, keeping up a constant circulation of characters in the population. In addition, Darwin believed that there was a source of new variations as a result of the disturbing influence of a changed environment upon the reproductive system. This action might occasionally produce adaptive characters, as in the Lamarckian effect, but more often it simply upset the natural copying process to produce random variations. Darwin's lifelong commitment to a limited amount of Lamarckism and to what was later called blending inheritance (the mixing of parental characters) were integral parts of his worldview.

Animal Breeding

The C notebook shows Darwin taking an important step in the search for information—he began to study the work of animal breeders, where changes could actually be observed within a species (Secord 1981; Vorzimmer 1969b). No other naturalist of the time took this step, not even Alfred Russel Wallace. Darwin was impressed with the enormously different breeds produced by pigeon fanciers in a relatively short time. He joined their clubs and circulated questionnaires to gather their views on how new breeds were formed. He later came to believe that he observed the breeders applying a process of artificial selection and then saw how this model could be transferred to natural evolution. Several historians have explored this apparent analogy between artificial and natural selection because—as Darwin himself later realized—it offers such a neat model for how natural selection operates (Cornell 1984; L. Evans 1984; Ruse 1975a). The breeder sees many minute differences between the young animals in his artificially isolated population; he identifies a few of these variant characters as useful for his purposes and selects those individuals to breed the next generation. The desired character is enhanced over a series of generations through a process of artificial selection. But although the process of selection seems clearly illustrated in such cases, Darwin's notebooks show that there was no direct borrowing of the selection model from the breeders. Throughout the C and D notebooks there is evidence that he remained convinced that adaptive variations must somehow be elicited automatically in a changed environment (Lamarckism), and that he did not see the breeders' work as a useful analogy. He knew that the breeders worked by picking out useful variants to breed from, but did not see their method as in any way equivalent to what happened in nature (Herbert 1971; Limoges 1970; R. Richards 1997). Instead he made another crucial step which allowed him to create the idea of natural selection, and only then realized that there might be a parallel with what he had observed in the breeders' work.

Malthus

Darwin's new insight came from his reading of Thomas Malthus's work on the principle of population (see chap. 4). This occurred in September 1838 as Darwin was nearing the end of his D notebook. Malthus's sense that the constant pressure of an expanding population on resources generated a struggle for existence provided Darwin with an agent capable of producing a natural form of selection. Yet Darwin did not simply apply the logic of Malthus's view of human society to nature. Malthus had written his book to show that any hope of progress was illusory, and he made no effort to argue that the richest individuals are the winners in some kind of struggle for scarce resources. When he introduced the phrase "struggle for existence," he was discussing warfare among the primitive tribes of central Asia. In fact Darwin's first use of the population principle in the notebooks suggests a similar interpretation: a passage in which he portrays nature as a multitude of wedges being driven against one another seems to refer to competition between rival species each trying to drive the others to extinction: "One may say there is a force like a hundred thousand wedges trying to force every kind of adapted structure into the gaps in the oeconomy of Nature, or rather forming gaps by thrusting out weaker ones. The final cause of all this wedgings [sic], must be to sort out proper structure & adapt it to change" (D notebook: 135e).

In his E notebook Darwin recorded the realization that population pressure must entail a struggle between the individuals of the same species (Herbert 1971). Marrying this to the knowledge that animal breeders worked by exploiting the range of individual variation existing in any population, Darwin saw how the struggle for existence would eliminate any individual who had variant characters making it less fit to cope with the environment. By contrast, the better-adapted individuals would survive and breed, thus increasing the proportion of the next generation with the "fitter" character. He suspected there would be less variation in wild species, because they seldom were exposed to drastic changes in their environment, but he knew there would be some variation when geological forces began to modify the environment. The result would be some individuals with characters which were by chance better adapted to the new conditions, and others who would be disadvantaged. In the contest of life, he wrote, "a grain of sand turns the balance," and thus "if a seed were produced with an infinitesimal advantage it would have a better chance of being propagated" (E notebook: 115, 137). This insight confirmed Darwin's long-standing suspicion that death plays a creative role in the world, along with sex (Kohn 1980). Only now did he see that there was a parallel with the way the animal breeder picked out those individuals with a favored character to breed from.

The fact that Darwin acknowledged a debt to Malthus has generated a major debate over the extent to which his theory drew on the competitive ethos of nineteenth-century capitalism. Some accounts minimize the influence, treating Malthus as only a catalyst who helped Darwin put together insights already gained from his observations of nature (de Beer 1963). Others imply that Darwin simply applied the logic of Malthus's principle to nature, so that his theory was modeled directly on the ideology of free-enterprise individualism (Desmond and Moore 1991; Vorzimmer 1969a; Young 1969, 1985). But the connection is not as direct as is sometimes implied: Darwin's concept of the struggle for existence does not appear in Malthus's work, and on this point Darwin had to think creatively with the insight provided by the population principle (Bowler 1976b). Malthus did not anticipate the logic of what became known as social Darwinism, and thus could not provide the whole model for the selection theory. Darwin had to transform the concept of the struggle for existence by recognizing that it operated between individuals, and that as a result the better adapted would survive and breed.

The breadth of Darwin's reading reveals many ways in which the ideology of the time could have influenced his thinking (Manier 1978; Schweber 1977). He read David Brewster's review of Comte's positivist philosophy, which argued for the need to base all theories on mathematical foundations—exactly what was provided by the arithmetical logic of the population principle. His reading of Malthus was not an accident: it was part of a major program of investigation undertaken to throw light on the human implications of evolutionism. He read the work of Adam Smith, whose analysis of laissez-faire economics showed how individual reactions could work together to give an apparently purposeful overall effect. Looking for a way of measuring variation, he turned to the work of the Belgian anthropologist Lambert Quetelet (translation 1842), who pioneered the application of statistics to the human population. For any characteristic, Quetelet showed that there was a range of variation between two extremes, with most of the population clustered around the center of the range—the phenomenon later represented by the bell-shaped distribution curve. Here was a good example of population thinking, with the human race depicted as a group of diverse individuals rather than a fixed type. Quetelet actually discussed Malthus and may have triggered Darwin's reading of the population principle. All these factors contributed to provide a social foundation for the model of nature that Darwin would construct. His theory was not a slavish copy of Malthus's political philosophy; nevertheless, it may have been steeped in the ideology of the time. As Desmond and Moore (1991) stress, this was a time of social conflict in Britain, so Darwin could see on the streets the struggle that the political economists were trying to understand. Perhaps this real-life demonstration of struggle encouraged Darwin to apply the logic of individualism in a new way.

The M and N notebooks show that Darwin had already adopted a materialist perspective, at least on how his theory would apply to humankind (H. Gruber 1974; Herbert 1974–77). These notes anticipate many of the topics later articulated in the *Descent of Man*. Darwin saw no room for the traditional notion of a soul existing on a purely spiritual plane: the mind was a product of the material activity of the brain, just as the phrenologists held. He suspected that much of our unconscious behavior might be instinctive, programmed into our brains by the effect of evolution on our ancestors. The ways in which we expressed our emotions revealed our animal ancestry, as in the case of snarling to express anger. He was convinced that evolution would throw light on our moral values by showing how certain forms of social behavior had been programmed into us by natural selection. Morality was merely a rationalization of these social instincts. Such ideas would undermine the whole traditional view of human nature.

Yet Darwin did not want to reduce the whole universe to a chapter of accidents, with the human race as a product of mere chance. He still felt that the laws of nature were instituted by a wise and perhaps even benevolent God. Far from recognizing the full horror of a worldview based on struggle, he stressed that the end result of evolution was to keep species well adapted to their environments in an ever changing world. In this sense he transformed rather than destroyed the old natural theology of Paley (W. Cannon 1961a; Young 1985). Even struggle and death had a positive value in the divine plan.

Historians also debate the extent to which Darwin retained a role for the idea of progress: is the human race the intended product of the ceaseless operation of natural laws? Modern Darwinism dismisses the claim that evolution is automatically progressive and undermines the hierarchy by which progress could be measured. Darwin himself recognized these problems. As early as the B notebook he insisted, "It is absurd to talk of one animal being higher than another," noting that each great branch of the tree of evolution has specialized for different characters (74). We think great intellectual capacity is a sign of progress, but the insects are remarkably successful through the development of instinct. Yet Darwin could still fall into a progressionist way of thinking: "If all men were dead then monkeys make men—Men make angels" (B notebook: 169). Robert Richards (1991) argues that historians influenced by modern Darwinism have failed to appreciate that Darwin still saw evolution as a steady pressure to mount the scale of organization. By contrast, Desmond and Moore (1991) suggest that, in the harsh social climate of the 1830s, a more pessimistic reading of evolution was the most obvious. Darwin soon came to see natural selection as a force that would tend to raise the standard of organization whenever the circumstances were appropriate. But he was also aware that the advance could take place in a number of directions, not merely along a single line leading toward humanity.

DEVELOPMENT OF THE THEORY, 1840–1859

By the end of 1839, Darwin had conceived the outline of his theory. In 1842 he wrote out a short sketch of his ideas, followed two years later by a substantial essay (both reprinted in Darwin and Wallace 1958). The essay was meant to be published in the case of his premature death, but Darwin had no intention of putting his ideas before the public yet. He was well aware of their controversial nature—1844 was the year in which Robert Chambers's *Vestiges of the Natural History of Creation* appeared, and Darwin wanted to avoid a similarly poor reception for his own theory (Egerton 1970b). He also knew that much work would be needed to turn his provisional hypothesis into a fully articulated theory that would be accepted by the scientific community. A few close friends gradually were let in on the secret and asked to comment on it, although many others were sounded out on individual topics. He also took on research projects of his own, all intended to test one aspect or another of the theory.

Darwin's personal circumstances now changed substantially. In 1839 he married his cousin Emma Wedgwood, and in 1842 they moved to Down House in Kent, a large house in the country to the south of London. Darwin was already showing signs of the illness which would dog him for the rest of his life, forcing him to avoid all forms of excitement and eventually leaving him prostrated for months at a stretch. Many historians have attempted to identify this illness. At one time it was thought that Darwin had contracted a nervous ailment transmitted by the bite of a South American insect. Some have suggested that he was poisoning himself with patent medi-

19. Charles Darwin. This portrait of Darwin in middle age is from the frontispiece to his *Life and Letters* (F. Darwin 1887, vol. 1).

cine (Winslow 1971). A more plausible explanation is that the symptoms were the result of psychological stress brought about by the controversial nature of his theory (Bowlby 1990; Colp 1977). Emma, a deeply religious woman, was shocked when she realized the extent to which her husband's ideas challenged conventional beliefs.

Darwin could no longer stand the excitement of public debates, but he was not a complete recluse. He still went up to London regularly, and Down was accessible enough for friends and colleagues to visit. In addition, he had built up a vast correspondence network to provide him with information on subjects relevant to his theorizing (Darwin 1984–). He was trying to build up a community of scientists who would speak the new language of evolution (Manier 1980). The first to be let in on the secret was the botanist Joseph Dalton Hooker, who was given a copy of the essay in 1847 and who argued with Darwin for years before accepting the idea (Colp 1986; on Hooker, see

Allan 1967; L. Huxley 1918; and Turrill 1963). He was the son of William Hooker, the director of the Royal Botanical Garden at Kew, a position to which he eventually would succeed. Hooker had already made an exploratory voyage to the Antarctic and would spend the years 1847–1850 in India, so his knowledge of plant distribution was particularly important to Darwin.

The 1844 Essay

Darwin's 1844 essay provides a snapshot of his thinking in the years following the conception of the theory. In many respects it looks forward to the *Origin of Species*, adopting the same technique of leading the reader into the argument by explaining first how artificial selection works and then showing that there is an equivalent process at work in nature that adapts species to their environment. Darwin gives the following imaginary example to illustrate the action of natural selection:

> Let the organization of a canine animal become slightly plastic, which animal preyed chiefly on rabbits, but sometimes on hares; let these same changes cause the number of rabbits very slowly to decrease and the number of hares to increase; the effect of this would be that the fox or dog would be driven to try to catch more hares, and his numbers would tend to decrease; his organization, however, being slightly plastic, those individuals with the lightest forms, longest limbs and best eyesight (though perhaps with less cunning or scent) would be slightly favoured, let the difference be ever so small, and would tend to live longer and to survive during that time of the year when food was shortest; they would also rear more young, which young would tend to inherit these slight peculiarities. The less fleet ones would be rigidly destroyed. I can see no more reason to doubt but that these causes in a thousand generations would produce a marked effect, and adapt the form of the fox to catching hares instead of rabbits, than that greyhounds can be improved by selection and careful breeding. (Darwin and Wallace 1958: 120)

Significantly, Darwin imagines this process taking place on an island, where a small isolated population would be forced to adapt to the changing conditions. The changes would first turn the population into a distinct local variety of the original species from which it was derived, but eventually the differences would become so great that a new species, incapable of breeding with the parent form, would have evolved (see fig. 20). The process of splitting, or speciation, was a gradual one, and so there would be no obvious cutoff point at which a variety of an old species suddenly became a new species in its own right. This explained why naturalists were often unable to agree on whether a particular form was merely a well-marked variety or a distinct

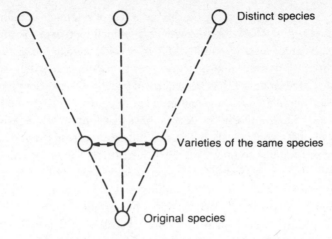

20. The relationship between varieties and species. Darwin supposed that varieties were "incipient species" because, if they survived and continued to change, they eventually would become distinct species in their own right. The diagram shows a single species diverging into three later ones because the original population has been subdivided, probably by geographical barriers. In the early stages of the process, the three separate populations are still so closely related that they could interbreed if they came together. At a later stage, the differences are so great the populations could no longer interbreed, and they are then distinct species. But since there is no clear dividing line, there is an intermediate phase when it is very difficult to determine whether there are three very strongly marked varieties of a single species or three separate species.

species. But Darwin was never tempted by Lamarck's claim that differentiation into species did not occur. Once a gap had been opened by divergent evolution, the two or more populations were separate and could be recognized as distinct entities. In practice, species existed but were not fixed by some underlying type (Kottler 1978). This was an important step toward a more populational view of species, although Darwin was unable to shake off completely the legacy of the old typological way of thinking (Beatty 1982).

The rest of the 1844 essay explores the implications of this model for natural history. Evolution must be a branching process, with groups of related species derived from a recent common ancestor. This explained how naturalists were able to classify species into ever widening categories, indicating the different levels at which they shared a common ancestor. It may be significant that philologists studying the historical development of languages had already begun to develop the model of a branching tree of relationships, so to some extent Darwin and his followers could invoke the analogy between the evolution of languages and species for support (Alter 1999). A group of related languages had been derived from a single ancient language, which was fragmented as the population speaking it was broken up by geographical barriers. Similarly, the geographical distribution of species was explained by observing the possibilities of migration across barriers and the resulting adaptations to newly occupied territories. The Galàpagos species were classic illustrations of what happened when separate founder populations were established in isolated locations, each able to adapt to its new environment in its own way. Rudimentary or vestigial organs (such as the human appendix) were explained as relics of structures that were useful in the past but were now being reduced in size because they no longer served a function in the modern organism. Selection would slowly eliminate as wasteful a structure that no longer served a purpose, although Darwin was also inclined to accept the inherited effect of disuse via the Lamarckian mechanism.

Breeders and Barnacles

Over the next two decades Darwin undertook researches to demonstrate applications of his theory (Ghiselin 1969). He also modified the theory in a number of ways as he began to appreciate the limitations of his early ideas. By the time Darwin came to write the *Origin of Species,* his thinking had become more sophisticated on a number of fronts. His understanding of the theory's fundamental implications had also changed significantly.

Darwin made further studies of pigeon breeding, showing the extent of the structural modifications made by artificial selection. He also worked on geographical distribution, investigating the mechanisms that could transport animals and plants to different locales. He studied how seeds could become stuck in mud on birds' feet, and how well the seeds could stand immersion in seawater, all with the aim of showing how oceanic islands gained their vegetation. He argued endlessly with Hooker on the application of the theory to the facts of geographical distribution. Whereas Darwin favored accidental means of dispersal, such as birds being propelled across oceans by storms, Hooker thought that geology might show there had once been land bridges between now-separate continents and islands.

The most extensive project was a major study of barnacles, a little-known group which had only recently been recognized as a subclass of degenerate crustaceans. Some unusual specimens had been discovered on the *Beagle* voyage, and as he began to investigate them, Darwin was drawn into a project to provide a complete description and classification of all the known barnacles (Darwin 1851–53). He was persuaded by taunts from Hooker that no one had the right to pontificate about species until they had undertaken a systematic study of a major group. He soon realized that the barnacles provided a useful way of testing many aspects of his theory (Ghiselin 1969; Ghiselin and Jaffe 1973). Most barnacles are hermaphrodites, although some species have dwarf males, often so small they appear to be parasites on the much larger hermaphrodite. Darwin realized that the group's ancestors had probably been hermaphrodites but had changed their reproductive strategy and developed the small males through a process of degeneration. In the parasitic males, all structures except the reproductive organs had become rudimentary.

The barnacles had been linked to the crustaceans through a study of their larvae. This illustrated the important point that evolution affected mainly the adult form of the organism, often leaving the embryonic stages unchanged so that they reflected the ancestry of the group to which it belonged. Darwin's theory thus incorporated an element of what would later be called the recapitulation theory, although the extent to which he can be associated with the full-blown version of this theory is controversial. Robert Richards (1991) has argued that Darwin adopted the law of parallelism (see chap. 4). But this form of recapitulation theory implies a linear model of evolution, with the lower animals being treated as immature versions of the more perfect human form. Darwin was committed to the rival vision of embryological development as a process of specialization, although at first he did not know of von Baer's work (Oppenheimer 1967). The law of parallelism reemerged in the recapitulation theory of Ernst Haeckel in the 1860s (Gould 1977b), but many historians are suspicious of Richards's effort to tie Darwin in with this school of thought. Darwin's vision of evolution as a branching process required him to see each branch as evolving in its own direction, so that the end product of one branch could never be treated as the ancestral or immature stage of the end product of another. This did not mean that the embryo was without significance for the recognition of ancestry: it could show the embryonic state of the ancestor but not its adult form. Larval barnacles revealed their ancestry among the crustaceans in their resemblance to that group's larval structure, not by passing though an adult crustacean stage.

Divergence and Struggle

This point can be illustrated by an important change which took place in Darwin's thinking in the 1850s. It has often been assumed that as soon as he formulated the selection theory, he must have recognized that it implied a vision of nature as a scene of unrelenting struggle and suffering. But Dov Ospovat (1979, 1981) argues that the early form of the theory still revealed its roots in Paley's vision of natural theology. In the 1844 essay, Darwin implied that species normally existed in state of perfect adaptation, with little or no individual variation. Only when the environment changed did variations appear, so that natural selection could operate to adapt the species to the new conditions. Such a view implied that natural selection was an episodic process interrupting long periods when species lived comfortably in a stable environment.

At one point the essay describes natural selection as an omniscient Being picking out those that are useful. Admittedly, this comes only a few pages before the naturalistic account of selection quoted above and may have been a rhetorical device intended to lead the reader from artificial to natural selection. But this terminology reinforces the impression of nature as a benevolent power superintending the operation of the various laws. Over the next decade or more, Darwin realized that this compromise was untenable. The principle of population implied that the struggle for existence continued even in a stable environment. Darwin's worldview became far more pessimistic, in part because of the death of his beloved daughter Anne in 1851. While at first he had hoped to retain belief in a God who created the universe for a purpose, he was increasingly plagued with doubts (F. Brown 1986; Gillespie 1979; Kohn 1989; Moore 1989).

One factor which triggered this recognition of selection's relentless power was Darwin's realization that the original theory did not explain a phenomenon that paleontologists now were discovering in the fossil record. Natural selection driven by isolation could explain the small-scale branching of species, as in the case of the Galàpagos finches. But the fossil record was revealing a much broader trend toward divergence: each new class began with a small number of very generalized species and then exhibited a pattern of branching toward a multitude of different specializations. Both Richard Owen and William Benjamin Carpenter recognized this trend by comparing the fossil sequences with the process of embryological specialization observed by von Baer (Bowler 1976a; D. Ospovat 1981). In both evolution and embryological development, specialized characters appeared only in the later stages.

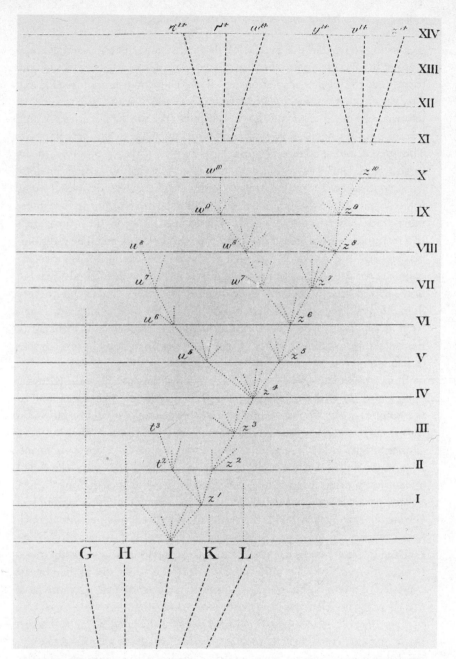

21. The branching tree from Darwin's *Origin of Species*. This example is part of
the treelike diagram used by Darwin to illustrate how species would diverge from
one another in the course of time. Note that, at each stage in the process, each
species throws out numerous small branches (varieties), most of which do not
survive. But once two separate forms are established, they continue to diverge
away from each other.

Seeking to explain this process of divergence and specialization, Darwin realized in 1854 that natural selection was a constant force that would tend to improve the level of adaptation even in a stable environment (see fig. 21). Any variant that was more specialized for the species' way of life would give its owner an advantage, and thus a character like the giraffe's neck would continue to be enhanced long after the initial switch to the new lifestyle. This pressure to specialize generated the divergent trends observed in the fossil record. Darwin's insight may have been inspired by another analogy from political economy (Limoges 1970; D. Ospovat 1981; Schweber 1980; for a contrary view, see Tammone 1995). Adam Smith had shown that it was more profitable for a number of workers to specialize for the various steps in a manufacturing process than for each to do the whole job. The physiologist Henri Milne-Edwards used this notion of the division of labor to explain the specialized functions of the different organs in the body. Darwin now realized that, by the same logic, a piece of territory could support more living things if they diversified to exploit the resources in as many different ways as possible. This would explain why natural selection constantly tended to improve the efficiency of adaptation by increasing the level of specialization.

Biogeography also transformed Darwin's thinking on this issue (Browne 1980; Kohn 1985). He had begun to examine the relative size of genera (i.e., the number of species they contain) and their degree of diversity. He found that small genera fell into two distinct categories: those containing very similar species, and those with highly divergent or even aberrant forms. Darwin concluded that the two types represent the beginning and the end of a historical process that affected all groups in the course of their evolution. When speciation first created a genus, all the component species were still very similar to one another. If the genus survived, it diversified into a much larger number of more specialized species. Eventually, however, after being replaced by later, more successful genera, the individual species were driven to extinction one by one until, shortly before the whole genus died out, it consisted of a few, highly specialized forms. Here again, Darwin came face-to-face with the relentless pressure of natural selection.

The effect of these insights was not always for the good, as far as modern Darwinians are concerned. By focusing on the pressure to specialize, Darwin was led to marginalize his early emphasis on the role of geographical isolation in speciation, as in the Galàpagos (Sulloway 1979a). It now appeared to him that the location most likely to ensure divergence was a large and crowded area where there were no isolated populations. Here, the pressure to specialize might divide a population into two without requiring a geographical barrier, because the animals at each end of the species' range might move toward different ways of making a living. Adaptive pressure itself would pull the population apart without the need for geographical separation. Darwin thus converted to what is now called sympatric speciation (speciation without isolation), although he still recognized that isolation could be a factor in circumstances, such as on oceanic islands. Most modern evolutionists think speciation is always allopatric (involving the intervention of a physical barrier between populations at least at the initial stage of separation). The problem with the concept of sympatric speciation is that, if interbreeding across the whole territory is possible, it is difficult to see what prevents a constant blending of the characters being favored at each end of the species' range.

By 1856 Darwin had created a much more sophisticated theory and had amassed a vast amount of supportive evidence. The climate of opinion also had begun to change, as more scientists became willing at least to consider the possibility of creation by natural law rather than by miracle. Lyell and Hooker encouraged him to publish, and he began what he called his "big book" on natural selection (partly published as Darwin 1975; see also Hodge 1977). The editor of the modern edition of the manuscript, Robert Stauffer, estimates that had Darwin continued, he would have published a large two-volume study sometime in the early 1860s. In fact, the big book was never finished, because Darwin was interrupted; he began to write the much shorter account of the theory we know as *Origin of Species*.

WALLACE AND PUBLICATION OF THE THEORY

The stimulus which prompted Darwin to publish in haste was the arrival in 1858 of Alfred Russel Wallace's paper on natural selection (on Wallace's life, see Beddall 1968; Fichman 1981; George 1964; Marchant 1916; McKinney 1972; Raby 2001; Williams-Ellis 1966). Most accounts of this event credit Wallace as the codiscoverer of the selection theory, and many imply that history has been unjust to him. They ask why the theory should be known as Darwinism when it was also discovered independently by Wallace. There are dark hints that Darwin tried to play down Wallace's role in order to boost his own status. An even more extreme version of this view implies that Darwin plagiarized his principle of divergence from Wallace (Brackman 1980; Brooks 1983; for a more balanced account of their ideas on divergence, see Beddall 1988).

This last argument rests on a bizarre manipulation of the evidence, but there are other good reasons for doubting that a conspiracy of silence deprived Wallace of credit. We know that Darwin developed his basic idea of natural selection two decades earlier than Wallace. Wallace hit on the idea of divergent evolution in the mid-1850s but did not conceive the idea of natural selection until 1858, by which time Darwin was well into the writing of his "big book." Wallace's 1858 paper was brief and hastily written, and we know that, even when published in conjunction with material by Darwin, it had little influence. It would have taken years for Wallace to prepare a book-length account of the theory and its applications. The 1858 paper did not, in any case, contain the essence of Darwin's theory—Wallace's conception of how natural selection works was different from Darwin's and may not have recognized the role of competition between individuals. Darwin was misled by a partial overlap between their ideas and was panicked into thinking that he might lose priority for the whole theory.

Unlike Darwin, Wallace came from a poor background and eventually became a professional natural history collector, supporting himself by the sale of exotic specimens. His first expedition was to South America in the years 1848 to 1852, accompanied by Henry Walter Bates (Beddall 1969). Here, he gained some insight into the problems of biogeography and ecology which had first started Darwin thinking about the origin of species. The ship carrying his collection home for sale was destroyed by fire, but fortunately he was insured; he soon set out on another expedition to the Malay Archipelago, the islands which now constitute Indonesia.

Like Darwin, Wallace took Lyell's *Principles of Geology* with him to South America. He adopted the theory of gradual geological change but soon began to question Lyell's assumption that species were fixed. He became interested in how geographical barriers define a species' territory, arriving at conclusions similar to those Darwin reached in the Galàpagos. Wallace began to feel that creationism required too many arbitrary assumptions to explain the facts of distribution. He was impressed by Chambers's *Vestiges* and used the book to focus his attention on the problem of transmutation. During the 1850s he became convinced that branching evolution occurred, but had no idea what the mechanism of change might be. He began to plan a book on the species question, and in 1855 published an important article stating that "every species comes into existence coincident in time and space with a previously existing closely allied species" (Wallace 1855, reprinted in Wallace 1870, 1891). He began to correspond with Darwin, who expressed high regard for this paper, although Darwin did not disclose that he had already thought of a mechanism of evolution.

Wallace was impressed by Lyell's discussion of the "war of nature" between species, and he approached the question of the struggle from this direction. He read Malthus and—like Darwin—encountered the idea that population pressure generated an inevitable struggle. But he had also read the critiques of Malthus's work by socialists such as Robert Owen, which would have given him a perspective different from that of Darwin (Jones 2002). Significantly, Malthus used the term *struggle for existence* when discussing the competition between rival tribes, and Moore (1997) suggests that Wallace himself had observed firsthand the competition between rival ethnic groups. During a bout of fever, Wallace conceived his own idea of natural selection—apparently on the little-known island of Gilolo, not, as he subsequently claimed, the famous spice island of Ternate (McKinney 1972). He wrote up his idea in a short paper and mailed it to the naturalist he thought best able to evaluate it: Darwin (for the text of the paper, see Darwin and Wallace 1958; Loewenberg 1959; Wallace 1870, 1891).

Despite the widespread assumption that Wallace independently formulated the same theory as Darwin, there are considerable differences in the way they presented their ideas. Wallace's concept of natural selection may have differed significantly from Darwin's (for an evaluation of this claim, see Kottler 1985). A. J. Nicholson (1960) points out that Wallace tended to think of the environment as setting absolute standards of fitness against which the species is measured. Since most individuals normally measured up to this standard, there would be no evolution unless the environment changed. This would be a much less severe concept of selection, similar to Darwin's early model, which still allowed for perfect adaptation. More seriously, it can be argued that Wallace was not thinking of the key Darwinian mechanism—the struggle for existence between individuals within a population—but focused instead on competition between distinct varieties or subspecies (Bowler 1976c). This possibility centers on Wallace's use of the term *variety*, normally interpreted to mean "individual variant." But most of his references to variety seem to imply that he meant the well-marked local varieties known to exist within most species. Given Wallace's early interest in the struggle between rival ethnic groups, it is quite possible that he was envisioning a mechanism of selection based on competition between subspecies, not between the individuals making up each population. This was certainly a part of Darwin's theory and served as the driving force of divergence, but it is not the mechanism of individual selection upon which his reputation rests.

This interpretation of Wallace's idea is made more plausible by a striking difference between the ways in which he and Darwin made their discoveries. A major feature of Darwin's research was his study of animal breeding, and from a very early stage he stressed the parallel between artificial and natural selection. This focused his attention on factors operating within a single population, whether or not it gave him the idea of selection. Wallace made no attempt to study animal breeding and always denied that there was any similarity between artificial and natural selection. Without that model, it is much more likely that his understanding of competition focused on the relationship between the local varieties or subspecies with which he was familiar through his work on biogeography, rather than on the level of individual competition, which was the key to Darwin's mechanism of natural selection.

Whatever the differences between their ideas, Wallace's paper described a process of selection that was close enough to his own to convince Darwin that he had been anticipated in the publication of his twenty years of work. Had he been the dishonest person implied by some critical accounts, he would have destroyed the paper or delayed its publication. Instead, he turned to Lyell and Hooker, who arranged for Wallace's paper to be read to the Linnaean Society of London along with two short extracts from Darwin's writings. One extract was from a letter originally written to the American botanist Asa Gray, confirming Darwin's priority. When the papers were published, there was little comment, perhaps because they were seen merely as a contribution to the ongoing debate about the relationship between varieties and species (England 1997). The momentous topic of evolution could not be brought to public attention by short expositions in a technical journal. But Darwin now realized that he had held back from publication for too long, and that it might be dangerous to wait until he had finished his big book. He began to write the one-volume account of his theory, which was published at the end of 1859 as the *Origin of Species*.

6   The Reception of Darwin's Theory

When the *Origin of Species* was published on November 24, 1859, the 1,250 copies of the first edition were snapped up by booksellers on the first day. There was some support from scientists, but much of the initial response was negative. One clergyman declared Darwin to be the most dangerous man in England. We should not be surprised by the reaction of conservative forces. If evolutionism implied that humans were merely improved apes, and nature only a senseless round of struggle and death, where was the divine source of moral values? And if those traditional values were threatened, so was the established order of society, of which the Church was still a central pillar. Those who favored the theory tended to come from less privileged social backgrounds and were anxious to replace the old hierarchy with a new order favoring the middle classes. This group included the new generation of professional scientists who—unlike Darwin—depended on an income obtained by offering their services to the state or to industry. From the start, the theory was a religious, philosophical, and ideological battleground, and the scientific debates can be understood only in this context. Yet there were technical arguments both for and against the theory, and the debate over these arguments in the scientific community would determine the theory's fate. If it were discredited by the scientists, it could hardly be used to revolutionary effect in society at large.

Historians no longer accept that science and religion must inevitably come into conflict (Brooke 1991; Moore 1979). For every radical such as T. H. Huxley seeking to discredit natural theology there was another figure looking for a reconciliation between faith and the new knowledge. There was no sharp line between the scientific and the broader debates. Many of those who sought to discredit evolutionism did so because they found the philosophical and theological implications unacceptable. But we must not assume that because there was a nonscientific motive involved, the scientific arguments were lacking in substance. Darwin had built his theory on foundations provided by the contemporary understanding of nature, and in some areas—most obviously in the study of heredity—those foundations would soon be undermined. There were scientific arguments against the theory that seemed plausible at the time, even if those arguments no longer apply to our modernized version of Darwinism. There was thus genuine room for scientific debate, although to understand the structure of that debate we have to be aware of the underlying tensions. The issue would be decided by a combination of evidence, rhetoric, and the politics of the scientific community.

David Hull's collection of reviews of the *Origin* reveals the wide spread of opinion within the scientific community (1973b; for collections of primary sources, see Appleman 1970; and Lynch 2001). Even among those who objected to the theory, there was no unanimity. Natural theology had never been a unified doctrine, and those who came to the debate from different backgrounds inevitably focused on different problems. There was an even greater spread of opinion among the general public, and here the survey of the periodical press by Alvar Ellegård (1958) offers a useful barometer of support. Not surprisingly, conservative organs were more likely to object to the theory than were radical ones. But in the decade following the *Origin*'s publication, even the conservatives became more willing to accept the general theory of evolution. This was Darwin's great success: he precipitated a change in public (and hence in scientific) opinion by persuading most educated people that the world was subject to natural law and, by implication, that the origin of species must also be a law-governed process.

Ellegård's analysis also shows, however, that the theory of natural selection was much less successful. Evolution was gradually accepted, but there was widespread suspicion of the selection theory even among scientists. The Victorians did not accept evolution because they were convinced that Darwin had found the correct explanation of how it worked. They accepted it despite having major reservations about natural selection—and this means that they would not have been persuaded by the arguments now used to support the modern Darwinian theory. The transition to an evolutionary worldview must have involved something more than the sheer weight of evidence in its favor. Thus we must be aware of the rhetorical and political strategies of those who were promoting the theory.

James Secord's account of the response to Chambers's *Vestiges* suggests that, in some respects, the Darwinian debate of the 1860s was really only the concluding episode of a confrontation that had begun decades earlier (2000).

Michael Ruse (1996) argues that the evidence for evolution was never strong enough to persuade anyone by itself: people converted to Darwinism because it underpinned their faith in progress, and those who opposed the theory tended to do so because they rejected that faith, perhaps on religious grounds. These are important points which reinforce both the need to involve nonscientific factors in our understanding of the scientific debates and our sense that much early evolutionism was non-Darwinian in character. Yet we must be careful not to lose sight of the controversy within science itself. The elite of the scientific community had not been impressed by Chambers's book, in part because it did not offer a genuinely naturalistic theory which could be used as a guide to research. For this reason, Darwin's initiative must still be taken as crucial, because he forced scientists both to reassess the technical evidence for evolution and to confront the prospect of a truly naturalistic approach to the origin of living things. *Vestiges* paved the way for the emergence of a theory of progressive evolution, but we should not dismiss the debate over the *Origin of Species* as merely an epilogue to a transformation that was already almost complete. Within science there was a genuine debate which Darwin might have lost—and if he had lost, the idea of evolution would have remained a vague notion that might have had ideological appeal but could not have drawn support from the claim that it was founded on a scientific analysis of the natural world.

To analyze the scientific debate in the context of these broader developments, we must make distinctions not immediately obvious to the modern reader. For example, the meaning of the term *Darwinism* has changed over the years. By the 1870s, evolutionism was triumphant, and because everyone associated the theory with Darwin it became known as Darwinism. Darwinism meant little more than evolutionism—the term did not denote a commitment to the Darwinian theory of natural selection. Many of Darwin's leading supporters, including T. H. Huxley, remained suspicious of the selection theory. Even Darwin himself was not a Darwinist by modern standards, because he still admitted a role for Lamarckism. To understand the first generation of evolutionists requires recognizing that they were Darwinists only in the loosest sense. The metaphor of the struggle for existence was often applied in ways that do not correspond to the modern idea of natural selection. Non-Darwinian mechanisms of evolution and a preference for the idea of progress were rampant even among scientists. This is why I have challenged the conventional term *Darwinian revolution* (Bowler 1988): the initial conversion to evolutionism had to be supplemented by a second change of paradigm around 1900, which ushered in the age of modern Darwinism.

THE FOUNDATIONS OF DARWINISM

A few naturalists immediately saw that, whatever the ultimate fate of his theory, Darwin had opened up the question of the origin of species to scientific investigation. Without their support, Darwinism would have been swamped by the attacks of its critics. For a few years the situation hung in the balance, but by the 1870s most scientists had conceded that some form of evolutionism was more plausible than special creation, although the mechanism of change was still a subject of debate. The *Origin* showed that a number of well-known phenomena could be explained if it were assumed that new species were produced by divergence from a common ancestor. An increasing number of naturalists were thereby persuaded to adopt Darwinism as a working hypothesis. By the mid-1870s, the Darwinians were the dominant force in the British scientific community, although many gave natural selection a less prominent position than did Darwin himself. In most other countries as well, with the exception of France, evolutionism triumphed.

The Argument of the Origin

The *Origin* does not begin, as one might expect, with the wider case for transmutation. Darwin starts from the argument for natural selection as a powerful mechanism of adaptive evolution capable of bringing about the diversity of life. The reasoning behind this strategy was that Darwin knew his scientific readers were already aware of the more general arguments for a lawlike process of change. They knew that special creation was suspect as a scientific hypothesis but refused to adopt evolutionism because no plausible mechanism of change had been suggested. Lamarckism had been discredited in the earlier debates (although it soon would make a comeback) and the most radical scientists dismissed the vague law of development advocated in *Vestiges*. Darwin had to break the wall of suspicion by showing that he had devised a scientifically testable mechanism of evolution. He was not dogmatic about the role of selection and always admitted a limited role for Lamarckism. Personally, he thought selection was the dominant mechanism, but his main purpose was to convince his readers that—whatever the ultimate verdict—selection was a hypothesis worthy of evaluation. Science had the right to investigate the question of the origin of species, even if it eventually were to show that other mechanisms were more plausible.

The strategy worked, and the scientific community gradually adopted a more favorable attitude to evolution. Many objections were raised, and Darwin constantly revised the *Origin* to defend his position. There were six editions, the last in 1872 (for the variorum text, see Darwin 1959). In some areas, Darwin conceded the force of his opponents' arguments, but he never wavered from his main position. Some historians have portrayed him as full of inconsistencies, rushing from one point to another in a desperate effort to shore up a fundamentally unsound structure (Barzun 1958; Himmelfarb 1959). The account by Peter Vorzimmer (1970) is more sympathetic but presents a similar picture of him tackling difficulties piecemeal and often weakening one point in order to strengthen another. Some historians imply that Darwin eventually abandoned natural selection in favor of Lamarckism (Eiseley 1958; Himmelfarb 1959). Although he did concede a greater role for nonselective mechanisms, this is certainly not the case.

Some scholars take a more positive view of Darwin's efforts to defend the theory (Ghiselin 1969; Hull 1973b; Ruse 1979a). Given the limitations he faced, his efforts to maintain the integrity of his vision can be seen as heroic rather than pathetic—and of course, his general case for evolutionism was accepted. The differing interpretations offered by historians reflect rival opinions on the validity of the selection theory, but they also remind us to think carefully about what was actually going on. The fact that evolution was accepted while selection remained under suspicion shows that we must evaluate developments in the scientific community at different levels. There was a change in scientists' attitude to the fundamental question of whether they had the right to investigate this area. There was no simple conversion to a monolithic theory brought about by the weight of evidence. We must take into account the social forces at work in the scientific community and in society at large (Moore 1991).

Many of the best arguments for evolutionism were indirect; they supported Darwin's vision of branching, adaptive evolution without confirming that selection was the mechanism at work. These arguments fill the later chapters of the *Origin*, and they shaped the development of late-nineteenth-century biology in many areas where a detailed understanding of the selection theory was not required. Some of Darwin's followers extended the argument in directions that Darwin himself had avoided. This is especially true of efforts to reconstruct the history of life on earth from fossil and other evidence (Bowler 1996), efforts which generated much support for non-Darwinian mechanisms of evolution. Ernst Mayr (1991) argues that Darwin's great triumph was the wide acceptance of the theory of common ancestry, the assumption that many phenomena studied by anatomists, field naturalists, and paleontologists could be explained by the hypothesis of branching evolution. This point must be qualified, though, in the case of some explicitly non-Darwinian evolutionists.

The later chapters of the *Origin* explore all the arguments Darwin had outlined in 1844 (see chap. 5 above). There are two chapters on geographical distribution and one on the evidence from comparative anatomy, classification, and embryology. One area which turned out to be crucial was the fossil record, and here Darwin had to be cautious. He knew that new forms of life often seemed to appear suddenly in the record—the evidence that creationists took (as they still do) as confirmation of miraculous origin. Darwin borrowed Lyell's argument on the imperfection of the geological record, noting that the chances of most species leaving fossil remains were very slight (because fossils were laid down only in certain circumstances). This meant that there must be many "gaps" in the record, which would appear as sudden transitions from one population to another. In fact, two apparently contiguous populations might have been separated by millions of years in which no fossils were laid down. Because he stressed the imperfection of the record, Darwin could not argue that it could be used to reconstruct the evolutionary ancestries of modern species. But the theory predicted that new discoveries would occasionally fill in gaps, and in fact the discoveries of the next few decades encouraged paleontologists to hope that they could reconstruct the whole history of life.

In a second chapter on this theme, Darwin argues that once allowances were made for the gaps, the outline of the history of life was roughly as one might expect if his theory were true. He notes the evidence brought forward by Richard Owen that the earliest members of each class were generalized in structure, the subsequent development consisting of many divergent branches, each moving toward a different specialization. Darwin was anxious to distance his theory from earlier speculations such as Lamarck's, which had assumed an inevitable trend toward progress. He knew that in many cases quite lowly organized forms could be found surviving over vast periods of geological time, the so-called living fossils. This lack of change was quite understandable on Darwin's theory, because once a species had become well adapted to a particular habitat, it would survive unchanged as long as the habitat itself continued to exist. But highly specialized forms were vulnerable to geological changes gradually disrupting their environment, and Darwin's theory was perfectly compatible with the evidence that such species often go extinct in the course of geological time.

Darwin himself had little interest in trying to reconstruct the evolutionary history of particular groups, although in his final chapter he notes that embryology could be used as a clue to ancestral relationships. Earlier stages of development often remained unchanged, so that even distantly related forms could be identified by comparing their embryos, as Darwin had found in the case of the barnacles. But this point does not necessarily support Robert J. Richards's (1991) claim that Darwin endorsed the recapitulation theory. He certainly did not want to set up a ladder of evolution in which lower animals were regarded merely as immature versions of higher forms. The most one could expect was that the ancestry of a group might be identified—not the whole sequence of its evolution.

The *Origin* concludes on an upbeat note. Darwin predicts that naturalists will be converted once they see how the theory can illuminate so many otherwise incomprehensible facts. He also seeks to reassure those who fear the wider implications of the theory. The possible application to human ancestry is conceded in a single sentence. But the stress is on nature as the expression of the Creator's power, with Darwin arguing that it is better to think of Him as governing the world through law rather than by miracle. In only one area is the possibility of miracle implied, when Darwin speaks of life being "breathed" into the first organic beings (1859: 484, 490). Privately, he conceded to Hooker that he hated to have "truckled" to public opinion by using this biblical language, but it was important that he distanced his theory from the controversial topic of the spontaneous generation of life. To offset the negative image of a theory based on struggle and suffering, he suggests in the *Origin* that "as natural selection works solely by and for the good of each being, all corporeal and mental endowments will tend toward perfection" (488–89). Without implying a single hierarchy of evolution leading to humankind, he concludes by allowing his readers to suppose that the process is generally progressive:

> Thus, from the war of nature, from famine and death, the most exalted object which we are capable of conceiving, namely, the production of the higher animals, directly follows. There is a grandeur in this view of life, with its several powers, having been originally breathed into a few forms or into one; and that, whilst this planet has gone cycling on according to the fixed law of gravity, from so simple a beginning endless forms most beautiful and most wonderful have been, and are being, evolved. (490)

The last word is the only occasion in the book in which Darwin uses a derivative of the term *evolution*.

The First Darwinians

Some of the earliest support for Darwin came from field naturalists who appreciated his arguments based on biogeography and the complex interactions between species and their environments. The botanists Joseph Dalton Hooker and Asa Gray had already come to accept the theory as a result of their earlier contacts with Darwin, and both now came out in open support. Alfred Russel Wallace soon returned from his trip to the Malay Archipelago and began writing in defense of the theory. Unlike many of the early Darwinists, however, Wallace endorsed the detailed theory of natural selection, refusing to accept any alternative mechanisms.

Darwin's most flamboyant convert was not, however, a field naturalist but a young morphologist originally specializing in the structure of invertebrates, Thomas Henry Huxley (Barr 1997; Desmond 1994, 1997; Di Gregorio 1984; L. Huxley 1900; Irvine 1955; Lyons 1999). Darwin could not stand the excitement of public debate, but Huxley was willing to take on any opponent in defense of free thought. He was a gifted lecturer and writer (Jensen 1991). Unlike Darwin, he had risen to the middle class by his own efforts as a professional scientist, and his challenge to religion was based on the desire to present science as a source of authority to supplant the Church. He too had been around the world as the naturalist on a British naval vessel, but he had little interest in biogeography and ecology, making his reputation instead from the description and classification of marine invertebrates. Although dissatisfied with natural theology, he had been unable to support transmutation while there seemed no plausible mechanism, and he had ridiculed the notion of creation by law put forth in *Vestiges*. He saw that the *Origin* opened up the topic to scientific investigation, although he could never accept natural selection as a complete explanation. His favorable review of Darwin's book in the London *Times* on December 26, 1859, ensured that the theory was not howled down by the opposition. He also wrote a longer analysis for the *Westminster Review* (reprinted in T. H. Huxley 1893–94: vol. 2).

Huxley would not go all the way with Darwin. He called for a breeding program to attempt to produce a new species artificially and criticized Darwin's commitment to gradualism, suggesting instead that new species sometimes appear by saltations or large-scale mutations. His own work in comparative anatomy and paleontology was at first little affected even by the general idea of evolution (Bartholomew 1975; Desmond 1982, 1994; Di Gregorio 1984). Eventually he did make use of the idea, after the German Darwinist Ernst Haeckel convinced him that the search for phylogenies, or evolutionary genealogies, was reasonable (see below). Whatever his reservations, though, Huxley was determined to ensure that Darwin got a fair hearing.

Huxley has gone down in legend as the scientist who stood up to Bishop Samuel Wilberforce at the famous meeting of the British Association for the Advancement of Science at Oxford in 1860. Wilberforce, who was known as "Soapy Sam" because of the smoothness of his oratory, ridiculed Darwin's theory and its implication that humans were descended from apes. The meeting degenerated into uproar when Huxley declared that he would rather be descended from an ape than from a man who misused his position to attack a theory he did not understand. Modern research has exposed the traditional account of the Huxley-Wilberforce debate as a myth constructed by triumphant Darwinians in later years (Jensen 1988; Lucas 1979). In fact, Huxley's speech was not the most effective: Hooker actually made more impact on many of those present.

Huxley's real triumph was more subtle: he engineered a gradual takeover of the scientific community by those sympathetic to Darwin. Given that the details of the theory were controversial, the outcome of the debate would be determined not only by the evidence but also by the rhetorical and organizational skills of the rival parties. As a member of the new generation of professional scientists, Huxley was determined to wrest intellectual authority away from its traditional sources. Evolution was useful because it demonstrated that science could now determine the truth in an area once claimed by theology (Fichman 1984; Turner 1978). Huxley became a leading public figure, serving as a scientific expert on several government commissions. He was also a member of the informal "X club," an influential group whose behind-the-scenes activity shaped much of late Victorian science (Barton 1998; MacLeod 1970). By exploiting their position in this network, Huxley and his friends ensured that Darwinism had come to stay (Ruse 1979a). They controlled the scientific journals—the journal *Nature* was founded in part to promote the campaign—and manipulated academic appointments. Hull (1978) has stressed how important these rhetorical and political skills were in creating a scientific revolution. The Darwinists adopted a flexible approach which deflected opposition, minimized infighting among themselves, and made it easy for others to join their campaign. Many, like Huxley himself, were not rigidly committed to the theory of natural selection; they were simply anxious to promote the case for evolution. By the mid-1870s most scientists had accepted some form of evolutionism. T. S. Kuhn's prediction that younger scientists would be quicker to accept a revolutionary theory does not, however, seem to be borne out in this case (Hull, Tessner, and Diamond 1978).

The Darwinists' success points to a change of attitude within the scientific community. There were massive debates over the cause of evolution, but few now accepted that it was legitimate to invoke miracles as the source of new species. For Huxley and his fellow opponents of religion, this meant that the argument from design was dead. But others saw evolutionism as compatible with theism, because it was God who instituted the laws which govern the process, and His purpose might be visible in the end products. Conservative scientists such as Richard Owen and St. George Mivart tried to create a non-Darwinian evolutionism in which the guiding hand of the Creator was visible in the purposeful trends exhibited by the history of life (see below). To these scientists, the element of "random" variation built into natural selection seemed to violate the rule of law—Sir J. F. W. Herschel is said to have dismissed selection as "the law of higgledy-piggledy." Here the comprehensive nature of the Darwinian synthesis broke down. David Hull (1985) argues that Owen and Mivart were ostracized from the Darwinian camp for personal reasons—they were both rude to Darwin himself—but in fact their efforts to see a divine plan in the laws of evolution were also unacceptable to Huxley and his followers. The implication that evolution was guided in certain directions by supernatural forces was too obviously a compromise with the argument from design. In the short term, Owen's and Mivart's influence was limited, in part by the clumsiness of their tactics (Bowler 1985; Desmond 1982), but their teleological evolutionism was the forerunner of the openly anti-Darwinian schools of evolutionism which emerged later in the century.

Darwinism Abroad

Darwinism's reception by naturalists outside Britain varied from country to country (Glick 1974; Kohn 1985; Numbers and Stenhouse 1999). Much attention has focused on its reception in the United States (Daniels 1968; Loewenberg 1969; Pfeifer 1974; Russett 1976), where some naturalists with an idealist viewpoint took a very negative stance, including Jeffries Wyman (Appel 1988). The most active critic was Louis Agassiz, by now the country's most eminent naturalist. His idealism presented each species as a distinct unit in the mind of the Creator and left no room for transmutation, least of all by random variations (Lurie 1960; Mayr 1959a; P. Morris 1997; Winsor 1979). But Agassiz weakened his position by arguing that every minutely different form was a distinct creation, including those that many others considered to be merely local varieties. His own students found it impossible to accept this form of creationism, and many of them accepted some version of evolutionism. But they preferred non-Darwinian mechanisms that retained a sense or order and purpose in natural development, and several of them became founder members of the neo-Lamarckian school which flourished in later decades.

The American botanist Asa Gray bore the brunt of Agassiz's attack (Dupree 1959). Like Hooker, Gray was ideally suited to appreciate the arguments from geographical distribution and reinforced the case for evolution with studies of North American plants. His collected papers (1876) combine scientific arguments with a serious effort to show that Darwinism was compatible with religious belief. Scientific support also came from the paleontologist Othniel C. Marsh (Schuchert and Levene 1940). Like Huxley, Marsh played down the details of how evolution worked and got on with the job of showing how the fossil record supported the general idea of transmutation.

French scientists at first expressed little interest and only gradually converted to evolutionism later in the century (Conry 1974; Farley 1974; Stebbins 1974). The selection mechanism held little attraction, and French scientists became active in the neo-Lamarckian movement. Their negative attitude to Darwinism was in part the legacy of Cuvier: French biologists approached the question through comparative anatomy and saw transmutation—if it occurred at all—as a purely structural change. There was no tradition of field studies of the kind that led Darwin and Wallace to the problems of geographical distribution and ecological interaction. Deeper still, the rationalist tradition stretching back to Descartes prevented French thinkers from taking seriously anything so haphazard as a mechanism based on random variation.

The German response was much more positive, although German thinkers, too, evaded the implications of Darwin's selection theory. German biologists had a long tradition of working in comparative anatomy and morphology (the study of biological forms). They welcomed Darwin's book as a springboard from which to develop an evolutionary view of the way living forms had unfolded through time. There were social reasons, though, why the more radical implications of Darwinism were attractive (if not the details of natural selection). Some German scientists, of whom Ernst Haeckel was the most active, were political radicals who saw Darwin's rejection of design as a weapon in their fight against conservatism (Gasman 1971; Montgomery 1974; Weindling 1989a). It has been said that while Darwinism was born in England, it found its true home in Germany (Nordenskiöld 1946; Radl 1930). But this was not Darwin's Darwinism—Haeckel (translation 1876) openly proclaimed his intention of synthesizing the ideas of Darwin, Lamarck, and Goethe, and his work still reflected the spirit of Naturphilosophie.

THE SCIENTIFIC DEBATE

The first stage in the Darwinian revolution cannot be explained as a simple process in which the objective evidence for evolution convinced all scientists that the theory was correct. Evidence for and against evolution or natural selection was evaluated differently by scientists from different backgrounds, and some were more effective than others at promoting their position.

The first phase of evolutionary biology was based not on the selection theory but on the reconstruction of the history of life on earth (Bowler 1996). There were important discoveries of "missing links" in the fossil record. Soon, enthusiastic evolutionists were trying to reconstruct the detailed course of evolution, something Darwin himself had not encouraged. Ruse (1996) has sought to minimize the extent to which Huxley and his colleagues used the theory of evolution in their detailed science, and has dismissed the efforts made to reconstruct the history of life as scientifically worthless. This position is a reflection of hindsight, however, one that evaluates evolutionary biology of the late nineteenth century by modern standards. Reconstructing the history of life proved difficult with the techniques then available, and by the end of the century it was abandoned by the most innovative scientists (although paleontologists still work on ancestries, and their work forms the basis of most popular perceptions of evolutionary biology). But a whole generation of dedicated biologists took this project seriously, and to dismiss their work as insignificant because it failed to anticipate the revolutions of the early twentieth century leaves us no way of understanding why evolutionism was taken up with such enthusiasm by some of Darwin's contemporaries.

Recognizing the premodern character of much late-nineteenth-century evolutionism forces us to ask why scientists accepted the theory. We should not assume that their attention focused on the issues which are crucial today. Too many historical accounts of the nineteenth-century debates have concentrated on the critique of the selection theory as a prelude to the eventual synthesis of Darwinism and genetics. There is certainly an important story to be told about how Darwin's version of selectionism was transformed into modern Darwinism (see Depew and Weber 1995; Gayon 1998; Mayr 1982: chap. 11). But this is the story of Darwinism told from the perspective of modern science: it is not a complete history of evolutionism because it fails to acknowledge that the first generation of evolutionists were driven by other priorities. The historian of science has a duty to reconstruct the steps by which we came to our modern level of understanding. But if we want to understand what drove people to react the way they did in the past, we must be prepared to pay some attention to movements that now appear to have been blind alleys.

Biogeography and the Origin of Species

Among Darwin's earliest supporters were two botanists with a strong interest in biogeography. Joseph Hooker's introductory essay to his flora of Tasmania (1860) was one of the first scientific works to offer open support (Allan 1967; L. Huxley 1918; Turrill 1963). The leading American Darwinist was Asa Gray, who interpreted the distribution of North American plants in evolutionary terms (1876; Dupree 1959). On his return to England, Alfred Russel Wallace provided evidence from his studies of the distribution of animals (Fichman 1981; George 1964; Marchant 1916; McKinney 1972; Williams-Ellis 1966). His 1864 paper on Malayan butterflies (reprinted in Wallace 1870) showed that local varieties were often found on islands, and pointed out the impossibility of drawing a sharp line between varieties and species. He established the boundary still known as Wallace's line, between the Asian and Australian faunas in the Indonesian islands (Camerini 1994; Fichman 1977; Mayr 1954). The boundary was defined by the deep water between Bali and Lombok, which had prevented a land connection from being established even in periods when the sea level was much lower than at present. Wallace went on to coordinate a major project to explain the geographical distribution of animals in terms of migration and adaptation to new environments (Wallace 1876).

Wallace's onetime traveling companion Henry Walter Bates provided a new line of evidence through his investigation of mimicry in the brightly colored insects of the Amazonian forests (1862, 1863; Beddall 1969; Woodcock 1969; for a different interpretation of Bates's work, see Blaisdell 1982). Many insects camouflage themselves to escape detection by predators, but others have acquired a taste which makes them unpalatable, and proclaim the fact with bright warning colors. Bates found that some edible species mimic the warning color of an unrelated inedible form, thus gaining the same degree of protection. This evidence for adaptive coloration favors natural selection rather than Lamarckism, because individual insects cannot alter the colors of their wings.

Field studies showed that species were not the neatly designed units postulated by the creationists, but the studies also revealed confusion among the Darwinians about how species actually formed. Darwin's Galàpagos experiences originally had led him to assume that populations had to be geographically isolated if they were to diverge and become distinct species. But by the time he wrote the *Origin*, he had abandoned this position because he felt that small, isolated populations might not contain enough variation to allow rapid evolution. He now argued that the best conditions for speciation occurred when a large population was spread over a continuous area covering a range of physical environments. Specialization for the different environments would fragment the population and allow the separate subpopulations to diverge into distinct species. When the German biologist Moritz Wagner (translation 1873) insisted that geographical separation was essential to allow the formation of distinct populations and hence new species, Darwin rejected the claim—mistakenly, in the view of most later biologists (Mayr 1959b; Sulloway 1979a).

The History of Life on Earth

Wallace's biogeography project included the use of fossils to trace the point of origin for newly evolved groups, but most applications of evolutionism to reconstruct the history of life focused on morphology and paleontology. Fossil discoveries sometimes revealed the missing links in hypothetical evolutionary genealogies. Although Darwin himself was reluctant to engage in such speculations, his more enthusiastic followers could not wait to begin a complete reconstruction of the development of life on earth, even where no fossils were available. Morphologists were used to working out relationships between groups, and now those relationships could be interpreted not as idealized components in a divine plan but as real connections based on common ancestry. It has been argued that evolutionism did not make much difference to this program of research, but in fact it introduced new constraints because all the intermediate stages had to be real animals capable of surviving in a particular environment.

To the opponents of evolution, the apparent gaps in the record were genuine evidence for the miraculous origin of new forms. They insisted that all the intermediate stages predicted by evolutionists ought to be visible among the known fossils. The fossil record thus became a vital area of debate despite Darwin's efforts to sideline the issue. Some crucial fossil links were discovered just in time to influence the debate. Many gaps were left unfilled, but the opponents of evolution were put on the defensive because they could never predict what new discoveries might turn up.

The project to reconstruct the evolution of life on earth took off first in Germany, where morphology was already established in the universities (Bowler 1996; Coleman 1976; Nyhart 1995; E. Russell 1916). Carl Gegenbaur translated the concept of the archetype into a search for common ancestries (Di Gregorio 1995). He suggested how the first terrestrial vertebrates could have evolved from fish. Even more active was Ernst Haeckel, who coined the term *phylogeny* to denote the hypothetical line of ancestry leading toward a known form (Bölsche 1906; Di Gregorio 1992; R. Richards 1991; Weindling 1989a). Although Haeckel proclaimed himself a Darwinist, his real motivation derived from the radical gloss he put on the idealists' search for the underlying structural relationships. He favored the theory of inheritance of acquired characteristics and called in selection only to eliminate the less successful species produced by the Lamarckian process. Haeckel was a convinced progressionist and portrayed the tree of evolution with a central trunk running up to its crowning product, the human species.

Haeckel's first comprehensive work was a reassessment of morphology on evolutionary lines (1866). His more popular books reached a wide audience in Germany and in translation (1876, 1879). But his hypothetical reconstructions of ancestries were based on serious evaluations of anatomical and embryological evidence that would influence a whole generation of biologists. Haeckel created the wave of enthusiasm for the theory that the embryological development of the individual organism (ontogeny) recapitulates the evolutionary ancestry of its species (phylogeny). The conditions under which recapitulation could take place were specified by Fritz Müller (translation 1869), who warned that not all forms of variation would allow ancestral stages to be preserved (see fig. 22). Haeckel simply assumed that in most cases variation would proceed by adding on new stages to individual development. He went far beyond Darwin in his enthusiasm for using embryology as a model for the history of life on earth. Recapitulation allowed Lamarckism a major role because it saw progressive evolution as the addition of extra stages. It also highlighted a progressionist vision in which lower animals were merely stages in nature's ascent toward the fully mature form of the human species (Gould 1977b; but see also R. Richards 1991, which stresses Darwin's links to this program).

One problem with Haeckel's almost linear model for the ascent of life was that it encouraged the idea of what later would be called living fossils. Every lower animal was a stage in the process by which life matured toward the human form (see fig. 23), and it was thus possible to use living species as illustrations of what the missing fossils must look like. This assumption violates the most important implication of Darwin's branching-tree model, in which ancestral forms are seldom preserved unchanged in later periods. Using anatomical and embryological evidence, Haeckel proposed hypothetical reconstructions of every stage in the ascent of life from the origin of

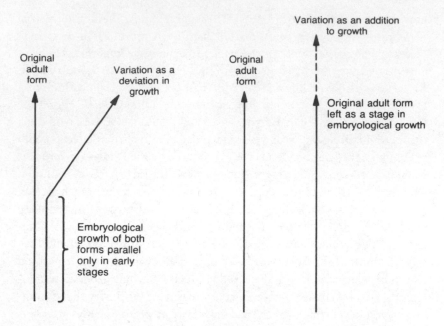

22. Variation and the recapitulation theory. Fritz Müller (translation 1869) outlined two modes of variation defining the circumstances in which recapitulation could take place. If a new character is produced by a deviation in the original pattern of development (left), the old adult form disappears and the new course of development cannot recapitulate the recent history of the species. But if the new character appears by adding a stage to the original pattern of development (right), the old adult form is preserved as a stage in the development of the new species, and recapitulation is observed. This kind of variation is postulated more readily in a Lamarckian scheme, where acquired characters are seen as being added onto the existing adult form and then incorporated into the pattern of embryological development, so that they become hereditary. The Darwinian concept of random variation fits much more easily into the model of deviation.

multicellular creatures up through the ancestry of the vertebrates and of the successive vertebrate classes. His ideas would be debated throughout the rest of the century (see chap. 7).

Other evolutionists became enthusiastic contributors to Haeckel's project to reconstruct the history of life. There was much embarrassment when T. H. Huxley assumed that an apparently organic slime discovered on the deep ocean bed was a relic of the simplest undifferentiated protoplasm from which all higher life had evolved, only to have it soon dismissed as an artificial product of the collecting process (Rehbock 1975). A more extended controversy centered on the discovery in Canada in 1865 of what was

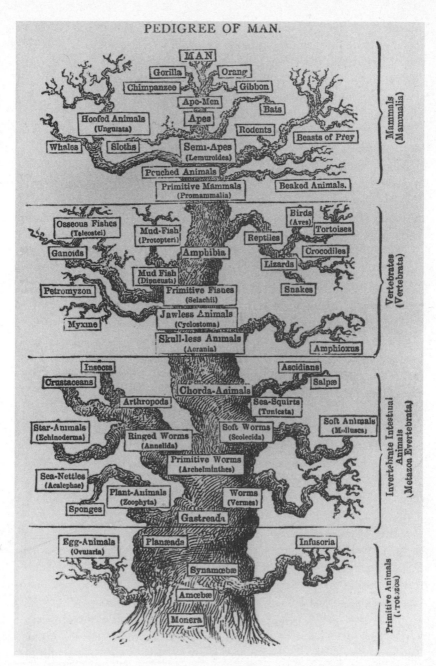

23. Ernst Haeckel's tree of life. Note that Haeckel's tree has a main trunk leading to the human race—all divergent forms are depicted as side branches. From Haeckel's *History of Creation* (1876, vol. 2, facing p. 188).

interpreted as the first evidence for Precambrian life, thought to be a giant foraminiferan and named *Eozoön canadense* (O'Brien 1970). The evolutionists rejoiced at what seemed the first real evidence that the sudden explosion of life in the Cambrian period was preceded by earlier phases of development. Unfortunately, it was soon demonstrated that the "fossil" was the result of purely mineral action within these ancient rocks—it would not be until well into the twentieth century that genuine evidence of Precambrian life was found.

More promising discoveries filled in some of the later gaps in the record of evolution. Attention soon began to focus on the link between reptiles and birds. Huxley first suggested that some small dinosaurs had legs and feet almost indistinguishable from those of modern birds (their fossilized footprints at first had been attributed to birds). This, he claimed, was the first sign of the missing link between the two classes (Bowler 1976a: chap. 6; Desmond 1982; Rudwick 1972: chap. 5). The first specimens of *Archaeopteryx* added further evidence of an intermediate stage: here was a creature with reptilian features, including a mouth with teeth, yet the fossils showed the clear imprint of feathers surrounding the bones. In America, O. C. Marsh discovered more toothed birds (Marsh 1880). These links were too fragmentary to allow a detailed reconstruction of how birds evolved, but they were intermediates which no creationist would have predicted. They showed that the clear distinctions between the modern classes did not extend into the distant past.

A number of more complete evolutionary sequences also came to light, the most famous being the ancestry of the modern horse (see fig. 24). The horse is specialized for grazing and for running on the open plains, its broad hoof being in effect the enlarged nail of a single toe. If the evolutionists were correct, it must have evolved from a less specialized mammal with the normal five digits on each limb. Huxley at first tried to link the horse with certain European fossils—no one would have thought of an American ancestry, since horses were unknown there until introduced by Europeans. But Marsh soon convinced him that the horse must have evolved in America before migrating to Eurasia and then becoming extinct in its original homeland. A whole series of fossils was discovered in the American West which eventually linked the modern horse back to a small, multiple-toed ancestor Marsh named *Eohippus*, the "dawn horse." In a popular lecture, Huxley later proclaimed this sequence to be "demonstrative evidence of evolution" (1888), although the true story of horse evolution has turned out to be more complex that he imagined.

By the later decades of the century, Sir J. W. Dawson of Montreal was

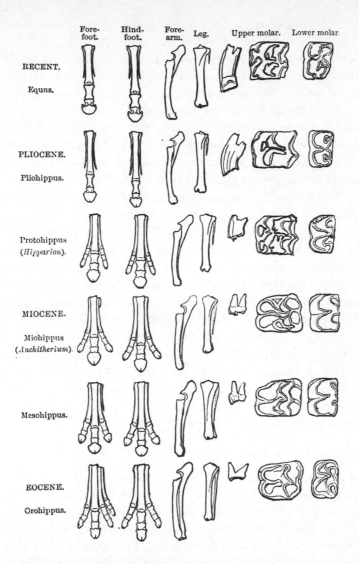

24. Evolution of the horse family. This diagram shows the modifications to the teeth and lower parts of the limbs of the horse, from the four-toed *Orohippus* of the Eocene to the modern horse. The sequence was constructed by O. C. Marsh from American fossils and regarded by T. H. Huxley and others as clear evidence of evolution. It was widely reprinted in the late nineteenth century— this example is from A. R. Wallace's *Darwinism* (1889: 388). Note that, by arranging the fossils into a linear sequence, the diagram gives the impression that horse evolution proceeded steadily in the direction of increasing specialization. Later discoveries have shown that the process is much better represented as a tree with many extinct branches.

almost the only major paleontologist still defending what we would now call creationism (see J. Dawson 1890; O'Brien 1971; Sheets-Pyenson 1996). The fossils seemed to endorse evolutionism, but they were not directly supportive of the Darwinian selection mechanism. The simple fossil trends such as those Marsh constructed for the horse family seemed to offer better evidence for a more directed mechanism than anything based on random variation. Evolutionists who wanted to create a non-Darwinian evolutionism that would preserve some aspects of the old, teleological viewpoint seized upon the fossil record. In some cases they argued that many lines evolved in parallel toward the same goal. In Britain, conservative naturalists such as Richard Owen and St. George Mivart made use of these effects to argue that natural selection was an illusion: evolution was the unfolding of a preordained pattern defined in advance by the Creator (Desmond 1982). Although he had once been an opponent of transmutation and had written a critical review of the *Origin* (reprinted in Hull 1973b), Owen now conceded that the development of life unfolded in regular patterns without the need for miraculous creation (MacLeod 1965; Owen 1866–68; E. Richards 1987; Rupke 1994). He was an opponent of Darwinism without being an opponent of evolutionism. In America, the paleontologists Alpheus Hyatt and Edward Drinker Cope founded a neo-Lamarckian school of evolutionism which also stressed the existence of almost linear patterns of parallel evolution in the fossil record (see chap. 7).

The Debate over Natural Selection

By the 1870s most biologists were convinced that evolution did occur, but distrusted Darwin's claim that natural selection was the chief cause. Wallace was one of the few to accept that selection was indeed paramount (although he made an exception in the case of human origins), but even he disagreed with Darwin on a number of technical points. The two differed over how the sterility between two distinct species is brought about: Darwin believed that it was a by-product of structural diversity, while Wallace thought that natural selection actually favored the buildup of mechanisms to prevent hybridization. The two naturalists also disagreed on the efficacy of Darwin's subsidiary mechanism of sexual selection, described at length in his *Descent of Man* of 1871. Darwin introduced this mechanism to account for characters such as the bright plumage of many male birds, which seemed unlikely to have an adaptive advantage and hence could not have been developed by natural selection. Darwin argued that such characters were an advantage in the competition for mates: once a certain color or structure became associated with courtship, those males in which it was well enhanced would attract more females and leave more offspring. The same argument applied to weapons such as the horns of male deer, used more in competition for females than in defense against predators. Wallace rejected the idea that the choice exercised by females during courtship displays could have a role in developing male characters (Cronin 1991; Kottler 1980).

The disagreements between Darwin and Wallace always took place in a friendly atmosphere, but there were other scientists determined to reject natural selection altogether. Their opposition to the selection theory was often motivated by religious or moral objections to the idea that competition weeded out the poorly adapted products of random variation. But this does not mean their objections can be dismissed as expressions of mere prejudice, and in some cases the critics identified genuine weaknesses in the way Darwin had formulated his ideas.

One line of objections centered on Darwin's methodology (Hull 1973a; Ruse 1975b, 1979a). Had he employed the true scientific method to arrive at his theory, or had he indulged in unwarranted speculation beyond what the evidence allowed? Important figures such as Sir J. F. W. Herschel conceded that there was a role for hypotheses in scientific research, and Darwin self-consciously used what we now call the hypothetico-deductive method (Ghiselin 1969). But many scientists thought that the testing process must be able to establish the absolute truth of the hypothesis—something no modern philosopher of science would accept. Darwin knew that to demand demonstrative proof of his theory was unreasonable—by such a standard the origin of species would never be accessible to science. Even Huxley fell into the trap by demanding that two mutually infertile species should be produced by artificial selection before the theory was proved. At least Huxley realized that the theory was truly scientific because it could be tested, and for Darwin himself the great strength of his theory was that it made sense of a wide range of otherwise inexplicable facts. Opponents who demanded absolute proof and then dismissed the theory as speculation because their standards of proof could not be met were adopting an oversimplified view of the scientific method.

The critics insisted that natural selection was not a *vera causa*, or true cause, of the kind acceptable to science. The Darwinians favored the philosophy of John Stuart Mill, whose *System of Logic* of 1843 (reprinted in Mill 1981–91: vols. 7 and 8) upheld the tradition of British empiricism according to which all knowledge is derived from the senses, and "cause" simply refers to the constant conjunction between events. Natural selection was a plausible cause, even though its actions could not be observed directly, because all the component effects (variation, heredity, population pressure) were empirically verifiable. Yet Mill himself, like Huxley, fell into the trap of demanding absolute proof before the theory could be fully accepted, although both accepted that science could legitimately investigate it.

The opponents of Darwinism aligned themselves with the rival idealist school of thought represented by William Whewell's *Philosophy of the Inductive Sciences* (1847b, 1989; see also Fisch and Schaffer 1990; Yeo 1993). They held that it was possible to intuit truths about nature ahead of empirical investigation, and they believed that the conventional view of species as fixed objects created by design was such a legitimate intuition. Darwin's complex speculative framework was far too fragile to overturn such accepted truths and thus could not be a *vera causa*. Idealists also felt certain of the existence of the First Cause, God, and of the ability of that Cause directly to influence the events we observe. When the secondary causes investigated by science are inadequate to explain things (as they seemed to be in the case of the origin of species), it was perfectly acceptable to invoke the miraculous action of the First Cause. Even if creation were a lawlike process, as many were now prepared to admit, the law of creation would have to accommodate the element of purpose and goal-directedness we see in the world. The fact that Darwin could not provide an explanation of variation showed that there was room for such an element of design to be incorporated into evolution.

Such methodological points were reinforced by detailed arguments intended to show that natural selection could not explain all the phenomena we observe (Gayon 1998; Vorzimmer 1970). In 1864 the botanist Carl von Nägeli insisted that many species exhibit characters so trivial that they could have no adaptive advantage and, hence, could not have been developed by selection. Darwin insisted that Nägeli had exaggerated the extent of such nonadaptive characters because naturalists did not know enough about most species' lifestyles to say with any confidence what was of use to them. But he took the objection seriously and conceded that there were some characters that selection could not produce directly. Darwin invoked the "correlation of growth," which he believed would sometimes link two characters— the equivalent in modern terms of a single gene influencing two separate characters. For selection to promote one character that was an advantage to the organism, it would have to boost the other as a by-product, even if the latter conferred no benefit of its own.

This point was taken up by one of Darwin's most persistent critics, the Catholic anatomist St. George Mivart (J. Gruber 1960). Mivart's *Genesis of Species* (1871) provided a cornucopia of arguments against the selection theory, some of which are still used by modern creationists. He argued that even advantageous characters must have passed through incipient stages when they would have been too poorly developed to be of any practical use—so how had selection passed through this incipient stage? The Darwinists countered with the principle of the change of function enunciated most clearly by Anton Dohrn in 1875 (translation 1993): most adaptive structures were not built from nothing but were modifications of a structure originally used for something else. Even here, Mivart had a counterargument: how could a limb function adequately if it were in a halfway state between, say, a leg and a wing? Surely it would be of no use for walking or flying. Darwin and Dohrn could, however, point to structures that did seem to function adequately in two different ways.

Another objection derived from similarities in structures developed in entirely different branches of evolution. Why, asked Mivart, was the eye of a cephalopod (squid or octopus) almost identical to the human eye? Surely the selection of random variation could not have produced such an entity in two so widely separated groups. Mivart suggested instead that there must be predispositions built into life directing evolution along fixed channels, a position reinforced by the alleged discovery of lines of parallel evolution in the fossil record. Darwin thought this hypothesis unnecessary: the "independent invention" of a similar structure for the eye was not all that unlikely if there were only a limited number of ways in which a functioning eye could be constructed.

The Problem of Heredity

An equally persistent line of objection centered on the internal working of the selection mechanism. Darwin's theory of heredity shared the conventional view of the time that the characters of the two parents would be blended in the offspring—if their hair color, for instance, was different, the offspring would have hair of an intermediate shade. It was accepted that some characters were inherited on an all-or-nothing basis (most obviously sex), but these were thought to be exceptions. Darwin explained heredity with his theory of pangenesis, published in his *Variation of Animals and Plants under Domestication* (1868; see also Farley 1982; Geison 1969; Hodge 1985; Robinson 1979; Vorzimmer 1963). This supposed that each part of the body budded off particles, or gemmules, responsible for shaping the equivalent part in the offspring. The gemmules were transported around the body, perhaps in the bloodstream, and were concentrated in the reproductive organs. Because each parent normally contributed some gemmules for each character, blending would be the general rule. Modern genetics regards the case of all-or-nothing inheritance as a better illustration of the true nature of heredity. But because most characters are influenced by a number of different genes, the effects studied by Mendel are seldom apparent, and most of Darwin's contemporaries shared his view that blending was the normal process. Darwin's views were grounded in a completely pre-Mendelian view of heredity, which made it impossible for him to think in terms of unit characters transmitted from parent to offspring.

Blending heredity had important consequences for Darwin's view of how natural selection would work, and led to a much discussed claim by the engineer Fleeming Jenkin that the selection theory was implausible if blending took place (Jenkin 1867; reprinted in Hull 1973b). Some later commentators, especially Loren Eiseley (1958), have implied that Jenkin's review destroyed Darwin's confidence in the selection theory and led him to abandon it in favor of Lamarckism. Pangenesis does allow for a Lamarckian effect, because changes in the parents' bodies (acquired characters) would be reflected in the gemmules they produced. But there is no evidence that Darwin gave up natural selection, although Jenkin forced him to think seriously about its operations. Vorzimmer (1970) goes to the opposite extreme by minimizing the effect of Jenkin's review. A more reasonable position is that the review forced Darwin to change the way in which he described the operations of selection without requiring him to give up the theory (Bowler 1974b; Gayon 1998: chap. 3). The debate highlighted problems that would be solved with the advent of Mendelian genetics, but the claim that Darwin's theory was unworkable without genetics is an artifact of hindsight. It was perfectly possible to construct a theory of natural selection on the basis of blending heredity.

The objection by Jenkin that has caught historians' attention is his claim that, if selection works on large-scale variations, or "sports of nature," then blending will render it ineffective because the influence of a single sport will be swamped by intermixture with the unchanged bulk of the population. A common analogy is that of putting a single spot of white paint into a bucket of black—stir it up and the effect of the white is vanishingly small. Jenkin was referring to what a later generation of biologists would call hopeful monsters, individuals with a major favorable variation. He insisted that if the hopeful monster had to breed with unchanged individuals, the effect of its new character would be halved in each generation and would soon become negligible. There was a case of a significant monstrosity (we would call it a mutation) creating a new breed under artificial selection. This was the Ancon sheep, produced from a single ram with abnormally short legs. The farmer had realized that the new character was useful to him because it stopped the sheep from jumping over fences, so he bred only from the one ram and succeeded in fixing the new character in his flock. But Jenkin pointed out that this worked only because the farmer was able to force the offspring of the sport to interbreed among themselves. In the wild, the sport's offspring would have interbred freely with the rest of the population and the character would have been swamped.

Darwin's letters show that he was disturbed by Jenkin's review. Yet it is difficult at first sight to see why—he had always insisted that natural selection worked through small individual differences, not through large-scale sports. But he thought of even small favorable variants as single individuals, probably quite rare. That is why he abandoned the idea that evolution worked best in isolated populations, because on this model there might not be any favorable variants at all in a small population. Darwin had still not acquired the modern way of thinking about variation in a whole population. Wallace pointed out to him that there was no need to treat variations as individuals: in any population most characters would exhibit a continuous range of variation. Think of height in the human population, for instance: there is a continuous range from the very tall to the very short, with most people bunched around the mean, represented by the bell-shaped distribution curve. In this situation, half the population is above and half below the mean value. If being tall turns out to be advantageous, half the population—which is, by definition, of above average height to some extent—will gain some benefit. In these circumstances blending is not a problem because no swamping is possible. Half the population is favored, half disadvantaged, and the effects of their differential breeding will be significant.

Jenkin himself conceded this point but claimed that selection acting on such a continuous range of variation would change the population only up to a fixed limit defined by the current maximum. No further evolution would be possible, just as artificial selection seemed to reach a limit and could never turn the species into something really new. Darwin rejected this notion of a fixed limit, holding that over a long period of time new variations would always appear and evolution would continue. When this belief was linked to Wallace's view of continuous variation, it was possible to construct a plausible mechanism of selection based on blending heredity.

It may seem odd that an engineer like Jenkin should attack Darwin's theory, but Jenkin was in touch with the leading physicist of the day, William Thomson (Lord Kelvin), who was disturbed by the implications of Darwinism. Kelvin did not oppose evolution, but he rejected any materialistic application of the idea and shared the view that something more purposeful than natural selection must be involved. His work in thermodynamics offered an indirect way of attacking Darwin because it could be used to undermine Lyell's estimates for the vast age of the earth on which Darwin relied. Because Darwin assumed that natural selection was an immensely slow process, he required a long timescale for evolution, and in the *Origin* he included an estimate of a particular case of erosion (of the Weald in southern England) as having taken up to three hundred million years. Kelvin knew that such estimates were inspired by Lyell's steady state view of earth history, and by the early 1860s he was already challenging the uniformitarian theory by insisting that it was inconsistent with the new science of energy (Smith and Wise 1989: chaps. 16, 17). The earth was hot inside, so it must cool down—it could not maintain itself in a geologically active state indefinitely. He also estimated the time it would take for the planet to cool from a molten state, concluding that the total age might be only a hundred million years. Lyell's theory, and along with it the theory of natural selection, was thus unacceptable by the standards of the latest physical theories. The impact of Kelvin's attack grew over time, especially as he gradually reduced his estimates of the time available (Burchfield 1975). By the time Darwin died in 1882, he was seriously worried by this attack. The much shorter time span of earth history accepted by physicists and geologists in the later nineteenth century played an important role in stimulating the introduction of non-Darwinian theories of evolution.

DARWINISM AND DESIGN

Surprising as it may seem in today's world of revived biblical literalism, there was little opposition to Darwin's book on the grounds that it challenged the Genesis account of creation. The geological controversies in the early decades of the century had convinced most educated people that the text of Genesis must be understood in a nonliteral way that would be consistent with the development of the earth over a vast period of time. The challenge of Darwinism was that it turned nature into an amoral chaos showing no evidence of design by a creator. The argument from design was central to the pre-Darwinian worldview, at least in the English-speaking world, and its defense was a leading theme in the debates of the 1860s.

Naturalistic theories of evolution played an important role in the process by which nineteenth-century thought became secularized (Chadwick 1975; Mandelbaum 1971; Young 1970b). But Darwinism was not the only issue, and many in the Church were more concerned about other threats (Chadwick 1966; Symondson 1970). The evangelical movement was urging a return to a religion of personal salvation, while some Anglicans' concern for ancient tradition led them back to Rome. Even more disturbing was the debate over the "higher criticism" sparked by the translation of D. F. Straus's *Life of Jesus*. This treated the Bible as merely another ancient document and encouraged some liberals to doubt the veracity of many of the gospel stories, especially those involving miracles. When a book called *Essays and Reviews* offered some British contributions to this movement in 1860, the debate in some quarters overshadowed that concerning the *Origin of Species* (Brock and MacLeod 1976; Ellis 1980; Willey 1956).

Although the Darwinian debate was important in this context, it was no simple clash between science and religion. Those who portrayed the debate in such simplistic terms were active participants in the conflict who wished to conceal the existence of a middle ground (Draper 1875; White 1896). Many of the scientists who raised objections against the selection theory were motivated by moral or religious concerns. A few continued to reject evolution altogether, including Louis Agassiz (Lurie 1960) and J. W. Dawson (Cornell 1983; O'Brien 1971; Sheets-Pyenson 1996). At the opposite extreme, T. H. Huxley's radical scientific naturalism offered a powerful challenge to orthodoxy and was strenuously resisted by those with more conservative views. But there were many liberal Christians willing to accept the theory of evolution provided it could be reconciled with the argument from design (Durant 1985; J. Greene 1961; Moore 1979). Some strict Calvinists welcomed the Darwinian scheme because they saw that natural selection was not a source of inevitable progress and thus would allow humanity to be portrayed as a fallen race in need of salvation (Moore 1979).

Most of those who worked for compromise tried to evade the more materialistic implications of Darwinism so that evolution could be seen as a purposeful process (Elder 1996). Clergymen such as Charles Kingsley and Frederick Temple worked actively to promote a liberal theology in which creation was seen as an indirect process controlled by the laws God had instituted. The term *Darwinisticism* has been suggested for this teleological view of evolution (Peckham 1959). But this clumsy term seems redundant, given that many doubted that there was room for natural selection within the compromise position. Darwin's thinking had roots in the utilitarianism of Paley's argument from design, but he had come to realize how difficult it was to reconcile the theory of natural selection with goodness of God, given the theory's reliance on struggle and a relentlessly selfish nature which permitted the evolution of parasites whose adaptation caused suffering in their hosts (Gillespie 1979; Moore 1989). Darwin doubted that any natural theologian would want to make God even indirectly responsible for such suffering. To many, it seemed that evolution could be made acceptable only by replacing natural selection with something more purposeful and humane. But was it possible to have a genuinely scientific theory in which the course of evolution was guided by supernatural forces? This was the question on which Chambers's *Vestiges* had foundered, as far as many scientists were concerned, and it would remain problematic so long as there were efforts to explain the course of evolution in terms of future goals rather than mechanistic laws. Many of the non-Darwinian mechanisms of evolution that would be suggested later in the century would emerge from the attempt to grapple with this question.

The notion of a law of creation was implicit in the argument of *Vestiges* and had been endorsed by respectable figures such as Baden Powell (1855) before Darwin published. The widespread acceptance of evolutionism in the 1860s made it essential for those who still thought in terms of design to see evolution as the unfolding of the Creator's plan or purpose. But it seemed obvious that natural selection's reliance on random variation made it incompatible with the argument from design: how could God control a process which incorporated chance rather than law at every step? In 1861 Sir J. F. W. Herschel wrote of the evolutionary process: "An intelligence, guided by a purpose, must be continually in action to bias the direction of the steps of change—to regulate their amount—to limit their divergence—and to continue them in a definite course. . . . On the other hand, we do not mean to deny that such intelligence may act according to law (that is to say on a preconceived and definite plan)" (Herschel 1861: 12). The belief that evolution was actually shaped by the Creator's will has been called theistic evolutionism, and it became a popular compromise in the 1860s. Richard Owen's theory of derivation (1866–68: vol. 3) proposed an innate tendency for species to change in directions which manifested creative purpose in the variety and beauty of its results. St. George Mivart's assault on the selection theory also was intended to provide room for divine purpose in the course of evolution. Ellegård's survey of the popular press (1958) reveals the popularity of such notions outside the scientific community.

Scientists such as Owen and Mivart insisted that there were regularities and patterns in the history of life which were too orderly to be explained as products of a process governed by random variation and local adaptation. This search for an underlying order or pattern in evolution represents what has been called the "idealistic" version of the argument from design (Bowler 1977, 1983). God's designing hand was seen not in the benefits of adaptation conferred on individual species but in the overall harmonies of the plan of nature. The physiologist William Benjamin Carpenter (1888: chap. 15) expounded this viewpoint by demonstrating orderly patterns of evolution in the beautiful shells of minute sea creatures, the Foraminifera. Many advocates of this position conceded that everyday variation might be random as Darwin supposed, but they proposed the existence of a different kind of variation, perhaps one involving sudden saltations by which new species were formed in accordance with some preordained plan. These arguments illustrate the difficulty that many scientists—to say nothing of religious thinkers—had in accepting Darwin's completely naturalistic theory. Theistic evolutionism was a halfway house which tried to admit evolution and the rule of natural law while retaining the hope that the laws were directed by a supernatural intelligence which provided visible proofs of its activity.

Was there any hope of seeing natural selection as the expression of divine purpose? The most notable effort to incorporate Darwinism into the teleological viewpoint came from Asa Gray, a deeply religious man who felt it necessary to show that his acceptance of Darwinism did not threaten his faith (Dupree 1959; and more generally, Persons 1956). Gray's collected essays (1876) offered scientific support for Darwin's theory but also sought to defend it against the charge that natural selection implied atheism because it negated the argument from design. Natural selection explained adaptation by law rather than by miracle, and the theist always has assumed that the laws of nature are expressions of God's will. No one accused Newton of atheism because he showed that the planets are maintained in their orbits by natural law, so why should anyone accuse Darwin for showing that species are maintained in a state of adaptation by an equivalent process? Gray used the analogy of a billiards game: is the ball's course determined by natural law or by design? Obviously, by both—the player uses the laws of physics to carry out the plan he had in mind when he made the shot. The scientist merely shows how the universe works, leaving the theologian to seek the ultimate purpose.

The argument implies that any mechanism for maintaining adaptation is compatible with design. But this view is difficult to sustain in the case of natural selection, where the element of chance variation seems to make the outcome unpredictable and where death is the driving force of change. Natural theologians always had had to explain away the existence of suffering, and Gray tried to turn this dilemma to his advantage by suggesting that natural selection at last gives us an explanation: the suffering of the unfit was a regrettable but necessary by-product of the mechanism God had chosen to achieve the greater good of keeping species well adapted. In the end, however, Gray conceded that there were problems with this suggestion and implied that perhaps variation was not random after all, so there were really no unfit individuals (the "scum of creation" one critic had called them) to be eliminated: "Wherefore, so long as gradatory, orderly, and adapted forms in Nature argue design, and at least while the physical cause of variation is utterly unknown and mysterious, we should advise Mr. Darwin to assume, in the philosophy of his hypothesis, that variation has been led along certain beneficial lines" (Gray 1876: 148). Here Gray implied that the Creator has established laws of variation which can somehow anticipate the future adaptive requirements of the species.

Darwin responded in the conclusion to his *Variation of Animals and Plants under Domestication*. All the evidence from breeders, stated Darwin, suggested that variation was not directed along a predetermined course—that explained why selection was necessary for consistent change to be produced. More seriously, a real conceptual problem faced the scientist confronted with Gray's suggestion. We do not think of the laws of nature as having been set up to anticipate the future needs of living things. We would be most surprised if the rocks falling off a cliff face were all nicely shaped for us to use in house building. We do not expect nature to load the dice in this way, and to suggest that in the case of variation the dice *are* loaded is to imply that a supernatural influence pervades the whole of nature. Scientific investigation would be unable to explore how the element of teleology was incorporated into the laws.

This point became clear in a suggestion made by the duke of Argyll, who wrote a number of popular critiques of Darwinism (1867, 1898). Like Gray, Argyll wanted to defend the argument from design without rejecting evolution. He suggested that rudimentary organs might be structures being prepared for use by the species in some future environment. To Darwin, these structures were vestiges of organs which had once been useful, but which had dropped out of use because the species had changed its lifestyle. But Argyll supposed that the laws of variation could somehow see into the future, recognize what the species will need, and then begin to prepare new organs which were not at first of any value. This is to build an element of teleology into the evolutionary process that most scientists would find unacceptable, and Wallace challenged Argyll on this point.

Argyll also illustrates another line of support for theistic evolutionism based on the assumption that many characters were not actually useful to the organism. His *Reign of Law* (1867) argued that the brightly colored plumage of many birds could have no adaptive benefit and suggested that the Creator had established the laws of nature to generate structures which we would find beautiful. Darwin could not accept the claim that one species could evolve characters for the benefit of another, and his theory of sexual selection was developed to provide an alternative explanation of these colors. Argyll thought the colors were so bright that they went beyond anything that would be of use in the search for a mate. They provided evidence that the laws of evolution were designed to achieve a purpose beyond mere utility. The scientific community eventually realized that the kind of open invocation of God's designing hand reflected in the writings of Gray and Argyll was unacceptable. Lamarckism and other non-Darwinian theories began to gain ground in the later decades of the century precisely because they provided mechanisms of evolution which seemed more compatible with the belief that they had been instituted by a divine lawgiver.

HUMAN ORIGINS

Darwin knew that the application of his theory to human origins would be its most controversial aspect. The reaction to Chambers's *Vestiges* had confirmed the strength of conservative opinion on this issue and ensured that everyone was aware of the materialist implications of evolutionism. To avoid further inflaming the controversy, Darwin decided not to deal with the emergence of humans in the *Origin*. But he felt that it would be dishonorable to conceal his opinions, and he inserted a single sentence at the end hinting that "light will be thrown on the origin of man and his history" (1859: 488). The question of human evolution almost immediately became central to the debate, even though Darwin himself did not publish his views until he brought out *The Descent of Man* (1871).

The problem arose from evolution's challenge to the traditional view of the status of humanity's mental and moral qualities. These always had been assigned to the spiritual world as characteristics of the soul, not the body. Christians had assumed that animals had no spiritual faculties, so there was a complete gulf between the animal and human kingdoms. Evolution would bridge this gulf and imply that all human characters are parts of nature, subject to natural law. To preserve the uniqueness of humankind, evolution would have to be rejected outright or qualified by the assumption that something very special, and presumably of supernatural origin, had happened on the one branch of evolution leading toward humans. Alternatively, the human situation would have to be reassessed so that the meaning for our lives came from our position at the head of the panorama of evolutionary progress. For this to be plausible, evolution had to be presented as an essentially purposeful process. Darwin and the evolutionists also had to explain how nature had generated the higher faculties of the human mind (R. Richards 1987).

Humans and Apes

Conservative thinkers were appalled at the claim that we might be no more than superior apes. The latest discoveries from Africa reported the gorilla as a ferocious animal, hardly what one would choose for an ancestor. Even some of Darwin's supporters found this aspect of the theory hard to swallow, and at first the evolutionists had to struggle hard to gain a hearing for their natural account of human origins. Since Darwin himself did not contribute to the debate, it was left to his followers to make the case. The Huxley-Wilberforce debate of 1860 came to a head over the question of human origins, when Wilberforce asked if Huxley claimed descent from an ape on his grandmother's or his grandfather's side and received an appropriately sharp reply.

Huxley was already working to establish a close similarity between human and ape anatomy, especially in the structure of the brain. He had clashed with Owen on this issue, and at the 1860 British Association meeting the two fell out in public. Huxley's work on the topic was collected in his book *Man's Place in Nature* (1863, reprinted in Huxley 1893–94: vol. 8; on the debate, see Cosans 1994; Rupke 1993: chap. 6; and L. Wilson 1996). Owen defended the traditional classification by Cuvier and Blumenbach, in which apes and humans were assigned to two different orders in the system of classification. The apes were the Quadrumana ("four-handed," because their feet were anatomically similar to hands), while humans were the only species in the order Bimana (two-handed). Huxley argued that the differences were so trivial that apes and humans belonged in a single order, the Primates. There was also a highly technical disagreement on the structure of the brain: Owen claimed that the human brain contained a structure, the hippocampus minor, not possessed by apes. The traditional interpretation has been that Owen's science was unsatisfactory, perhaps because he relied on preserved specimens to dissect. But Huxley cleverly manipulated his presentation of the evidence in order to discredit Owen. He convinced many of his readers that humans cannot be separated from the apes by any large gulf based on their physical structure (see fig. 25).

Showing that humans were closely related to apes reduced the gulf between them but did not prove an evolutionary link. Fossil evidence of an intermediate form, popularly known as the missing link, was needed. In fact, there would be a whole series of intermediates, assuming the process of evo-

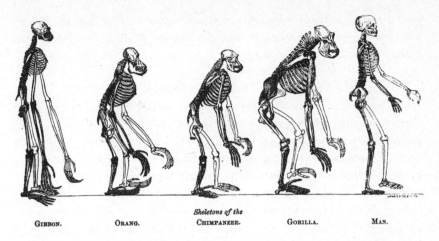

Skeletons of the
GIBBON.   ORANG.   CHIMPANZEE.   GORILLA.   MAN.

25. Frontispiece to T. H. Huxley's *Man's Place in Nature* (1863). Huxley lined up the skeletons of the gibbon (shown at a scale twice that of the others), orangutan, chimpanzee, gorilla, and human to show that the basic structure of the skeleton is the same in humans and apes.

lution was slow and gradual. Darwin's branching model of evolution meant that there would be no simple ladder of ascent from a living ape (chimpanzee, gorilla, or orangutan) to the human species. Humans and the modern apes had diverged from a common ancestor, and although that ancestor might be classified as an ape it would not be identical to any of its descendants, because all had changed to some extent. Humans had changed more than the other Primates, but the great apes had become more specialized for living in the forest. However complex the process, though, the case for evolution would be immensely strengthened if even a single intermediate fossil could be discovered. Huxley confronted this problem head on by examining in detail the only two ancient human specimens available to him. (On hominid fossil discoveries, see Bowler 1986; Reader 1981.)

The Engis skull from Belgium was ancient but anatomically fully human. More promising was the Neanderthal skull from Germany, discovered in 1856. It had thick brow ridges and a receding forehead, much like an ape, and was thus a candidate for the missing link. It had been dismissed by some authorities as the skull of an idiot or perhaps a pathological throwback, or atavism. Huxley insisted that the skull was not pathological (and indeed many similar remains were subsequently discovered, confirming that there once had been a Neanderthal race or species living in Europe). It was one of the most apelike human skulls he had ever seen, yet its cranial capacity was fully modern. Surely the missing link would have had a brain intermediate in size between that of an ape and a modern human. Huxley was honest enough to admit that, despite its apelike appearance, the Neanderthal skull could not be seen as an evolutionary link between humans and our hypothetical ape ancestors.

In Germany, Ernst Haeckel produced his book *The Evolution of Man* (translation 1879), which sought to reconstruct the stages of development from anatomical and embryological evidence. Haeckel's use of the recapitulation theory encouraged him to think in terms of a more or less linear ascent of life toward the human form (although he admitted many side branches). He also tended to neglect Darwin's warning that the common ancestor would not be exactly like a modern ape, presenting the gorilla as the last stage in evolution before the emergence of the first primitive humans. He coined the name *Pithecanthropus alalus* (ape-man lacking speech) for the hypothetical intermediate he hoped would be found in the fossil record.

Archaeology and Human Antiquity

The absence of a fossil link between humans and apes was not surprising to anyone who accepted Darwin's warnings about the imperfection of the fossil record. Indirectly, however, the evolutionists benefited from another line of evidence which had only recently become available: the archaeological record indicating that Stone Age humans had been in existence on the earth for long periods before the dawn of recorded history (Bowler 1989a; Daniel 1975; Grayson 1983; Hammond 1980; Trigger 1989; Van Riper 1993).

In the early nineteenth century it was still accepted that, despite the great age of the earth indicated by geology, humans had existed for only a few thousand years. The biblical chronology was still valid for human history. This belief had been threatened in the 1840s, when Jacques Boucher des Perthes discovered stone tools in gravel beds in the Somme valley, in northern France, which also contained the bones of extinct animals. The beds were geologically recent but ancient in human terms, and no one accepted the finds as genuine. In the late 1850s a group of British geologists began to find similar evidence in caves at Brixham, Devon, and elsewhere. They visited Boucher des Perthes's sites and confirmed that his discoveries were genuine. There was a sudden revolution around 1860 in which the antiquity of our species was extended back into the latest stages of the geological record. Charles Lyell's *Geological Evidences of the Antiquity of Man* (1863) summed up the latest developments, although he added little of substance to the discoveries (Bynum 1984).

The archaeological evidence was more compatible with an evolutionary viewpoint than with the old version of human history. If humans had evolved gradually from an ape ancestry, one hardly would expect our ancestors to appear suddenly with a fully developed civilization only a few thousand years ago. Brain size and intelligence would have had to expand very slowly beyond the ape level, and at some point in this process the purely human attributes of material technology, language, and culture would have begun to emerge. The archaeological discoveries confirmed not only the vast span of human history but also the primitive level of ancient technology. The evidence comprised mostly stone tools and weapons consistent with a hunter-gatherer level of subsistence. The notion of a stone age soon became popular, and the evidence increasingly suggested that there was a progressive sequence of technological development. The evolutionist John Lubbock coined the terms *Paleolithic* (Old Stone Age) and *Neolithic* (New Stone Age) to denote the era of primitive chipped stone tools and the later period when more sophisticated polished stone tools became available (Lubbock 1865). The Stone Age fitted into a sequence already established in which stone was eventually replaced by bronze and then by iron.

Like many evolutionists, Lubbock tried to understand the nature of our Stone Age ancestors by studying the culture of modern "savages" such as the Australian Aborigines, who had still been using stone tools when first encountered by Europeans. Savages were now increasingly seen as "primitives"—peoples who had retained a lifestyle and culture typical of an earlier period of human history. Somehow, they had been left out of the great march of progress leading to modern industrial civilization (on the evolutionary model in anthropology, see chap. 8). For those who also accepted the theory of biological evolution, including Lubbock, there was an overwhelming temptation to assume that a primitive level of culture had been retained by people who had not advanced intellectually as fast as the triumphant Europeans. Evolutionism thus contributed to a growing sense of racial superiority among whites, in which other races were perceived as relics of earlier stages in human evolution. In effect, modern "primitives" became a model for the missing link—indeed, it was almost as though the link were not missing at all, because it was preserved by these living fossils even though the actual fossils were not available.

Darwin on Human Origins

In his *Descent of Man* (1871), Darwin himself promoted this progressionist model of human mental and cultural evolution—in fact, he cited evidence reported by physical anthropologists claiming that nonwhite races had smaller brains and more apelike features. But unlike many of his fellow evolutionists, Darwin wanted to understand the actual processes which had allowed one branch of evolution to develop such an expanded level of mental faculties. Cultural evolutionism simply assumed that progress was inevitable—the real problem was to explain why some races had lagged behind. It favored a linear model of development, with Europeans at the top and "lower" races illustrating what European ancestors had looked like. Darwin, however, knew that evolution was a branching process. Even if one assumed that mental and cultural progress represented the means by which the human branch had adapted to its environment, there was still the question of why our ancestors had adopted a specialization different from that of the apes. If more intelligence was an advantage, why had not the apes chosen this course too? Darwin accepted that once the human branch was established, intellectual progress was inevitable, but he was enough of a biological evolutionist to ask about the process of divergence from a common ancestor. This was a question that many of his contemporaries ignored in the rush to create the linear model of progress.

Darwin realized that he had to minimize the alleged gulf between human and animal mental faculties. He tried to show that the differences were only in degree, not in kind: there were no totally unique human characters because all were possessed to at least some slight extent by the higher animals. Humans were superior to the animals, but not separated from them by an unbridgeable gulf. He argued that animals displayed true intelligence as well as instinctive behavior. They communicated with each other through a primitive form of language and they had social instincts which led them to cooperate with other members of their group and thus served as the basis for self-sacrifice and morality. He told the story of a monkey which risked its own life to save that of a friendly zookeeper who had been attacked by a fierce baboon. Darwin also argued that animals had the full range of emotions experienced by humans, devoting another book (1872) to arguing that human facial expressions could be explained by noting the equivalents in the higher animals.

Darwin went too far in his desire to "humanize" animal behavior. He accepted unverified travelers' tales which exaggerated animals' intelligence and foresight and their capacity for cooperation and unselfishness. He identified warning cries with the far more complex process of communication involved in human language—and dismissed "primitive" human languages as clicks and grunts similar to those made by animals. He was quite right to note the degree of cooperative behavior in some highly social species, although his critics argued that he had not provided evidence of anything equivalent to the human awareness of moral values.

Darwin felt he had made a case for breaking down the gulf between animals and humans. He also wanted to explain how a process of purely natural evolution could have produced the massive extension of our mental and moral faculties. In addition, a few physical characters were difficult to account for: even Wallace thought that the absence of body hair was difficult to explain, given that all our close relatives had dense coats. Here, Darwin invoked his theory of sexual selection, arguing that less hair had become a sexually attractive character and thus had been favored over many generations because those with less hair had mated more frequently. For this reason, *The Descent of Man* contains his most extensive discussion of sexual selection.

To explain human intelligence, Darwin could have called on the progressionists' assumption that this was an advantage that would be boosted by natural selection. But he realized that this assumption begged an important question: why had the human branch of evolution been able to exploit this advantage while the apes had not? He theorized that the difference was brought about by an adaptive change involving a transition from a forest environment to the open plains. Our ancestors had moved out onto the plains and had become fully upright because this was now the best means of locomotion. This had freed their hands from playing a role in locomotion among the trees; thus the hand, with its fully opposable thumb, had developed a much more precise ability to grasp objects, and this consequence of the adjustment to bipedalism indirectly had favored selection for increased intelligence. Our ancestors had developed bigger brains because their hands allowed them to make use of sticks and stones as primitive tools. The apes had stayed in the trees, where their hands continued to be used mainly for grasping branches, and thus had not had the need for additional intelligence.

The details of Darwin's theory were highly speculative, but he had hit upon an important point ignored by almost all his contemporaries. His theory implied that the divergence of humans and apes from a common ancestor required an adaptive explanation. The higher human faculties were only a by-product of this change in the mode of locomotion by one group of African apes. He predicted that our ancestors had stood upright before they began to develop bigger brains, something that was confirmed by the fossil record only in the following century. Most of Darwin's contemporaries ignored this aspect of his thinking because they assumed that humans were the goal of a triumphant progression through the animal kingdom, not an unexpected consequence of a small group of Primates changing their habitat.

A more serious problem for the theological critics was the origin of the moral faculties. Darwin argued that moral values were merely the intellectual expression of social instincts implanted in us by evolution because our ancestors lived in social groups. We were willing to recognize the rights of others, and even to sacrifice our own interests for those of the group, because the instinct to behave in this way was useful to creatures living a highly social lifestyle and thus could be developed by evolution. Darwin thought that early humans had been willing to work for the good of their own tribe alone (as, he claimed, did many "savages" today), but our increasing intelligence led us to generalize this instinct to the whole human race, generating universal moral values and the religious sanctions used to enforce them.

At first sight, it would seem unlikely that natural selection could enhance an instinct which left the individual at a disadvantage with respect to others: surely any individual who behaved this way would be at risk in the struggle for existence. Here, Darwin cited an idea first suggested by Wallace corresponding to what later evolutionists called group selection. He supposed that the struggle for existence took place not only among individuals but also among groups. A group composed of individuals who instinctively cooperated with one another would do better than another whose members constantly fought among themselves, so the latter would be eliminated. He also invoked Lamarckism, arguing that we learned the advantages of living in and cooperating with a group, and that such learned habits eventually might be converted into inherited instincts.

In this area Darwin's thinking was in tune with the attitudes of his time (R. Richards 1987). Herbert Spencer had proposed a Lamarckian theory of mental evolution in which human nature was shaped by evolution. The mind was not a blank slate because the results of experience were gradually embedded in the structure of the mind by evolution, creating instinctive patterns of thought and behavior which predisposed our mental lives to move along predetermined channels. We could still learn to cope with new situations, of course, and if the situation became a permanent part of our species' lifestyle, the learned habits would eventually become new instincts. Spencer's theory assumed that, as the species' interaction with its environment became more complex, so its mental faculties would become more advanced. When human language and culture first emerged, they provided an even more stimulating environment in which the mind could develop.

Defending Human Uniqueness

None of this convinced the skeptics, who claimed that human moral faculties were completely different from anything possessed by even the higher animals. Some of those who accepted the basic idea of evolution for the human body nevertheless thought that something quite exceptional must have happened to create the human mind. The critics maximized the gulf between animal and human mental powers and insisted that we had unique faculties that could not be seen as extensions of animal mental powers. Such a view underpinned Owen's opposition to Huxley and was also taken up by Darwin's most active scientific critic, Mivart. The duke of Argyll made a telling point against Lubbock and the archaeologists in his *Primeval Man* (1868; see also Gillespie 1977). Although Lubbock had easily disposed of critics who tried to defend the old view that the earliest humans were fully civilized, he was on much shakier ground when he attempted to reconstruct the "primitive" culture and morality of our distant ancestors. They might not have had the time to develop an advanced technology, argued Argyll, because that required a long sequence of individual discoveries—but their moral standards could have been as highly developed as ours. Pointing to stone tools and making a comparison with the crudest descriptions of modern "savages" was no proof that these people were not fully human.

Even some of Darwin's closest supporters could not go the whole way with him on this issue. In his *Antiquity of Man* (1863), Lyell accepted a progressionist view of the fossil record and conceded some plausibility to Darwin's theory, but he speculated that the human species could have been formed by a sudden leap that introduced a new level of awareness into the world. He had long been disturbed by the possibility that evolutionism would destroy the status of humankind, and was now able to accept the basic idea of evolution only with this exception (Bartholomew 1973). Darwin wrote that this part of Lyell's book made him "groan"—he could see no point in making the origin of humanity a separate issue.

Wallace, too, became increasingly disturbed on this point. At first he had gone along with Darwin and had even made the important suggestion that the differentiation of the human species into distinct racial groups might have taken place before the final emergence of our full mental powers. But by the late 1860s, he was having second thoughts, which seem to have coincided with his growing interest in spiritualism (Kottler 1974; R. Smith 1972; Turner 1974: chap. 4). If the spirit could survive the death of the body, then something had emerged that transcended the physical evolution of the brain. In his paper "The Limits of Natural Selection applied to Man" (1870, 1891), Wallace expounded the view that a supernatural agency had influenced the final stages of human evolution. He now suspected that even some of our physical characters had no natural explanation, including the loss of body hair, which Darwin attributed to sexual selection. More important was a whole range of mental powers which seemed to have no value for humans living in a primitive society. What was the advantage in developing the ability to appreciate music or to perform abstract mathematical calculations? These were hardly likely to be employed by our distant ancestors, so how could natural selection have developed them? Wallace did not believe that modern "savages" had a lower level of mental ability than Europeans, but he saw no sign that they actually used the higher mental powers—so why should our ancestors, living a similar lifestyle, have gained any benefit from them? The answer was that such powers had not been developed naturally. They had been shaped by a supernatural interference with the last stage of evolution and were designed to achieve full expression only in a civilized state.

The doubts of Lyell and Wallace show that the evolutionary account of human origins was not accepted by everyone, even among those who saw the scientific strength of the argument. Here was a major conceptual problem, one which has remained and become the source of much modern opposition to evolution by creationists. In the short term, the opposition was overcome by the Darwinists insisting on the progressive nature of the evolutionary process, thus giving those who preferred a theologically acceptable form of the theory a way out of their dilemma. If evolution was designed to produce ever higher mental and moral powers, indeed if the human spirit were a predictable outcome of the process, then it would be possible to argue that our nature is divinely created after all. We retain our unique status by being the intended goal of spiritual progress. This assumption was to become the mainstay of liberal theological acceptance of evolutionism in the late nineteenth century.

EVOLUTION AND PROGRESS

In the early stages of the debate, however, the opponents of the traditional worldview made the most use of the idea of progress. For Huxley and his supporters, it was vital to stress that progress was the result of natural law operating without any visible sign of supernatural guidance. Darwin's theory incorporated elements of liberal ideology, including Malthus's principle, and was increasingly absorbed into a liberal progressionism which saw the social and economic improvement of the nation as depending on the efforts of able individuals, who needed freedom from the traditional restrictions designed to maintain the influence of the aristocracy (Desmond and Moore 1991). To those who endorsed this policy, the idea of progress was vital to underpinning the claim that social betterment was the inevitable continuation of nature's own law-bound activities (Bowler 1989a; Ruse 1996). But where Chambers's *Vestiges* merely invoked a vague law of progress, Darwin actually showed how the activities of individuals going about their daily business could result, in the long run, in benefit to the population as a whole. Small wonder that the middle classes would welcome such a theory, once they could overcome their scruples about its apparent materialism.

Working-class radicals had long been atheists and materialists and had used Lamarckian evolutionism to discredit the traditional worldview. But they called for outright revolution, which was anathema to the successful middle classes who wanted to enjoy the fruits of their newfound prosperity and thus favored steady reform. This latter group made Darwinism its own, thereby associating the theory with an ideology which preserved a role for the elite. What had changed was that the elite was now expected to demonstrate its superior ability in order to confirm its position as the guardian of progress.

Huxley and Scientific Naturalism

T. H. Huxley was a leading exponent of this new social policy. He worked hard to convince the working classes that the real message of Darwinism was that they should abandon the hope of revolution and put their faith in a process of gradual improvement whose benefits would eventually trickle down to their level of society (Desmond 1994, 1997). But Huxley himself typified a new element in the liberal movement. He was not a businessman and saw no benefit in the kind of unrestrained free-enterprise capitalism favored by the commercial elite. Rather, he was a professional scientist who earned a salary from a government-funded educational institution, the Royal School of Mines. He wanted state involvement in science and education to allow bright people from unprivileged backgrounds to make their way in the world. He welcomed evolutionism because it undermined the old ideology of a perfectly designed but static world in which the traditional ruling classes maintained their position by divine right. Darwinism challenged the conservative science of the gentlemanly specialists at ancient institutions such as Oxford, Cambridge, and the Royal College of Surgeons (Owen's home base). At the same time, however, Huxley was able to make common cause with wealthy liberals such as John Lubbock, who did their science for love rather than for money, because they shared his vision of an elite which governed both the scientific community and society at large. The shift from the amateur to the professional scientist was thus by no means complete (Desmond 2001).

Huxley accepted Darwin's theory—despite his many technical misgivings—because it was a weapon in his battle to professionalize science and science education (Fichman 1984; Turner 1978). But although he wanted to exploit the idea of progress in the call for more general reform, his interests did not completely coincide with those of the commercial classes who favored free-enterprise individualism and begrudged every penny spent by the government, even on education and scientific research. Darwin's theory articulated far better with the ideology of this latter group, because it stressed that evolution was the cumulative effect of individual activity. His ideas increasingly were identified with the evolutionary philosophy of Herbert Spencer, now achieving international recognition as the spokesman of liberalism and industrial progress. Huxley did not share Spencer's commitment to laissez-faire, and the two eventually fell out precisely because Huxley became concerned about the extent to which Spencer's philosophy was being used to undermine the values of social cohesion. But in the 1860s they could make common cause against the old Tory establishment.

These differences help to explain why Huxley was an enthusiastic supporter of Darwinism without his being convinced that natural selection was the chief driving force of change. He welcomed Darwin's theory because it offered a purely naturalistic explanation of the origin of species. Huxley's philosophy has been called scientific naturalism because he insisted that we explain the world solely in terms of the natural laws that science can reveal. There was no sign of design in the sense of an artificial pattern of creation or a concern by nature for the welfare of individuals. Nature was harsh and unforgiving, but if we understood nature's laws we could manipulate the world for our own benefit. Yet in the area of evolution, Huxley's imagination was constrained by his training as a morphologist: he had little interest in adaptation and tended to see the whole process in terms of an abstract transformation of organic forms. This is why his scientific work was inspired more by Haeckel's quasi-Romantic developmentalism than by the selection theory, and why he tended to think of evolution as a series of sudden transformations driving species along a path determined by internal biological factors. Only in the long term was he prepared to invoke the constraints imposed by the environment, remaining opposed to the almost idealistic model of development proposed by the theistic evolutionists in which order, harmony, and beauty play the dominant role.

We can see this aspect of Huxley's thought very clearly in the message he proclaimed in his *Man's Place in Nature*. Here, as in so many of his popular writings and lectures, Huxley was playing to the gallery, trying to convince the working classes that evolution was inevitable, even if we had to wait for the next step to occur. Evolution told us that progress must come:

"Thoughtful men, once escaped from the blinding influences of traditional prejudice, will find in the lowly stock whence man has sprung, the best evidence of the splendour of his capacities; and will discern in his long progress through the Past, a reasonable ground of faith in his attainment of a nobler future" (1863: 111). Yet the transformation of society into a better state would not come through the daily struggles of individuals gradually accumulating their gains. Earlier in his text Huxley had cited the embryological analogy, but had invoked it in the context of insects, where development is not gradual but episodic: "History shows us that the human mind, fed by constant accessions of knowledge, periodically grows too large for its theoretical coverings, and bursts them asunder to appear in new habiliments, as the feeding and growing grub, at intervals, casts aside its too narrow skin and assumes another, itself but temporary" (58). This might easily have been taken as a prediction of revolution, but Huxley's real aim was to convince the underprivileged that education was the route to a future liberalization of society in which they too would benefit.

Progress was governed by law, although not necessarily the law of natural selection. Evolution fitted into Huxley's ideological program along with other areas of science which showed that aspects of human nature once linked to the spiritual dimension could now be explained by natural law (Barr 1997; Paradis 1978). Huxley was active in promoting what his opponents claimed was a materialistic view of the mind. Animals were merely automata, complex robots, and even humans were subject to the same mechanistic laws of physiology. The conscious mind was an epiphenomenon, a by-product of the brain's activity with no power to alter the mechanistic operations of nature. This naturalistic program was also promoted by the physicist John Tyndall in his notorious Belfast address given to the British Association in 1874 (reprinted in Tyndall 1902). Yet Huxley insisted that he was not a materialist—the world of consciousness could not be explained in mechanistic terms, even though it depended completely on the material world. In this respect his philosophy was by no means as radical as that being promoted by materialist thinkers in Germany (Gregory 1977).

Huxley's attitude to materialism paralleled his approach to religion. Despite his hostility to organized religion and his claim that the advance of science routinely demolished the claims of theologians, he would not proclaim himself an atheist. The world gave no evidence of being designed by a wise and benevolent God, but on the question of the ultimate source of the law-bound universe, science had nothing to say. Huxley coined the term *agnosticism* to denote an active state of doubt on the question of the existence of a God, as opposed to an outright denial of His existence. He called some of his lectures to working men "Lay Sermons," and much of his writing has a moralistic tone that seems almost a call for a new religion.

Spencer's Social Evolutionism

In the 1860s Huxley made common cause with Herbert Spencer, even though the latter's philosophy was more in tune with the free-enterprise ideology of the commercial entrepreneurs. Although little read today, Spencer was one of the most highly regarded philosophers of the mid–nineteenth century. His "Synthetic Philosophy" was based on the idea of universal evolution, and Spencer popularized the term *evolution* and convinced everyone that it denoted an essentially progressive process (Bowler 1975). To many, Spencer stood at the heart of the evolutionary movement—Darwin merely provided the scientific details in one crucial area of the program. Yet Spencer had little patience with factual details, and there were significant differences between his thinking and Darwin's. That is why Darwin's theory could be exploited in entirely new ways by later scientists, whereas Spencer's philosophy declined in influence as the times changed. Darwin may have been a progressionist, but he was far more aware of the problems posed by the concept of progress.

Spencer gained his reputation as a philosopher because he used evolutionism to resolve a conflict which had divided thinkers on the question of human nature for several generations (on his life and work, see Duncan 1911; J. Greene 1959b; Kennedy 1978; Peel 1971; H. Spencer 1904). The utilitarian tradition had insisted that the mind was a blank slate written on by experience in the form of learned habits—it is a learning machine with no built-in structure. Kant and the idealists saw the mind as imposing categories defined by its own structure onto the external world. By the 1850s even the leading utilitarian, John Stuart Mill, had admitted that a moral philosophy could not rely on associations built up in the mind by pleasure and pain. Psychologists such as Alexander Bain were showing that the mind did have distinct faculties not based on experience. They hinted that these faculties might be localized in particular areas of the brain, just as the phrenologists had claimed (Greenaway 1973; R. Richards 1987; Young 1970a). Spencer realized that the faculties might be composed of the accumulated experience of many generations, built up through evolution. What one generation learns in order to cope with its environment, its descendants inherit as a built-in instinct. The mind has a structure, but it is not absolutely fixed, because individuals can learn new habits and influence future generations through evolution.

Darwin had his own ideas on how these instincts might be formed by natural selection. But Spencer had developed his ideas in the 1850s before Darwin published, and they were firmly rooted in the Lamarckian principle of the inheritance of acquired characteristics. Spencer had published an article in favor of Lamarckism as early as 1852 and his *Principles of Psychology* of 1855 invoked Lamarckism to explain how the faculties of the mind were constructed and modified. He accepted the theory of natural selection when Darwin published it, and he coined the phrase "survival of the fittest" (Paul 1988). Throughout the rest of his career, Spencer invoked both selection and Lamarckism. More generally, his evolutionism had been inspired by Karl Ernst von Baer's embryological model in which development proceeds from the most general to the more specialized structures. He invoked this as a universal process, responsible for everything from the formation of the solar system to the origin of life and the emergence of human societies. Although in principle an advocate of branching evolution, Spencer always treated the branch leading to the human species as the main line of development. This was the essence of his Synthetic Philosophy, outlined in an 1857 essay on progress (reprinted in H. Spencer 1883: vol. 1) and in his *First Principles* of 1862.

Spencer's popular reputation derived from his association of his philosophy with the ideology of free-enterprise capitalism. His *Social Statics* of 1851 had attacked the idea that the state could play any useful role in promoting the well-being of individuals—everything should be left to the individuals themselves, who must sort out their social interactions in terms of their own self-interest. It is this aspect of his thought, coupled with his introduction of the phrase "survival of the fittest," that led a later generation to dub Spencer a social Darwinist, although the Lamarckian element in his thought renders this attribution problematic (see chap. 8). Spencer still thought of himself as a moral philosopher, even though his critics accused him of rejecting the whole tradition of morality which had underpinned Christian society. In one sense the critics were right: for Spencer, whatever is successful is good. The world determines what our morality should be, not a conscience reflecting eternal moral values. Yet neither Darwin nor Spencer saw their ideas as promoting ruthlessness or immorality. Darwin protested when a newspaper described his theory as justifying the actions of Napoleon or tradesmen who cheat. Spencer accepted that people cooperate and thus subdue their own short-term interests for the benefit of living in an orderly society. He also accepted that sympathy was an important human emotion, and that individual (though not state-funded) charity was a legitimate response to the genuine misfortune of others. More seriously, his social philosophy was intended to promote the virtues of thrift, self-reliance, and initiative. These are very much the virtues of the old Protestant work ethic, and as James Moore (1985b) argues, some liberal Christians were able to see themselves as Spencerians because of this. In America, John Fiske's *Outline of Cosmic Philosophy* (1874) adapted Spencer's cosmic evolutionism to more traditional tastes by sidelining the element of struggle and presenting cooperation as the goal toward which evolution was working. Evolution thus became the expression of the Creator's plan to create morally aware beings such as ourselves—a view that would be developed by many liberal Christian thinkers later in the century.

A key feature of Spencer's thinking was his assumption that the same laws of evolution govern both the biological process by which humans have been created and the emergence and development of society. For him, the evolution of the mind and the evolution of society went hand in hand. There was no recognition, as in Darwin's *Descent of Man*, that there might have been some unique turning point in the history of the primates where one group separated itself decisively from the apes by adopting a new habitat and lifestyle. Spencer imagined a kind of feedback loop between mental and social evolution: the higher the mental powers, the greater the complexity of the society that the individuals could create; the more complex the society, the greater the stimulus it provided for further mental development. Everything cohered to make progress inevitable or to weed out those who did not keep up. These views would serve as a foundation for many later developments in evolutionary psychology and sociology, but their influence diverted attention away from the more radical aspects of Darwin's own thought.

This, then, was the optimistic message offered by the scientific naturalism of the more radical evolutionists. Nature was a great impersonal machine, but nature's laws had brought us this far and, if understood and controlled, offered the hope of unlimited technical, social, and (by the evolutionists' own standards) moral progress in the future. Even the more liberal Christians, taught to see humans as made in the image of God, had to rethink their priorities to take any part of this message on board. Yet the element of progressionism built into the evolutionists' position, coupled with their evident rectitude and concern for others, made the transition not an impossible one to make. Liberal theologians did indeed accept the message of evolution, although they preferred to regard the progress as a spiritual one intended by the Creator. When Darwin died in 1882, he was buried in Westminster Abbey (Moore 1982). This was certainly not because he had made things comfortable for the Anglican church, although by then the reconciliation of evolutionism with liberal Christianity was well under way. It was because he was perceived as a national hero, someone who had transformed the way almost everyone thought in a way that had created a new basis for morality. Only those more conservative Christians who still saw humanity as a fallen species in desperate need of redemption had refused to climb aboard the bandwagon, and they would begin to raise their voices in the next century.

7   The Eclipse of Darwinism
Scientific Evolutionism, 1875–1925

"The eclipse of Darwinism" is a phrase Julian Huxley used in his survey of evolution theory (1942) to describe the situation which existed before the "modern synthesis" of genetics and the selection theory. Darwin had faced mounting opposition to the theory of natural selection in his own lifetime, and its popularity had continued to decline until, by the end of the century, its opponents were convinced it would never recover. One attack, translated from the German, had the title *At the Deathbed of Darwinism* (Dennert 1904). Evolutionism remained unquestioned, but biologists had developed a variety of non-Darwinian theories, many of them based on hints already provided by the first generation of critics whom Darwin had confronted face-to-face (for contemporary surveys, see Delages and Goldsmith 1912; Kellogg 1907; Romanes 1892–97; note also the histories by Nordenskiöld [1946] and Radl [1930], biologists who still thought that Darwinism was dead).

Later historians at first paid little attention to this outburst of anti-Darwinism; their accounts jumped from the debate over the *Origin of Species* to the reemergence of Darwinism promoted by figures such as Julian Huxley. Loren Eiseley (1958) mentions Hugo De Vries's mutation theory as an alternative, but only because De Vries played a role in the "rediscovery" of Mendelism and the foundation of genetics. More recent histories have also tended to stress the reemergence of Darwinism in the mid–twentieth century without paying much attention to the preceding wave of opposition. For a leading Darwinist such as Mayr (1982), the story of evolution has a main line leading from Darwin to modern Darwinism, and there is little point in wasting time on theories that turned out to be blind alleys. This is a legitimate approach for a biologist whose main interest is the emergence of the theories which have dominated modern thinking. But to historians who take a more direct interest in the science of the late nineteenth and early twentieth centuries, such a rigid application of hindsight is inappropriate. We must learn how the theory of evolution worked at the time—what scientists found exciting and promising about it—even when their work turned out not to be as productive as they had hoped. This is necessary to explain both the reason the idea of evolution became popular and the context within which the later initiatives arose. This latter point is significant because a key component of the new developments (Mendelian genetics) was strongly influenced by the anti-Darwinian theory of evolution by sudden saltations. Cultural historians interested in the wider application of evolutionary ideas should be aware that those ideas did not all derive from the Darwinian selection theory, despite that theory's subsequent triumph.

A wide variety of alternatives to selectionism came under consideration (Bowler 1983, 1988). The selection theory was not completely abandoned, and August Weismann's neo-Darwinism proclaimed it to be the only acceptable mechanism of evolution. But Weismann's dogmatism alienated the many biologists already inclined to invoke at least some non-Darwinian processes. The Lamarckian theory of the inheritance of acquired characters, accepted as a subsidiary mechanism even by Darwin himself, now came to be regarded as a complete alternative to selectionism by the school of thought known as neo-Lamarckism. Orthogenesis explained nonadaptive trends in terms of forces from within the organism predisposing variation to take place along fixed paths. These theories were supported in part because they preserved an element of teleology that countered the apparent materialism of the Darwinian theory. Biologists reluctant to concede that evolution is a haphazard, trial-and-error process preferred to believe that development is predisposed to advance in purposeful or orderly directions. The anti-Darwinian movement thus exploited themes pioneered in the theistic evolutionism of earlier naturalists such as Mivart. Some non-Darwinian theories were derived from the recapitulation theory, in which the course of evolution was modeled on the embryological development of the individual organism.

This developmental model of evolution dominated efforts to reconstruct the history of life on earth, the centerpiece of the first evolutionary biology. It flourished within the morphological research tradition, which was being used to gain biology increased status as a professional discipline (Caron 1988; Coleman 1971; Maienschein 1991; Nyhart 1995). This program existed before evolutionism came to the fore, but it was transformed by the new theory to create the conceptual framework within which the first generation of evolutionary biologists worked. Non-Darwinian ideas flourished in part because they fitted in with the mind-set of the morphologists, many of whom self-consciously repudiated the fieldwork tradition within which Darwin had worked. At the same time, however, this new focus on professionalization has led many historians to marginalize the Darwinian revolution because it did not result in the creation of a new disciplinary structure within biology. This approach risks throwing the baby out with the bathwater, because it ignores the possibility that evolutionism's real influence may have been in providing bridges between disciplines and preventing a total fragmentation of biology. Biogeography and paleontology also played a role in reconstructing the history of life, and they were museum-based disciplines with much closer links to the old natural history.

The collapse of the developmental model of evolution came about in an unexpected way. Neo-Darwinists such as Karl Pearson and W. F. R. Weldon were trying to put the study of natural selection on a more scientific basis. But they had limited success, and by the end of the century, a number of biologists had begun to argue that adaptation (whether Lamarckian or Darwinian) was irrelevant to evolution. They believed that forces arising within individual development were paramount, and increasingly they were inclined to argue that these forces operated by occasional, quite sudden transformations, or saltations. This "mutation theory" helped foster the climate of opinion within which Gregor Mendel's laws of heredity were rediscovered in 1900 after decades of neglect. Many early geneticists were also saltationists who bitterly opposed the Darwinian selection theory. But they opposed Lamarckism too, because the unit characters transmitted by the genes did not allow for slow modification by the organism itself. The developmental viewpoint evaporated as it became clear that the new science of heredity undermined the recapitulation theory and anything associated with it. Lamarckism and Darwinism both were repudiated as expressions of an outdated science which had sought to reconstruct the history of life on earth rather than study the process by which new characters are actually produced. This move was characteristic of the revolt against morphology around 1900 which focused attention much more on the study of living processes than on the dissection of dead specimens (G. Allen 1975a). It would be several decades before the geneticists realized that mutations cannot create new species and reluctantly called in natural selection to explain why some mutations increased their proportion in the population while others declined (see chap. 9).

RECONSTRUCTING THE HISTORY OF LIFE

Despite Darwin's own reservations, many of his followers saw reconstruction of the course of life's development as the most promising area in which to apply the theory. The fossil record might be imperfect, but new discoveries were being made at an increasing rate, and some of the new fossils would fill in crucial gaps. Even where fossils were not available, anatomy and embryology offered clues about relationships that would allow the branches of the tree of life to be fitted together—and there was always the hope that fossils would turn up to test the hypotheses. Biogeography also became important as better knowledge of both the living and fossil inhabitants of different continents offered the hope of identifying the point of origin of each major group and its subsequent migrations.

From Morphology to Paleontology

The enthusiasm for reconstructing what Ernst Haeckel called "phylogenies" (evolutionary genealogies) was in part a consequence of the changing state of science in the mid–nineteenth century. The new generation of professional biologists favored a move from the field to the laboratory in order to enhance the scientific credentials of their discipline. One way to do this was to focus on morphology, the study of animal form, since the dissection of specimens to study structural relationships was done in a laboratory environment. Morphology was the key to T. H. Huxley's campaign to gain more support for scientific biology in the education system. In Germany, Haeckel was merely the best known of a group of morphologists struggling to create a niche for their subject in the university system (Nyhart 1995). Even field studies were being put on a more formal basis as the number of professionals based in natural history museums increased (Winsor 1991), and here, too, morphology was important because it provided the evidence upon which the classification of new species was based.

Haeckel led the way in showing how evolutionism could transform the morphologists' project to uncover the relationships linking the major groups of animals (see chap. 6; Di Gregorio 1992; R. Richards 1991; Weindling 1989a). Even Huxley saw the practical implications of the theory for his research only after reading Haeckel's first major work on evolutionary morphology (Haeckel 1866). Traditionally, morphologists looked for the archetype, or most basic structure, of each group. But for the evolutionist, the idealized archetype became the real, common ancestor from which the group had diverged (Bowler 1996: chap. 2; Coleman 1976). This placed more constraints on the hypothetical reconstructions because the ancestor had to be visualized as an actual creature with its own adaptations. The transformations which led from the common ancestor to its various descendants also had to be plausible in terms of the pressures of adaptation and internal coherence. More excitingly, where the archetypes of the major groups were seen as completely isolated from one another, evolutionism implied that transitions between them had to be possible. Wherever possible, fossils would be used to throw light on these ancient transformations, but even where fossils were not available anatomy and embryology would fill in the gaps.

Because the major animal types already had been differentiated before the known fossil record began, much of evolutionary morphology would remain hypothetical in the sense that testing by the fossil record was impossible. The genealogies based on morphological resemblances were often rather abstract because no direct information was available about the adaptive pressures on the organisms. As the range of fossils increased, emphasis gradually shifted to those areas where fossils were known or might be found (Bowler 1996; Buffetaut 1986; Desmond 1982; Rainger 1991; Rudwick 1972: chap. 5). This focused attention on the conditions under which the transformations occurred, and phylogenetic research began to make much more use of geological and geographical evidence about changing conditions which might have triggered evolutionary transformations and about the migration of species to new areas and environments. By the early twentieth century, surveys of the history of life on earth had begun to take on a modern appearance, with discussions of topics such as the causes of mass extinctions and evolutionary radiations in addition to the purely morphological transformations postulated in the immediately post-Darwinian era. This becomes obvious when comparing popular accounts written by specialists. Haeckel's survey (1876) was based largely on morphology and seems dated today, but the account by the American paleontologist Henry Fairfield Osborn (1917) would be understood by any modern reader.

The evolutionary morphologists' project ran into serious difficulties because its originators could not foresee the complexity of the processes they were studying. The recapitulation theory encouraged the assumption that the course of evolution would follow almost teleological patterns or trends in the history of life. If evolution was mirrored in the development of the embryo, as Haeckel claimed, it was natural to assume that it too would progress steadily toward some final goal. Morphologists often reconstructed the tree of life with a main trunk leading toward the human race, all other developments being dismissed as side branches. The "lower" modern animals were treated as living fossils—relics of earlier steps along the main line preserved almost unchanged today. The element of divergence stressed by Darwin was minimized in these early pictures of the history of life.

It soon became clear, however, that evolution was a far more complex process. Anton Dohrn (translation 1993) and E. Ray Lankester (1880, see Lester 1995) argued that evolution was not always progressive; when species took up a less challenging lifestyle, they often degenerated. This meant that many of the "lower" forms of life were not truly primitive; they were the degenerate descendants of forms that once had climbed much farther up the scale. They gave no real clues about the truly primitive stages of evolution and served only to confuse the search for relationships. In its most extreme form, this viewpoint led the embryologist E. W. MacBride (one of the last major supporters of the recapitulation theory) to treat all the invertebrate types as degenerate offshoots from the main stem of vertebrate evolution: "The invertebrates collectively represent those branches of the Vertebrate stock which, at various times, have deserted their high vocation and fallen into lowlier habits of life" (MacBride 1914: 662; see also Bowler 1984).

MacBride's assumption that degeneration would follow from the adoption of a less energetic lifestyle, such as creeping on the seabed, suggests why non-Darwinian ideas of evolution began to flourish at this time. Lamarck's theory had always been associated with the claim that the driving force of evolution was a change in the species' habits. It was assumed that the selection of random variations left individual animals at the mercy of the environment—the fact that a new habit alters the adaptive relationship between the species and its environment was not considered relevant to Darwinism. So, morphologists like MacBride favored Lamarckism, which was also more easily associated with the recapitulation theory. Paleontologists, too, thought they could trace linear adaptive trends in the fossil record which could be explained in the same way. The reconstruction of phylogenies became a focus for anti-Darwinian thinking, and the concentration of scientists' efforts in these areas created a climate of opinion in which Darwinism could become eclipsed.

In the end, though, it was evolutionary morphology which was eclipsed. Rival hypotheses were offered to explain key steps in the history of life, and morphology offered no way of deciding between them. The problem was created by the possibility of convergent evolution. Darwin assumed that homologies were a sign of common ancestry: if two different structures were modeled on the same foundation, they were modified descendants of the same ancestral form. The fact that the forelimbs of bats, whales, horses, and humans were all modeled on the same system of bones suggested that all mammals evolved from a single progenitor. If two very different species independently adapted the same habits and lifestyle, they might evolve similar structures, although at first it was assumed that such similarities would be only superficial. Owen distinguished between homologies and analogies (e.g., the fishlike structure of ichthyosaurs and porpoises), which do not indicate a true relationship. But if the process of convergence went beyond superficial analogy and generated detailed internal similarities, how would morphologists distinguish between these convergences and true homologies? They could not: one biologist's homology was another's convergence, and thus rival views of evolutionary relationships could be generated. When William Bateson (1894) recognized that he could provide no definitive evidence to support his theory of the origin of the vertebrates, he turned his back on morphology and began the work that led to his becoming one of the founders of modern genetics.

Bateson and his contemporaries were involved in what Garland E. Allen (1975a) calls the "revolt against morphology" at the end of the nineteenth century. Reconstructing phylogenies had turned out to be a waste of effort, and biology had to turn to a new, experimentalist, approach. Later generations have shared this view, and phylogenetic studies have played little role in conventional histories of evolutionism. Ruse (1996) argues that evolutionary morphology was indeed a waste of time, at best only second-rate science driven by the ideology of progress. But using hindsight to dismiss the morphologists' efforts leaves us with no way of understanding why the opportunity offered by evolutionism seemed so promising at the time. The morphologists could not anticipate the difficulties that would turn up, and the hope of reconstructing the whole course of the evolution of life on earth clearly exerted great fascination. Ruse is right to stress that much of the enthusiasm was generated by the ideology of progress, but the project was not quite as pointless as its frustrated critics made out. Phylogenetic studies continued into the twentieth century, and many of the problems eventually were solved, some having to wait until the advent of modern molecular biology. The popular view of evolutionary biology is still dominated by the books and museum displays which chart the course of evolution "up" to humankind. Whatever one's opinion of evolutionary morphology as science, as an expression of a whole generation's enthusiasm for the idea of evolution it cannot be ignored.

The Tree of Life

Haeckel (1876, 1879) presented an overview of the evolution of life from the spontaneous generation of the first living things to the appearance of humans, focusing on a main line of development through the emergence of the first vertebrates, the reptiles, the mammals, and finally the primates and the first primitive humans. Haeckel's books say little about fossils or adaptation to changing environments. For him, phylogenetic studies were essentially morphological; he used anatomical and embryological similarities to work out relationships. By the early twentieth century, though, the emphasis had shifted to areas of evolution where there was a significant fossil record, and where both geological and geographical factors were integrated into the reconstruction process (Rainger 1991). Instead of accepting the gradual progress up the scale depicted by Haeckel, scientists saw the evolution of life as an episodic process in which mass extinctions and sudden bursts of evolution were triggered by environmental challenges. In a curious way, the image presented is more Darwinian—even though few of the paleontologists endorsed the selection theory. The abstract progressionism of Haeckel was replaced by a more down-to-earth viewpoint in which much of evolution depended on the hazards of migration and the response to environmental change. The lesson spelled out by Lankester that evolution would stagnate without external stimulus was now accepted by all. Even before Darwinism began to revive in the 1920s and 1930s, phylogenetic science had begun to pave the way for the emergence of a view of evolution in which adaptation, rather than automatic progress, was the key (Bowler 1996).

Materialists such as Haeckel began their surveys of evolution with the assumption that some process of spontaneous generation served as the origin of life. Following Darwin's lead, however, most British evolutionists tended to avoid this issue. The theory of spontaneous generation was championed by Charles Bastian, who sought to associate it with Darwinism. But T. H. Huxley and his colleagues thought Bastian was unreliable, and they did not follow his lead (Strick 2000; and for a collection of primary sources, see 2001). Haeckel also attempted to reconstruct the process by which the first metazoans (multicelled animals) were formed by following the very earliest stages in the differentiation of the embryo. A great deal of morphologists' attention focused on the emergence of the major animal phyla— Cuvier's four main types, now much increased in number. Whereas preevolutionary morphologists had treated the types as utterly distinct, evolutionists now had to show how the most primitive (presumably wormlike) animal could have diverged into groups with such major structural differences. A long-standing debate centered on whether the main groups of arthropods (crustaceans, insects, and arachnids) had derived their characteristic segmented structure from a single common ancestor or had evolved it separately by convergent evolution.

Most attention focused on the origin of the vertebrates (Gee 1996). Haeckel followed the Russian morphologist Alexandr Kovalevskii, who had noted the similarity between the larval form of the sea squirt, or ascidian, and the immature form of some vertebrates such as the tadpole. They postulated that the ascidian larva once had been the adult form, and had evolved into two branches, progressively into the vertebrates and, by degeneration, into the modern ascidians. In 1875 Anton Dohrn (translation 1993) challenged this view by deriving vertebrates from annelid worms, arguing that the segmental structure which both types possessed was a homology indicating common descent (in Kovalevskii's theory, segmentation independently evolved in several different types). The debate raged through the rest of the century and was never satisfactorily resolved. William Bateson showed that the acorn worm *Balanoglossus* was a primitive Chordate and, hence, ancestral to the true vertebrates, but his frustration at being unable to refute rival hypotheses led him to abandon phylogenetic work altogether. The Lamarckian E. W. MacBride played an important role in establishing what became the most generally accepted view, demonstrating a link between vertebrates and the larval forms of echinoderms (sea urchins and starfish).

Controversy surrounded all the major steps in vertebrate evolution. Evolutionary morphologists originally assumed that the first vertebrates lacked a bony skeleton, so that lampreys and sharks corresponded to the main steps on the way toward bony fish. But growing fossil evidence suggested that bone had evolved at an early stage, the sharks being a degenerate (though in some respects successful) offshoot. The modern lungfish also were treated as "living fossils" offering a clue to the intermediate stage between fish and amphibians, but again the fossil evidence forced a change of emphasis as it became clear that another group of fish, the crossopterygians, had fins whose structure was almost preadapted to being modified into legs. By the 1930s scientists recognized that the "conquest of the land" was unlikely to be explained by some abstract improvement of the vertebrate type, and paleontologists such as A. S. Romer were suggesting that limbs evolved in fish desperately trying to find new pools in a world of increasing drought (1933: 105).

T. H. Huxley had noted the resemblance between some small dinosaurs and birds. He suggested that modern flightless birds such as the ostrich were relics of the first true birds, flight having evolved later (Di Gregorio 1984: chaps. 2, 3). Most authorities thought it more likely that wings had evolved in small, tree-dwelling reptiles, with the flightless birds being a degenerate form derived from flying ancestors. Huxley had also commented on the origin of the mammals, questioning the view that the egg-laying monotremes (e.g., the duck-billed platypus) were a stage in the evolution of higher mammals from reptiles. His suggestion that the mammals were polyphyletic—mammalian characters having evolved separately in several reptile groups—became widely accepted. Evolutionary morphology achieved a belated triumph in this area when Ernst Gaupp, in the early twentieth century, showed that several of the bones in the complex jaw of reptiles were homologous with the bones in the mammalian inner ear. At the same time, South African fossils revealed by Robert Broom began to fill in stages in exactly the kind of transformation Gaupp had predicted. Broom's book on the "mammal-like reptiles" (1932) summed up decades of work showing how the mammals had evolved and represented a triumph of phylogenetic research in which morphology had been transformed and extended through the use of the ever increasing fossil evidence. Significantly, however, Broom revealed in his conclusion that he believed evolution to be the unfolding of a divine plan aimed at producing humanity. One did not have to be a Darwinian to make major contributions in this area.

Early-twentieth-century paleontologists such as Osborn and Romer were far more concerned with relating evolution to environmental changes than Haeckel and Huxley had been. This change of emphasis seems to have been particularly effective in America, and it paralleled significant changes in the role played by museums and other research centers as biologists sought to modernize their discipline (Rainger, Benson, and Maienschein 1988). The purely morphological research tradition was transformed into a wider project to understand the factors shaping the history of life on earth. Paleontologists looked to the geological evidence of the conditions at different eras in the earth's history for evidence of physical transformations that could have affected the climate and thereby forced animals and plants to evolve. The idea that evolution was a cyclic rather than a continuous process became commonplace as evidence grew that major episodes were triggered by the great physical revolutions which defined the geological periods. This was not a revived catastrophism, but it did introduce an element of discontinuity repudiated by Darwin and his immediate followers.

Another factor which helped to focus attention on the role of the environment was biogeography. Alfred Russel Wallace and others had begun to explain how species expand their territory to reach boundaries defined by geographical barriers, most notably in the case of Wallace's line. Wallace's book on the geographical distribution of animals (1876) was a monumental synthesis which triggered an outburst of research as biologists tried to identify the original homeland of each newly evolved form and reconstruct its subsequent migrations and adaptations to new territories (Bowler 1996: chap. 8; Browne 1983). Wallace collated evidence for the distribution of modern species but also was able to use the fossil record, which was now improving for areas outside Europe and North America. He theorized that most new species had evolved in the harsher climate of Eurasia or North America, and had then expanded southward, exterminating the earlier inhabitants or driving them into isolated refuges in the southern hemisphere. By the early twentieth century, such patterns of evolution and migration were widely accepted, contributing once again to the growing belief that progressive evolution was always a response to environmental challenge.

THE AGE OF THE EARTH

Geology had initially been a source of major difficulty for the Darwinian theory. The physicist William Thomson (Lord Kelvin) used the new science of thermodynamics to undermine the credibility of Lyell's steady state view of earth history (Burchfield 1975; Smith and Wise 1989). Since Darwin assumed that natural selection was a very slow process, his theory required the vast amounts of geological time postulated by Lyell. By showing that the latest physical theories would not permit the earth to remain geologically active for an indefinite period of time, Kelvin shortened the age of the earth to an extent which made the selection theory untenable. Evolution was still possible, but it needed to work much more rapidly than Darwin had supposed. The shortened time span—which even the geologists conceded— put the pressure on biologists to think of evolutionary processes that would work more rapidly than natural selection.

The strength of Kelvin's position lay in its appeal to fundamental laws of physics. Thermodynamics was based on the principle that energy always becomes less available, and hence, that hot bodies always cool down. A steady state earth is, in effect, a perpetual-motion machine and is incompatible with this principle. But Kelvin drove home this point by trying to calculate exactly how long it would take a body the size of the earth to cool down from a molten state. Here, he was on weaker ground, because he had to estimate the internal temperature of the earth and the rate at which heat is conducted to the surface. But even if his estimates were wrong, the general principle was right, and he argued that his figures were unlikely to be so far wrong that the Lyellian timescale could be rescued. By 1868 he was arguing that the total age of the earth could not be much more than 100 million years, only a tiny fraction of what Darwin's theory required.

Darwin was certain that Kelvin was wrong, but could not put his finger on the reason why (Burchfield 1974). Other evolutionists, however, began to give ground. Wallace proposed that evolution would work more quickly in periods of environmental stress, while T. H. Huxley, after challenging Kelvin's figures, conceded that biology had to take its timescale from geology. As the century progressed, Kelvin reduced his estimates even further, until eventually even the geologists began to protest. But within this environment biologists began to explore other evolutionary mechanisms, including Lamarckism and saltationism, that might plausibly be argued would proceed faster than natural selection.

Geologists were also aware of the weakness of Lyell's principle of absolute gradualism. The massive discontinuities at certain points in the geological record could not all be explained away as gaps in our knowledge; they were evidence of relatively sudden transformations on the earth's surface and among living things. Catastrophism as such did not reemerge, but the possibility of earth movements and vulcanism on a scale much greater than anything observed in the course of human history had to be taken seriously. Such events might trigger waves of extinction and rapid evolution that would serve as punctuation marks in the history of life.

Kelvin also applied his calculations to the sun, which (in the absence of any internal source of energy) would also eventually cool down, leaving the earth without its source of light and heat. More generally, the physicist Rudolph Clausius extended the principles of thermodynamics to predict the "heat death" of the whole universe, when no energy sources would be left. Although the Victorian period is often depicted as obsessed with the ideology of progress, these applications of the latest developments in physics suggest that there was a darker side to the Victorians' worldview (Brush 1978; Gillispie 1960: chap. 9). Evolution might have progressed toward humanity, but in the long run, all life was doomed to decline into the darkness and the cold. Such pessimistic speculations link with Lankester's warning about evolutionary degeneration to show that the idea of progress did not go unchallenged.

Kelvin's estimate of the age of the earth was undermined by the revolution which took place in physics around 1900 (Dalrymple 1991). The discovery of radioactivity introduced an entirely new factor which he had been unable to include in his calculations. In 1903 Pierre Curie announced that the radioactive decay of elements such as radium liberated a slow but steady supply of heat. These elements are present in small quantities throughout the earth's crust and, presumably, in its core. By 1906 Lord Rayleigh had shown that the energy produced by the radioactive elements should more than compensate for the heat lost into space. The rate of decay of some of these elements is so slow that the balance could be maintained over billions of years. Within a few years, geologists such as Arthur Holmes were using the rates of decay to produce new estimates of the age of the earth, which soon reached figures of the same order of magnitude as the one accepted today, that is, over four billion years (Lewis 2000). The Cambrian period lay at least half a billion years in the past. Lyell was vindicated—but not completely, because the new ideas on radioactive heating suggested that there was so much heat for the earth to get rid of that occasional episodes of massive vulcanism or crustal rearrangement would have to occur. The absolutely gradualistic model of evolution would never return.

NEO-LAMARCKISM

In the late nineteenth century, the most popular alternative to natural selection was a revival of Lamarck's concept of the inheritance of acquired characters. In the earlier part of the century, this theory had been rejected as too materialistic. But natural selection was far worse in this respect, and biologists soon realized that the inheritance of acquired characters might allow evolutionism to take on a less harsh aspect which would preserve an element of teleology. If natural selection was beset with problems, why not replace it with an alternative mechanism of adaptation? Some early advocates of this position did not even realize that Lamarck had pioneered the idea, and only in the new century did the American biologist Alpheus Packard publish an account of Lamarck and his writings (1901). Most of Lamarck's original program was no longer acceptable; but, taken in isolation, the mechanism of the inheritance of acquired characteristics gained a new lease on life. In the 1890s the term *neo-Lamarckism* become popular— in his survey *Darwin and after Darwin* (1892–97), George John Romanes depicted a polarization of opinion between neo-Darwinism and neo-Lamarckism.

The Emergence of Lamarckism

Darwin himself accepted that the inheritance of acquired characters might supplement natural selection, and his view of heredity allowed for the effect to occur. In its simplest form, Lamarckism supposed that changes of structure produced by the activity of the adult organism (the acquired characters) might be transmitted at least in part to the offspring. When the organism changed its habits and used its body in different ways, the effects of use and disuse would be obvious—this is how the weightlifter acquires stronger muscles. If the new habit were kept up over many generations, and if the effect were transmitted, however slightly, from parent to offspring, then the modification would build up over time and eventually produce a permanent change. If the new habit were a useful one taken up by the whole population, then by definition the change would be adaptive or purposeful, as in the case of the giraffes which stretched their necks to reach a new source of food in the trees. Variation within the species was directed, not random, so there were no unfit individuals to be eliminated by struggle. But the question of whether acquired characters were really inherited now became crucial. Darwin assumed that they were, and the neo-Lamarckians built on this assumption to argue that natural selection was unnecessary. But, like Darwin, they were unable to create a satisfactory model of heredity to substantiate their theory, and as time went by, the lack of direct experimental evidence for the effect became more critical.

There was considerable support for Lamarckism in the late nineteenth century, but the movement was never a united one. Use-inheritance was the most popular aspect of the theory because of its apparent moral superiority over natural selection. But many naturalists, including some botanists, preferred to focus on the direct action of the environment upon the organism. Plants adapted to new conditions too—but here there was no conscious response via a change in habits. In America, a distinctive school of neo-Lamarckism emerged among naturalists influenced by Louis Agassiz's idealist view of nature. They stressed the orderly nature of adaptive trends and explained them in terms of the recapitulation theory. Several of the most influential Americans were paleontologists who drew their inspiration from the linear patterns they found in the fossil record. But their fascination with this effect led them to identify trends that led in nonadaptive as well as adaptive directions, thus forging a link between Lamarckism and orthogenesis (for contemporary surveys, see Kellogg 1907; Packard 1901; for a modern account, see Bowler 1983).

Behind this diversity of approaches lay a distaste for the Darwinian selection theory and the hope that Lamarckism would allow evolutionism to become a worldview based on order and purpose. Some of the most active neo-Lamarckians were philosophers and literary figures rather than scientists. Edward J. Pfeifer (1965) suggests that American neo-Lamarckism was a continuation of natural theology. Many Lamarckians saw the inheritance of acquired characters as the kind of mechanism a wise and benevolent God would institute to produce adaptation and progress. Lamarckism was a response to the problems that Asa Gray identified with the selection mechanism: its harshness and its apparent dependence on chance (see chap. 6). But not all Lamarckians were religious; indeed, one of the leading British supporters was the philosopher Herbert Spencer, widely regarded as a Darwinian. For many, the attraction was not theological but moral. As popularly understood, natural selection left the organism at the mercy of its environment—life or death depended on the luck of the draw in the process of random variation. The organism (and by implication the individual human) was little more than a puppet in the hands of forces it was powerless to control. By focusing on new habits as the driving force of evolution, Lamarckism allowed the organism to be seen as an active, creative agent in charge of its own and its species' destiny. Anyone seeking to emphasize that human life does indeed have a purpose would find this a more plausible account of how our higher characters were built up.

One indication of the complex nature of neo-Lamarckism is the important role that the evolutionary philosophy of Herbert Spencer assigned to the inheritance of acquired characters. Widely portrayed by later generations as a social Darwinist, Spencer nevertheless invoked both natural selection and Lamarckism to explain how evolution works, and remained convinced that the latter was indispensable. When August Weismann attacked Lamarckism and proclaimed the all-sufficiency of natural selection, Spencer responded by insisting that "either there has been the inheritance of acquired characters or there has been no evolution" (1893). He assailed the credibility of pure selectionism by using arguments that seemed quite plausible when judged in terms of a pregenetical theory of heredity (Ridley 1982). If a new structure were produced by random variation, he argued, the rest of the body would have to adjust to the change, and this would require a host of additional variations. To suppose that all of these would appear at exactly the right time by random variation was to invoke too great a coincidence. There was also the difficulty posed by the disappearance of organs that became useless because of changed habits. Selection might explain a reduction in size, but hardly a complete elimination, while if Lamarckism were valid, an unused organ would continue to diminish in size until it wasted away completely. Spencer was thus a vigorous defender of Lamarckism, and the process of individual self-improvement played a central role in his social philosophy.

Spencer was written out of the history of Lamarckism by the advocates of a different line of moral argument based on the theory. An early advocate of this rival interpretation was the writer Samuel Butler, who clashed with Darwin and his followers in the 1880s (Pauly 1982; Willey 1960). Originally a Darwinian himself, Butler was alerted to the possibility of an alternative form of evolutionism by reading Mivart's *Genesis of Species.* Increasingly concerned to allow a sense of purpose in evolution, Butler saw Lamarckism as a means of retaining an indirect form of the design argument. Instead of designing species by miracles, God had transferred His creativity to living things, allowing them to be the agents of their own and their species' development. He soon realized that the mechanism based on self-development had already been proposed by Lamarck and Erasmus Darwin, and became angry that the younger Darwin had ignored their contribution. Butler's *Evolution, Old and New* (1879) was both a defense of Lamarckism and a personal attack on Darwin which succeeded in alienating him from the Darwinian community. A series of later books with titles such as *Luck, or Cunning?* (second edition 1920a) drove home the point that natural selection destroyed any sense of individual creativity in animals and humans. In an essay, "The Deadlock in Darwinism," Butler poured out his contempt for the selection theory: "To state this doctrine is to arouse instinctive loathing; it is my fortunate task to maintain that such a nightmare of waste and death is as baseless as it is repulsive" (1890: 308).

Butler's main interest was the mental life of animals and humans, although unlike most Lamarckians he did not regard consciousness as the highest form of mental activity. He argued that instinct was the most perfect mental operation, the end product formed when an acquired habit became an inherited mental predisposition (1916, 1920b). But his philosophy of evolution depended on conscious choice as the first step directing any new development toward a purposeful goal, and this was the message that a whole generation of Lamarckians drew from the theory. Although ostracized by the early Darwinians, Butler was increasingly cited by the next generation of evolutionists and was even on good terms with Darwin's son, Francis, an eminent botanist.

Another botanist, the clergyman George Henslow, applied Lamarckism to plant evolution by stressing the automatic adaptive response of plants to any change in their environment (1888, 1895). He showed that a plant is always capable of adjusting to environmental stress in a positive way, and much of his work was devoted to cataloging examples of plants grown in conditions that were hotter, drier, and so on than those they were used to. Typically, he did not appreciate the idea that, to confirm the Lamarckian effect, he would have to show that the acquired characters would be retained in the next generation, even if that generation were not exposed to the new conditions. Since each generation could acquire the adaptations for itself, showing that they appeared in offspring still exposed to the new conditions did not prove inheritance. This was a common failing of many Lamarckians:

proving the existence of acquired characters and then merely assuming that they were inherited.

Henslow also observed that the shapes of flowers were related to the ways that insects entered them. He argued that this illustrated a direct response by the plant to an external stimulus. In each generation the insects had exerted pressures on the flowers, and the accumulated effect of the distortions had shaped the evolution of the species so that, now, the structure was produced even before any insects reached the plant. This argument contains another logical fallacy common among the Lamarckians. The mere fact that a structure looks as though it is shaped as a direct response to external stimulus does not prove that the structure was evolved by the cumulative effect of such responses. If a particular shape is advantageous to the species because it encourages insects to pollinate it, for example, then natural selection will drive evolution toward that shape because individuals better adapted to the insects' requirements will leave more offspring in every generation.

Neo-Lamarckism was also strong in Europe. For all that he had called himself a Darwinian, Ernst Haeckel had incorporated a large element of Lamarckism into his theory, invoking selection mainly at the species level to eliminate the less successful products of direct adaptation. Later naturalists such as Theodor Eimer began to support use-inheritance at the expense of any role for selection in the production of adaptations (translation 1890). For Eimer, Lamarckism offered a direct challenge to Weismann's neo-Darwinism. Even the eminent pathologist Rudolph Virchow cited the inheritance of acquired characters (Churchill 1976). In France, too, Lamarck was belatedly hailed as the true founder of evolutionism. French biologists had never been enthusiastic about Darwinism, and as they reluctantly conceded ground to evolutionism, it was only natural for them to explore this homegrown alternative. Biologists such as the paleontologist Alfred Giard and the experimentalist Lucien Cuénot provided a variety of arguments for the Lamarckian position (Limoges 1976; Persell 1999).

American Neo-Lamarckism

The neo-Lamarckian movement in America had a number of different sources, some quite different from those in Europe (Cook 1999; Dexter 1979; Pfeifer 1965, 1974). There was support from field naturalists attempting to explain local adaptations in terms of the direct response of organisms to their environment. Alpheus Packard used evidence from a study of blind fish living in caves, which were thought to have lost their eyes through the cumulative effect of disuse (1889; see also Bocking 1988). But the most characteristic American line of support for the theory came from paleontologists who thought they could provide evidence for Lamarckism from the adaptive trends revealed by the fossil record. Their interpretation was derived from a morphological perspective which led them to seek orderly patterns in the transformation of structures through time, and was strongly based on the recapitulation theory (Gould 1977b: chap. 4; Rainger 1981). The leading exponents of this tradition were Edward Drinker Cope and Alpheus Hyatt.

Cope is best known for his dispute with O. C. Marsh over the opening up of the rich fossil beds of the American West (Lanham 1973; Plate 1964; Shor 1974; D. Wallace 1999; on Cope's life, see Osborn 1931). But the popular books on the Cope-Marsh feud concentrate on their personality clash, scarcely noting the immense difference in theoretical perspective created by Cope's neo-Lamarckism. Both Cope and Hyatt converted to evolutionism in the 1860s, but neither was happy with the Darwinian theory. Hyatt was a student of Agassiz, and Cope, too, was deeply influenced by the idealist philosophy of nature, which stressed the orderly arrangement of species rather than adaptation. Neither began as Lamarckians; their starting point was the desire to reconcile Agassiz's idealist vision of development modeled on embryology with the new evolutionism (Bowler 1983: chap. 6). The key to this was the recapitulation theory, the belief that the development of the embryo repeats the evolutionary history of its species. But where Agassiz had postulated a divinely instituted parallel between human ontogeny and the history of life on earth, his followers confined their search for developmental patterns to evolution of particular groups of organisms. Within each group, they sought not an irregularly branching process of adaptation but a linear pattern of development identical to that seen in the embryological development of later individuals. They stressed that many different species within the group all passed separately through the same basic pattern of evolution in parallel.

Cope and Hyatt both began their theorizing from a position that ignored the role of adaptation in evolution, only later conceding that a Lamarckian process of specialization would explain the trends they saw in the fossil record. Their law of acceleration of growth was published in Cope's "On the Origin of Genera" of 1867 (reprinted in Cope 1887) and in Hyatt's paper on fossil cephalopods (1866). According to this law, evolution proceeded by a series of sudden additions to the development of the individual. At certain points in time, every individual in the species underwent a new phase of growth which advanced all to the status of a new species. To make room for this addition, the old adult form was compressed back into an earlier stage of development, which was then accelerated to accommodate the addition. Cope insisted that, within each group, evolution was not a branching process. Every species advanced through the same linear pattern of growth stages, although not necessarily at the same rate, so that one species might exhibit the ancestral form of another. Moreover, the similarities which allowed the most closely related species to be grouped into genera were not the result of common descent but were merely an indication that the species had independently reached the same point in the growth cycle (this was the significance of Cope's title). The evolution of the group thus consisted of a number of lines advancing in parallel through the same sequence, the whole pattern being revealed in the ontogeny of those species which had reached the highest stage.

At this point, neither Cope nor Hyatt thought that the direction of advance was determined by adaptation. Cope explicitly denied any adaptive value for the new characters defining genera, claiming they were steps in a regular pattern designed by the Creator. He soon realized that theistic evolutionism begged the whole question of a natural cause for evolution. He abandoned his opposition to the principle of utility and conceded that most developments have an adaptive purpose. Cope and Hyatt both saw that the inheritance of acquired characters would serve as the guiding force they were seeking, because once a group had begun to develop habits for a particular way of life, continued exercise of the habits would drive all the species along the path of specialization (Cope 1887, 1896; Hyatt 1880, 1884, 1889). The Lamarckian process was well suited to the recapitulation concept, because it required new stages to be developed in the adult organism and then compressed back into embryological development so that they could be inherited. Cope postulated a growth force named *bathmism*, which became focused in those parts of the body most in use. He continued to insist, however, that the developments produced by this force were not continuous. The pressure for change built up over many generations until it was suddenly released at an "expression point," giving rise to an evolutionary leap or saltation. Neither Cope nor Hyatt had read Lamarck when they developed these ideas, and they came to realize only later that what they were proposing contained elements of Lamarck's thinking.

Cope was a deeply religious man who, like Butler, saw that Lamarckism allowed consciousness to be seen as the guiding force of evolution. Instead of being designed by an external creator, species designed themselves as consciousness gradually extended its manifestations in the organic world. New habits allowed the development of purposeful new structures. The combination of mental powers with the physically transforming effect of bathmism represented the Creator's powers transferred into nature. And because evolution gradually encouraged organisms to develop their mental powers, it could be seen as having a spiritual dimension: its ultimate goal was the emergence of the human soul (see Moore 1979).

Although Hyatt did not accept this "psycho-Lamarckism," he considered the theory's implication that living things are in control of their own destinies an important point in its favor. So compelling was this view that it became necessary for those who still accepted Darwinism to show that it was not merely the trial-and-error process caricatured by its opponents. Some Darwinists argued that the selection theory could also accommodate an element of conscious direction by the organisms. The mechanism of "organic selection," later known as the Baldwin effect, was suggested independently in the 1890s by the paleontologist Henry Fairfield Osborn and the psychologists Conwy Lloyd Morgan and James Mark Baldwin (Baldwin 1902; R. Richards 1987: chaps. 8, 10). Like Lamarckism, this process was supposed to start with the deliberate choice of new habits by animals faced with a changed environment. The new habits generated acquired physical characters, but there was no need to assume that these were inherited in order for the effect to determine the course of evolution. The acquired characters allowed the species to adjust in the short term, providing time for the process of random variation to come up with equivalent hereditary characters. Selection would then move the species in the direction mapped out by the new habit—after all, selection could hardly favor long necks for giraffes unless they had already changed their habits to feed off trees. As a pupil of Cope, Osborn hailed the mechanism as a compromise between selection and Lamarckism. More perceptively, Morgan and Baldwin realized that it gave the Darwinists the opportunity to steal the Lamarckians' best nonscientific argument.

The Baldwin effect illustrates the strength of feeling at the time over the question of selection's moral implications. But the most distinctive feature of the American school of neo-Lamarckism was its insistence on the regularity of the development within each branch of evolution. This was not a necessary component of Lamarckism, as illustrated by Packard's study of the purely local adaptation of blind cave fish. But Cope and Hyatt had begun from Agassiz's idealist perspective in which morphological change must follow an orderly pattern, and had merely applied Lamarckism to explain the trends they were already committed to. Cope insisted that the evolution of a vertebrate group such as horses had proceeded far too regularly in a single direction for it to be explained by random variation and selection. In Hyatt's case this fascination with the regularity of growth and evolution extended into the bizarre idea of "racial senility," in which the trends eventually extended themselves so far that the results became nonadaptive and contributed to the group's extinction. In this sense, neo-Lamarckism also incorporated an element of orthogenesis.

Toward the end of the century, the tide began to turn against Lamarckism. Its best evidence was indirect, and the continued discovery of new specimens threw doubt on the conclusions drawn by Cope and Hyatt from the fossil record. The history of families such as the horse family began to look more like an irregularly branching tree than a series of parallel lines (Bowler 1996). More serious still was the demand for experimental proof of the Lamarckian effect among biologists caught up in the revolt against morphology. The study of heredity was increasingly treated as an experimental problem, and if the inheritance of acquired characters could not be subject to direct testing, it would lose credibility. The French physiologist C. E. Brown-Séquard reported positive results with the inheritance of epilepsy induced by brain mutilations in guinea pigs. This was widely discussed (e.g., Romanes 1892–97: vol. 2), but few were convinced that the effect was truly hereditary, because of the possibility that a toxin had been transmitted to the young in the womb. In America, Osborn encouraged researchers to set up a major experimental program to test the Lamarckian effect, but gained little support (Cook 1999). The failure of this initiative highlights a polarization within the biological community: paleontologists and field naturalists continued to take Lamarckism seriously, but the experimentalists were hostile, especially after the new science of genetics became established. The concept of the unit gene transmitted unchanged from parent to offspring left no room for acquired characters to be incorporated. The split between the disciplines was not healed until the emergence of the modern Darwinian synthesis in the 1930s and 1940s (Mayr and Provine 1980). Only in Germany, where genetics developed in a less dogmatic form, was there an interaction between paleontologists and experimentalists searching for non-Darwinian mechanisms of evolution (Reif 1983, 1986; Rinard 1988).

Early-Twentieth-Century Lamarckism

Outside science, there was still a confidence that "Darwinism is dead," with writers such as George Bernard Shaw continuing to insist on the moral superiority of Lamarckism (see chap. 8). But the scientists who supported the theory grew more desperate as experimental evidence continued to elude them. This pressure was most acute in the English-speaking world, where genetics developed in a way that was particularly hostile to the idea that the environment could influence heredity. The antimaterialist psychologist William McDougall (1927) performed an experiment in which rats trained to run a maze seemed to pass the knowledge on to their offspring. Most of his colleagues thought the effect could be explained by the experimenter unconsciously selecting rats better at learning how to cope with mazes. By far the most controversial episode in the defense of Lamarckism centered on the experiments of the Austrian biologist Paul Kammerer, highlighted by Arthur Koestler in his book *The Case of the Midwife Toad* (1971). The scientific community has decided that Kammerer was a fraud, but Koestler—a prominent opponent of the Darwinian synthesis which emerged in the mid–twentieth century—argued that he had been the victim of a witch hunt.

Kammerer's experiments had been performed before World War I, when Lamarckism was not yet a lost cause. The work on which Koestler focused (only one of several investigations by Kammerer) involved the midwife toad *(Alytes obstetricans)*, a species which, unlike other toads, mates on dry land. The males of this species now lack the roughened mating pad on the forelimbs, which the males of other species use to grip the female while fertilizing her eggs in water. Amphibians are notoriously difficult to breed in captivity, but Kammerer was able to make his midwife toads breed in water rather than in dry conditions. The males developed mating pads and, according to Kammerer, the acquired character was then inherited. In fact, of course, this was not a new character but an old one which had been suppressed in this species. But Kammerer claimed the effect as experimental proof of Lamarckism.

Kammerer's career was disturbed by the poor economic conditions following the war, so, in the 1920s, he traveled to Britain and America to raise support for his work. Translations of his books and articles were published (1923, 1924), and there were headlines about his work and the implications of Lamarckism in the popular press. Kammerer was supported by the embryologist E. W. MacBride (1924), one of the last major British scientists still defending Lamarckism. The geneticists, led by William Bateson, were hostile to the Lamarckian theory and skeptical of Kammerer's experiments. No one could repeat the experiments because the toads were too difficult to breed in captivity, and the geneticists insisted that the preserved specimens left over from Kammerer's original work be reexamined by an impartial witness. The tests revealed that the mating pads were marked by india ink. Kammerer defended himself by suggesting that an assistant had injected the ink to preserve the marks during the war years, but his credibility was now destroyed, and shortly afterward he shot himself. Koestler hinted at a plot to discredit Kammerer because he had written against racism, and insisted that the work might have been valid. Most biologists think he was a fraud, while even those who think that no actual dishonesty was involved accept that Kammerer was mistaken in his interpretation of the results (Waddington 1975).

When Kammerer committed suicide in 1926, he was about to take up a research position in Soviet Russia. This was no coincidence, because in a few years' time Lamarckism would be catapulted to prominence in that country through the activities of T. D. Lysenko. The Soviets may have intended Kammerer to spearhead such a move (Zirkle 1949, 1959). There had been an earlier but abortive attempt to impose Lamarckism on Russian biology (Gaissinovitch 1980), but the Darwinian theory had never taken hold either (Rogers 1974; Todes 1989; Vucinich 1988), and now the Marxist philosophy of the revolutionary government was positively hostile to the capitalist implications of the selection theory. The Soviets also were opposed to the determinism of Mendelian genetics, which they took to imply that social reform would be ineffective in improving human character. Lamarckism was the obvious alternative, and in the course of the 1930s, Lysenko succeeded in having it incorporated into official communist philosophy. His rise to power has been portrayed as a crude attempt to impose ideological values onto science, but it also may have been a consequence of the political leverage he gained by promising an end to Russia's chronic wheat shortages (Joravsky 1970; Medvedev 1969; Soyfer 1994).

Lysenko's claim to fame was the discovery of the "vernalization" of wheat, a process in which the seeds are frozen so that they will germinate earlier the next spring. The effect had been known for some time, but Lysenko claimed that it could be inherited, as predicted by the Lamarckian theory. Thus, new breeds of wheat would be produced that were adapted to the short growing season in Russia. By promising to reform agriculture, Lysenko gained enough political support to begin making open attacks on Russian geneticists and the selection theory. Stalin allowed Lysenko to begin a purge of the Russian geneticists, and all research in that area was forbidden. Geneticists were forced to recant their support for their "bourgeois" science or were exiled to Siberia, where some simply disappeared.

The nightmare for Soviet geneticists ended only in the 1950s, when it became clear that Lysenko had failed to reduce the food shortages and had cut Russian agriculture off from improvements being made in the West using genetics. There is no shortage of critics arguing that the whole episode illustrates the disastrous consequences of political interference with the freedom of scientific inquiry. But did Lysenko really hope to create a Marxist science, or was he merely an opportunist who exploited Stalin's credulity? Marxist thinkers have since tried to develop a more subtle approach: they still believe that science can reflect social values, but they are warier of implying that ideology can rule out a certain theory whatever the empirical consequences (Lecourt 1977; Lewontin and Levins 1976). More generally, historians of science have begun to recognize that genetics itself was not a totally value-free science, so the hostility of Marxist thinkers may not have been altogether misplaced (Sapp 1987). It has also been noted that, whatever Lysenko's later excesses, his early work was in line with the kind of neo-Lamarckian work still going on in countries where genetics had not led to a hard-line determinist viewpoint (Roll-Hansen 1985). The episode certainly illustrates the dangers of allowing demagogues to impose their theories on the scientific community. But it should not lead us to suppose that modern Western science is value-neutral in ideological terms.

ORTHOGENESIS

Many Lamarckians were attracted to a related mechanism in which variation was driven not by functionally acquired modifications but by internally programmed forces generating characters unrelated to the organisms' needs. The term *orthogenesis* was popularized by Theodor Eimer to denote evolution which proceeded in a straight line, the term normally being restricted to those trends which had no adaptive purpose. On this theory, variation was not random but was constrained to move in a fixed direction. Natural selection was powerless to affect the outcome, except perhaps by causing the extinction of species which had developed positively harmful characters. The link between orthogenesis and neo-Lamarckism is particularly obvious in the American school, where a fascination with the regularity of development encouraged the search for linear trends in evolution, both adaptive and nonadaptive. It was possible to imagine that a trend that had begun with an adaptive purpose might ultimately become nonadaptive if, for instance, it continued to enlarge a structure beyond the optimum size needed for dealing with the environment.

Lamarckism was, at least, an adaptive mechanism and was not completely at variance with the Darwinian worldview. But orthogenesis struck at the heart of Darwin's vision by supposing that significant aspects of evolution were controlled by forces which produced change unrelated to the demands of the environment. The selection theory assumed that any harmful variation, however insignificant, would be wiped out in the struggle for existence.

Yet orthogenesis postulated variation trends that could produce totally non-adaptive characters, implying that the pressure of the environment was by no means as relentless as Darwin had claimed. The hostility displayed by the supporters of orthogenesis toward the whole utilitarian perspective of Darwinism, coupled with their desire to see evolution governed by predictable trends, illustrates the continued influence of an idealist perspective in biology (Bowler 1983: chaps. 3, 7).

Carl von Nägeli's theory of an "inner perfecting principle" was an example of what later became known as orthogenesis. In the 1860s Nägeli had opposed Darwin by stressing the nonadaptive character of many evolutionary developments, and his 1884 theory (translation 1898) postulated internally programmed variation trends that operated without reference to adaptation. The concept of orthogenesis was popularized in the 1890s by Theodor Eimer, who had begun as a Lamarckian (translation 1890) but who now began to stress the nonadaptive character of many trends (translation 1898). He studied the color variation among species of lizards and butterflies and became convinced that this variation had no adaptive significance. He arranged the species into linear patterns which he thought represented the course of the group's evolution, each series passing through the same sequence of color changes. This parallelism, he claimed, explained the similarities among unrelated forms which the Darwinians attributed to mimicry.

Eimer's sequences of living species provided no actual proof that a fixed trend operated through time; such proof could only come from a comparison of earlier and later forms in the fossil record. Paleontologists influenced by the idealist perspective of Agassiz were especially likely to search for such orderly trends and were by no means committed to explaining them in adaptive terms (Rainger 1981). Increasingly they looked for trends which, although they might have begun in an adaptive direction, led ultimately to the overdevelopment of structures in a way that might have contributed to the species' extinction. A much-discussed example was that of the Irish elk, which was supposed to have become extinct because its antlers eventually became too large as the result of an orthogenetic trend (Gould 1974b). It was argued that the trend that originally developed the antlers for use as weapons had somehow acquired a momentum of its own, so that it carried on affecting the species long after the antlers had reached the maximum size for effective use. This "overdevelopment" theory of extinction became widely popular among paleontologists in the late nineteenth and early twentieth centuries.

A widely cited exponent of this view was the American paleontologist

Alpheus Hyatt (1884, 1889), who argued that adaptive trends established by the Lamarckian effect almost always carried on beyond the point of maximum utility and drove the species inexorably toward "racial senility" and death. The analogy drawn with the life cycle of an individual organism was quite explicit and illustrates the power of the recapitulation theory. In the fossil record of the ammonoids, Hyatt thought he could distinguish a pattern through which all members of the group passed. Initially simple forms became gradually more complex as they responded to changes in the conditions, but then degeneration set in. Perhaps because they were unable to cope with conditions which had become unfavorable, the species began to lose their more advanced characters and regress toward the earlier, simpler forms. In other cases, they acquired bizarrely excessive characters. Whatever the form of the degeneration, the senile phase was always a prelude to extinction. Hyatt's theory presented the final stages of senility and death as predetermined by internal forces that the organisms could neither control nor reverse. The possibility that all branches of evolution might eventually run out of energy and degenerate to extinction presented a pessimistic view of evolution that chimed with the loss of confidence among many thinkers, especially in Europe, as the nineteenth century drew to a close (Bowler 1989a,c).

Orthogenesis remained popular among paleontologists into the early decades of the twentieth century (Bowler 1996: chap. 7). Robert Broom (1932) thought some trends could not be explained by any adaptive force. The Russian biologist Leo S. Berg (1926) provided an important example of support for the theory. Perhaps the most active exponent was the paleontologist Henry Fairfield Osborn, a wealthy and influential figure based at the American Museum of Natural History, who sought to block the rising tide of support for experimental biology (Rainger 1991). After playing a role in the discovery of organic selection, Osborn began to stress the role of predetermined trends equivalent to what had been known as orthogenesis, although he coined his own term, *aristogenesis* (Osborn 1908, 1912, 1917, 1929, 1934). He stressed that each major class underwent a period of rapid "adaptive radiation" at the start of its history, but once the various orders within the class were established, their subsequent evolution was a stable, linear process without the continuous small-scale branching postulated by the Darwinians. He saw the evolution of groups such as the titanotheres (giant ancient mammals) as a series of parallel lines advancing through the same pattern of development for horns and other features. Only in the middle phases of the pattern were the structures useful, so both the start of the trend and the final stages of overdevelopment had to be controlled by internal, nonadaptive forces. Osborn thought that outsized structures such as horns might play a role in making a species vulnerable to external threats such as sudden climatic changes.

The supporters of orthogenesis were often vague about the exact cause of the trends they claimed to see in the history of life. Broom openly invoked a divine plan, more or less as Mivart had done a generation earlier. Osborn, at least, was disturbed by the lack of a naturalistic explanation of predetermined variation and suggested that the interaction of energies within the living body might somehow allow the trend to become fixed in the germ plasm which controlled heredity. Some early geneticists wondered if there might be a predisposition for genes to mutate consistently in particular directions, although the experimental evidence soon eliminated this possibility. Osborn eventually conceded that his own explanation was unsatisfactory. He hinted that the evolutionist might have to rest content with the trends revealed by the fossil record and accept that no natural explanation was likely to be discovered. At this point, T. H. Morgan, whose work on mutations was laying the modern foundation of our understanding of variation (see G. Allen 1969a), accused him of toying with mysticism. Osborn's attitude reveals the growing incompatibility between the old morphological tradition, with its links to the idealist worldview, and the new experimental biology that would dominate twentieth-century biology.

Most of the fossil evidence used to support orthogenesis was rejected by the next generation of paleontologists, who helped to found the modern Darwinian synthesis (e.g., Romer 1933; Simpson 1944, 1953b). As more evidence was discovered, the neat linear trends linking specimens of successive periods dissolved into a welter of branches and subbranches. The trends were an artifact of insufficient evidence and existed more in the minds of paleontologists than in the record itself. The assumption that very large structures such as the antlers of the Irish elk were maladaptive has been challenged—even if too large to fight with, they may have been useful for display. In some cases, the founders of the modern Darwinian synthesis were prepared to admit nonadaptive trends as by-products of adaptive ones. Julian Huxley's concept of "allometry" (1932) noted that there were fixed ratios between the growth rates of certain structures, so if selection favored an overall size increase it might increase the size of an individual structure (such as the elk's antlers) at a rate much greater than would be required by adaptation. But such concessions are marginal, and the Darwinian synthesis has dismissed orthogenesis as the product of a worldview obsessed with abstract patterns at the expense of understanding the real world.

NEO-DARWINISM

Apart from the search for linear trends, Lamarckism could be treated as a mechanism of local adaptation, as Darwin himself accepted. Initially this flexibility allowed the emergence of a loosely defined Darwinian view of evolution, but toward the end of the century there was a noticeable polarization of views in which neo-Darwinism and neo-Lamarckism came to be seen as mutually exclusive. The split was precipitated largely by developments in the study of heredity. August Weismann's development of the concept of the germ plasm created a model of heredity in which, in principle, the substance responsible for transmitting characters from parent to offspring could not be affected by changes to the parents' bodies. The inheritance of acquired characters was impossible, and Weismann insisted that natural selection of random variation was the only mechanism of evolution. This was neo-Darwinism, but Weismann's ideas were at first highly controversial, and the Lamarckians continued to insist that his model of heredity was too materialistic and too narrow to explain effects they saw as an integral part of evolution.

Darwinism and Speciation

Within the general Darwinian framework, there was continued debate over how what we now call speciation occurred, that is, how a single species could become divided first into varieties and then into a group of distinct species. Darwin eventually abandoned his emphasis on geographical isolation and argued that ecological specialization could disrupt a continuous population (Mayr 1959b; Sulloway 1979a). When Moritz Wagner (translation 1873) insisted that isolation was necessary for speciation, Darwin and most of his followers objected, seeing Wagner's position as an alternative, rather than a supplement, to the selection theory. Wallace too accepted speciation without geographical isolation (what is now called sympatric speciation). The Darwinians had not yet fully appreciated the significance of what Mayr has called population thinking—they still tended to think of species as units defined by their unique morphological character. The crucial question should have been: What keeps the subpopulations from blending back together again by interbreeding? In 1886 George John Romanes suggested that natural selection would have to be supplemented by what he called physiological selection to explain speciation. Variations infertile with the original population might appear and, since they were unable to blend with the old species, eventually would evolve their own unique physical characters and become distinct species (see also Lesch 1975; Schwartz 1995). The problem with this suggestion was that, if variation by itself could explain infertility, why could it not produce all the other characters independently of natural selection?

Field naturalists eventually began to realize that something must prevent interbreeding from blending two potentially distinct populations back together again. If Romanes's suggestion was unsatisfactory, the only alternative was an initial period of geographical separation. Wagner's ideas were revived in the 1880s by John Thomas Gulick, whose work on Hawaiian land snails showed the correlation between varieties and distinct geographical locations (Gulick 1888; A. Gulick 1932; Lesch 1975). In Germany, Karl Jordan developed a populational view of species which also allowed him to appreciate the significance of geographical isolation (Mayr 1955). Jordan refused to be sidetracked by the debate between neo-Darwinians and neo-Lamarckians, realizing that for his purposes it was more important to clarify the basic nature of species. By the early twentieth century, the majority of field naturalists dealing with this issue had come to accept the importance of geographical isolation—and were lamenting the refusal of the new generation of experimental biologists to show any interest in their work. Eventually, these studies would play an important role in the creation of the modern Darwinian synthesis. Naturalists now recognize a range of isolating mechanisms that can prevent two varieties from blending back together again, including behavioral differences which may inhibit mating between them. Thus, two varieties can coexist in the same area, even though there is as yet no genetic barrier to interbreeding. But an initial phase of geographical separation is essential for such mechanisms to be established.

The field naturalists' complaints about the indifference of the experimentalists were articulated by Edward B. Poulton (1890, 1908). A student of animal coloration, Poulton was convinced that camouflage and mimicry had adaptive value—these colors could not be the product of internal forces, as Eimer claimed. But because most animals cannot control their colors (e.g., the colors of butterflies' wings), these adaptive effects cannot have been the result of Lamarckism; this left natural selection as the only viable explanation. Poulton pointed out that it was easy for the laboratory biologist to ignore the demands of adaptation, because animals kept in an artificial environment were not subject to the kind of threats that faced those living in the wild. As we shall see below, many early geneticists refused to accept that the demands of the environment placed any adaptive stress on the species. The fact that Poulton and other field naturalists had to speak out to defend the idea that adaptation played a role in evolution illustrates how far the anti-Darwinian trend in biology had developed by the turn of the century.

Weismann and the Germ Plasm

To some extent, the surge of anti-Darwinian thinking in the late nineteenth century was a reaction against the dogmatic selectionism of August Weismann. Neo-Darwinism emerged from Weismann's initiative in the study of heredity and variation, which was an attempt to resolve the problems and confusions in this area. Mayr (1985) regards Weismann as the most important nineteenth-century evolutionist after Darwin himself, and it is true that the concept of the germ plasm—the material substance responsible for transmitting characters from one generation to the next— helped to clarify the modern view of heredity and of natural selection. Weismann's attack on the problem of heredity arose very much as a continuation of Darwin's own program, in which the study of generation (reproduction) was an integral part of evolution theory (Hodge 1985). Yet in the end, he fragmented that program by creating a model of heredity in which the transmission of characters occurred independently of the process by which those characters were produced in the developing embryo. The key to this transition was his use of a new area of biology known as cytology, the study of cells, now increasingly seen as necessary for understanding all processes going on within the organism (Farley 1982; Robinson 1979).

Frederick Churchill (1986, 1999) has shown that Weismann's radical approach to heredity flowed originally from a traditional position inspired by the recapitulation theory. By applying this theory to his early work on the Hydrozoa, he became convinced that all organisms contained a potentially immortal kernel of reproductive material that could be transmitted to future generations (see fig. 26). Although failing eyesight forced him to give up microscopic work, Weismann felt that he had located this material in the rodlike chromosomes—so called because they are revealed by color staining—of the cell nucleus. This was an important insight which has been incorporated into the foundations of modern genetics. Indeed, the whole concept of a material substance that was somehow encoded with information that would be used to construct the body of the offspring was a crucial break with the past, although in this respect Weismann had been anticipated by Carl von Nägeli's concept of the "idioplasm" (Coleman 1965). In the course of the 1880s, Weismann refined the idea to create his model of the germ plasm as a substance that could be transmitted through the ovum or sperm to become the foundation for the next generation (translation 1891–92, 1893a; Romanes 1899).

In Weismann's mature theory, the germ plasm was totally isolated from the somatoplasm, which was the material constituting the rest of the body.

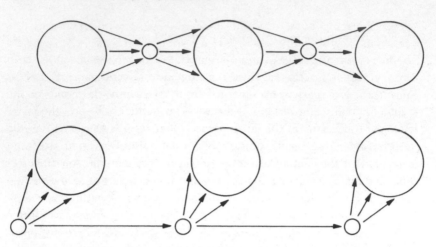

26.   Pangenesis and the germ plasm. These diagrams indicate the different
relationships between the body (large circle) and the hereditary substance
responsible for transmitting characters between generations (small circle). For simplicity, the combination of characters produced by sexual reproduction is ignored.
In Darwin's theory of pangenesis (top), the body actually manufactures the
gemmules which transmit its characters to its offspring. This implies that changes
to the adult body might be reflected in the gemmules it produces and thus could be
inherited. In August Weismann's theory of the germ plasm (bottom), the body
develops from germ plasm inherited from the parent but then merely transmits it
to the next generation. Since the body does not produce the germ plasm but
merely stores it, characters acquired by the adult body cannot be reflected in the
germ plasm and thus cannot be inherited.

The soma was constructed in the embryo using information coded in the
germ plasm, but it was constructed around the germ plasm, which it contained and protected but could not interact with. The flow of information
was one-way only, from germ plasm to soma. Once the soma was formed, it
acted as a host to preserve the germinal material so that it could be passed
on unchanged to the next generation. This claim encapsulates the essence of
what has come to be known as hard heredity, the belief that heredity cannot
incorporate the responses the body or soma makes to its environment.
Darwin's theory of pangenesis reflected the alternative, soft heredity, so
called because the body was supposed to manufacture its own hereditary
material (the gemmules), allowing changes in the body to modify the gemmules and thus be transmitted to the offspring. Weismann believed that the
germ plasm was composed of units he called determinants, each responsible
for producing a certain part of the body, although he did not anticipate the

Mendelian concept of discrete units of heredity. In sexual reproduction, determinants for each character derived from both parent are mixed to provide the information used to construct the embryo of the offspring.

Weismann's concept of the germ plasm made Lamarckism impossible. Since the body merely transmitted its germ plasm to the next generation, and did not actually produce it, there was no way that changes in the body due to use or disuse could be reflected in the process of heredity. Weismann dismissed the popular belief to the contrary as mere superstition. For him, variation within a population was due solely to the recombination of determinants as they were shuffled by sexual reproduction. Natural selection was the only possible mechanism of evolution, because it could favor those determinants responsible for producing useful structures and suppress those that were harmful. New variations could be produced only if the determinants' structure was distorted by an accident of duplication at the cellular level. And since the soma had no control over such accidents, this variation would be essentially random.

In a famous experiment, Weismann tried to prove his point by cutting off the tails of a group of mice and then continuing the process in their offspring over several generations. There was no indication that this orgy of mutilation produced mice with tails any shorter than normal. The Lamarckians insisted that this was not a fair test because accidental mutilations were not equivalent to the purposeful responses to the environment which they supposed to be inherited. Yet the experiment had a valid point: it showed that whatever was responsible for building the tails in embryonic mice was not produced by the tails of their parents. Mice without tails still carried the complete germ plasm for producing the structure in their offspring. Weismann could also point out that some of the alleged experimental demonstrations of the inheritance of acquired characters had used mutilations because these were so much easier to produce and check in the laboratory.

To those naturalists who had decided—often for other than experimental reasons—that use-inheritance must play a role in evolution, Weismann's theory symbolized the dogmatism of neo-Darwinism. If the Darwinian theory were to be narrowed down in this way, they would have none of it and would develop the alternatives to selection more vigorously than before. The germ plasm theory was a speculation based on inadequate evidence, and Weismann's insistence on the "all sufficiency of natural selection" (1893b) was misplaced. Herbert Spencer was one among many who responded by claiming that evolution depended absolutely on a Lamarckian component

(1893). In this respect, Weismann's dogmatism merely served to polarize opinions and thus promote the eclipse of Darwinism, despite its significance in the longer term.

Weismann never abandoned his opposition to Lamarckism, but he was forced to make concessions on other fronts. To explain how useless organs have been reduced to rudiments, he postulated a mechanism of "germinal selection" (1896), in which the determinants within the germ plasm competed for nutrients. Since germinal selection operated independently of natural selection, it could produce variation trends that went beyond anything that the Darwinian mechanism could accomplish. This might include the dismantling of a structure which was no longer of any benefit to the organism, but as Weismann himself later admitted, it might also lead to the production of new characters that were not adaptive. He insisted that this process could never get very far: only variations also favored by natural selection could get beyond a trivial level. But to his opponents, this concession revealed the speculative nature of Weismann's theorizing and the hollowness of his defense of Darwinism.

Biometry

Weismann had tackled the problem of heredity by setting up a model for the physical process by which characteristics are transmitted from one generation to the next. The school of biometry had its origins in a different line of attack on the same problem: the application of statistical techniques to analyze the range of variation within wild populations and the effect of selection on that range. The use of statistics in the study of the human population had been pioneered earlier in the century by Lambert Quetelet but only now was exploited on a larger scale. The biometrists applied the same techniques to animal populations in the wild in an attempt to provide direct evidence for the process that Darwin had postulated. Because the founders of the school were convinced of the power of heredity to predetermine character, their studies were conceived in terms of selection acting via hard heredity on the variability of the population (Depew and Weber 1995; Gayon 1998; Provine 1971).

The founder of the movement was Darwin's cousin, Francis Galton. Early in his career, Galton had performed an experiment involving blood transfusions between rabbits which seemed to disprove the theory of pangenesis (or at least that the gemmules were transported via the bloodstream). Following a trip to Africa, he had become convinced that human character was rigidly predetermined by racial origin and could not be improved by better conditions or education (Fancher 1983). Seeking a way of promoting the Darwinian cause, he decided to show that intellectual ability in humans was predetermined by heredity and thus could be affected by selection. His *Hereditary Genius* of 1869 (reprint 1892) used the pedigrees of eminent men to argue for the hereditary predetermination of ability but also made it clear that "isolated" genius was merely the extreme of a continuous range of variation in the population. Galton wanted to convince social thinkers that the only way to improve the human race was by restricting the breeding of individuals with inferior characters and by increasing the number of children among those more able. To gain support for this policy of racial improvement through what he termed *eugenics,* he began to encourage the study of the range of variation in the human population, and the application of similar techniques to animal populations, where the effects of selection might be more visible (Galton 1889, 1892; see Bulmer 1999; Cowan 1972a,b; De Marrais 1974; Forrest 1974; Froggatt and Nevin 1971a; Gilham 2001; Mackenzie 1982; Pearson 1914–30; Swinburne 1965; Waller 2002; Wilkie 1955; on eugenics, see chap. 8 below).

Biometrics pioneered the now-familiar method of depicting the range of variation by a frequency-distribution curve showing the proportion of the population occupying each part of the range. For most continuously variable characteristics, such as height in the human population, this follows the bell-shaped normal, or Gaussian, curve illustrated in fig. 27. But Galton and his followers wondered what would happen to the curve if the population were subject to selection over a number of generations. Selection ought to distort the curve, and its continued effect ought to become cumulative if it were reinforced by inheritance, because favored individuals not only survived preferentially, they also transmitted their character to the next generation. Galton wanted to know what proportion of an individual's unique character would be transmitted, bearing in mind that it would have to mate with a member of the opposite sex which might not share all the same features. He proposed a "law of ancestral inheritance" in which the individual derived half its inheritance from the two parents, one-quarter from the four grandparents, an eighth from the eight great-grandparents, and so on back through the generations. This bears no resemblance to the laws of Mendelian genetics, but it does encapsulate the basic idea of hard heredity, because Galton insisted that the environment could not influence how the package of ancestral inheritance could be expressed. Thus, Galton's followers were able to use it as the basis for an analysis of the effect of selection on a population.

Curiously, Galton himself thought that selection would be unable to produce a permanent change in the population because of what he called regression. Imagine a small sample taken from one extreme of the range of varia-

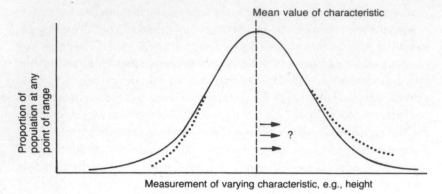

27. Continuous variation and selection. The solid curve represents the normal distribution for a continuously varying characteristic, such as height in the human population. The curve plots the proportion of the population occupying any point in the range against the measurement of the varying characteristic. Thus, the largest proportions are clustered around the mean value of the range (most people are of average height), while at the extreme ends of the range there are much smaller proportions (there are few very tall or very short persons). The crucial question for the Darwinists was: what is the permanent effect of selection upon such a distribution? If one extreme (e.g., tallness) is favored in the struggle for existence, those individuals at that end of the range will breed more than normal, while those at the other end of the range will breed less. The distribution will become skewed as shown by the dotted lines. But will this have a permanent effect in shifting the mean for future generations in the direction favored by selection? W. F. R. Weldon's method of gathering observations was designed to show that a permanent effect would be produced.

tion—for example, a group of unusually tall individuals. If these were to breed among themselves, Galton believed, the offspring all would tend to be shorter than their parents, so the next generation would regress toward the mean value for the species as a whole. After a number of generations without selection, the group would be indistinguishable from the main population. If this were the case, natural selection would not produce a permanent change in the population, because as soon as it was relaxed, its effectiveness would be undermined by regression toward the old mean. There would be a barrier to how far selection could change the population, and its effect could never be permanent, just as many of Darwin's critics had claimed. Galton's campaign to improve the human population by selective breeding was not intended to produce a new species of humanity, only to make the best of the heredity available. He believed that a saltation was needed to create a new species with a new mean value toward which regression would begin.

Galton's followers in the biometrical school thought he had misunderstood the statistical implications of his own theory of heredity. Two of them in particular became the leading defenders of the beleaguered Darwinian selection theory. The biologist W. F. R. Weldon was the driving force in the collection of data intended to demonstrate the effect of selection on wild populations, and the statistician Karl Pearson provided the mathematical techniques needed to analyze the information. Pearson was one of the founding fathers of modern statistics; he is also remembered for his work on the philosophy of science (1900). Pearson showed that even on the basis of Galton's law, selection ought to have a permanent effect, actually changing the mean toward which the species tended to regress (1896, 1898; see also Froggatt and Nevin 1971a; Magnello 1998; Norton 1973; and Provine 1971: chaps. 2, 3). Weldon used his fieldwork to show how this effect worked among populations of crabs and, later, of snails (1894–95, 1898, 1901; Magnello 1996). Although Pearson's and Weldon's early efforts were unconvincing, later work showed clear signs of a change in the population which could be attributed to selection. In the waters of Plymouth harbor, turned muddy by constant dredging operations, larger crabs survived more readily than small ones, and the character of the population in this area changed significantly.

Far from converting the scientific community to neo-Darwinism, the biometricians' observations precipitated a vicious controversy which distorted the initial reception of Mendelian genetics and may even have hastened Weldon's death in 1906. Their work demonstrated the effect of selection, but on so small a scale that it was easily dismissed by those who thought that other mechanisms such as saltations were needed to create new species. In addition, both Pearson and Weldon held advanced views on the philosophy of science which alienated them from many other biologists (Norton 1975a,b). Because it was hard to define adaptation in measurable terms, Weldon simply correlated death rates with an easily measurable characteristic such as the width of the crabs' shells, dismissing the need to explain why bigger crabs were at an advantage. Eventually he did suggest that bigger crabs were less likely to have their gills clogged in the muddy waters of Plymouth, but his later studies of snails again merely correlated easily measured features with death rates. Weldon's unwillingness to define adaptation fueled the suspicions of critics who were already convinced that adaptation was a less significant feature of evolution than the Darwinists believed. In America, Vernon L. Kellogg's similar program to test the effectiveness of selection was also inconclusive (Largent 1999).

The most important obstacle for the biometricians was their clash with the new Mendelian genetics. Even before the rediscovery of Mendel's laws, Pearson and Weldon had fallen out with William Bateson over his claim that discontinuous variations (saltations) were the real cause of evolution. Where the techniques of biometry were suited to the study of large populations, Bateson engaged in breeding programs with small populations in which he could crossbreed varieties with major character differences. He was convinced that such differences arose by saltations and dismissed continuous variation as trivial. He also argued that adaptation was irrelevant to the creation of new characters. When Mendel's laws were rediscovered in 1900, they fitted neatly into Bateson's program, and he became the leading British supporter of genetics. Although it has now been shown that Pearson did not reject Mendelism altogether (Magnello 1999), his distrust of Bateson prevented him from making a major effort to see if genetics could be reconciled with his Darwinian viewpoint.

The biometricians thus repudiated what we now see as the most important step forward in the study of heredity. The conflict between biometry and Mendelism was an ultimately futile debate which would be resolved only much later when it was realized that both sides had seen only part of the true picture (Cock 1973; Darden 1977; Olby 1989; Provine 1971). Although the biometricians rejected the concept of the unit character on which Mendelism was founded, they had realized that heredity and variation were not (as everyone, including Darwin, had thought) mutually antagonistic forces. The key to this insight was their focus on the population, not the individual. Variation in Darwin's sense is not a force affecting individuals, one constantly trying to disturb the exact copying of characters by heredity. It is a function of whole populations which exists because of the circulation of many different characters, each preserved by heredity, the whole complex being constantly reshuffled by sexual reproduction. Unfortunately, the biometricians were unable to appreciate that their opponents' laws of heredity (Mendel's laws) were compatible with the role for heredity that their model of selection required.

MENDELISM AND THE MUTATION THEORY

Although we now see genetics and Darwinism as complementary theories, the potential for synthesis was not apparent at first. The biometricians were the leading advocates of selection, but the early geneticists were strongly influenced by the rival theory of saltative evolution. Thus Mendelism emerged not as the savior of Darwinism but as yet another alternative to it, precipitating a new phase of the eclipse. The rediscovery of Mendel's laws of heredity in 1900 was prompted in part by an enthusiasm for evolution by saltation, and the first use of the term *mutation* was in the context of transformations so drastic that they would create new species instantaneously. The geneticists thus continued the tradition of assuming that the internal processes governing the production of discrete new characters were all that was needed to determine the course of evolution. Adaptation and selection were irrelevant. The new model of heredity reinvigorated the anti-Darwinian theory of saltative evolution, whose roots stretched back to Geoffroy Saint-Hilaire, a conviction that internal transformations of development were the key to the transformation of structure. Yet, at the same time, genetics was new—a product of the revolt against morphology that led early-twentieth-century biologists to focus on the controlled investigation of processes rather than the mere description of form (G. Allen 1975a; see Maienschein 1991 for the claim that older traditions were transformed rather than supplanted). The indifference of the Mendelians to adaptation was a product of their desire to work with small breeding populations kept under controlled conditions, a world apart from the field naturalists' studies of species in the wild.

The story of genetics begins, in principle, with Gregor Mendel's work on the inheritance of discrete characters in peas, performed in the 1860s. But Mendel's account of his laws was largely ignored, and thus does not figure directly in our story. Some revisionist accounts of the history of genetics even express doubt that he really intended to promote a complete new theory of heredity along the lines later attributed to him. The laws of inheritance which now bear his name were rediscovered independently in 1900, and they led to the creation of Mendelian genetics, providing Mendel with posthumous fame as the figurehead of the new science. Only in the 1920s did a new approach to the genetics of whole populations allow the creation of a genetical theory of natural selection. Far from undermining Darwinism, genetics had destroyed all the alternatives, leaving the selection mechanism as the only directing agent in evolution (see chap. 9; for general histories of genetics, see Bowler 1989b; Carlson 1966; Darden 1991; Dunn 1965; Olby 1966, second edition 1985; Stern and Sherwood 1966; Stubbe 1972; Sturtevant 1965).

Mendel's Laws

Gregor Johann Mendel was an unsuccessful student who received only a limited scientific education (for biographical studies, see Henig 2000; Iltis 1932; Orel 1995; Stern and Sherwood 1966). He did, however, have contacts with Franz Unger's school of botany at Vienna, and Unger was interested in both hybridization and transmutation (Gliboff 1999). Mendel entered the monastery at Brunn in Austrian Moravia (now Brno, Czech Republic), where he eventually rose to become abbot. He performed his work on the hybridization of peas in the monastery garden and published it in the journal of the local natural history society in 1865 (translation 1965; W. Bateson 1902; Stern and Sherwood 1966). He received no support from the scientific community and was positively discouraged by the Church. He died in 1884, worn out by his efforts to prevent the monastery from being taxed by the Austrian government.

When Mendel's laws were rediscovered in 1900, it was widely assumed that his account contained the essential principles upon which the new science of genetics would be based. But interpretation of Mendel's paper is complicated by the fact that some historians now believe that the rediscoverers read their own ideas into what he had written. Far from pioneering a new approach to heredity, he may have been trying to test the theory proposed by Linnaeus and others, in which new species might be produced by hybridization (Olby 1979, reprinted in 1985: appendix 5). For Mendel, this would have been an important alternative to Darwinism. His laws of heredity were important to him only because they implied the constancy of hybrids in founding new species. Nor is it clear that he thought in terms of paired particles in the germ plasm, the alleles of modern genetics. The mathematical rigor of his approach was certainly new, however, and matched exactly the insights his rediscoverers wished to promote. But their ideas were shaped by the post-Darwinian controversies over variation and by developments such as the cell theory and the theory of saltative evolution. They automatically interpreted Mendel's paper in this new light, and in this modernized form his laws were introduced to the scientific community at large.

Mendel's work on hybridization naturally led him to think of heredity in discontinuous terms, and what his experiments revealed was that discrete characters are inherited on an all-or-nothing basis—they do not blend. He crossed artificially bred varieties, each with a distinct character which was known to have bred true for many generations (a genetically pure line, as we would call it today). There was already a body of literature on hybridization, and discrete inheritance had been noted from time to time (Olby 1966, 1985; H. Roberts 1929; Zirkle 1951). But Mendel was the first to test whether the discrete character differences could be traced through the generations according to mathematical laws. He knew exactly what he was looking for and had picked out a suitable experimental subject in the garden pea, which had a convenient number of different characters and could be artificially fertilized to avoid contamination by foreign pollen. It has even been suggested that Mendel's results were a little too good to be true, perhaps "improved" by an assistant who had realized what he was looking for (Fisher 1936; Waerden 1968; Wright 1966).

Mendel picked out seven character differences—as it happens, this was the maximum number he could have chosen to give the results he was aiming at (because the relevant genes have to be on separate chromosomes). The characters included size, color of flower, whether the seed was wrinkled or smooth, and so on. In the example used here, Mendel used two varieties which differed in size—one was always tall, the other short. These were cross-fertilized so that Mendel could check how the two characters were transmitted to the hybrid offspring. There was no blending in any of the cases he studied: the hybrids showed only one of the two character states. They were all tall; none were of intermediate height, and the short character seemed to have disappeared altogether. The hybrids were then self-fertilized and the second hybrid generation revealed Mendel's famous three-to-one ratio. There was still no blending, but the short character, which seemed to have disappeared, reemerged in one-quarter of the plants in this generation, giving three tall plants for every short one.

The biologists who rediscovered Mendel's laws explained them by adopting a particulate model of heredity. A single unit, or particle, in the germ plasm had to be responsible for determining the character and hence for transmitting it from parent to offspring. For each character difference, there would have to be two forms of the relevant particle (one for short and one for tall), which we now would call the alleles of a single gene. To be consistent with the nature of sexual reproduction, the organism had to carry two units for each character. It had to have derived one of those units from each of its parents, and it had to transmit only one of the pair to its own offspring. That unit would combine with an equivalent unit from the other parent to make a pair for the offspring. To explain the phenomena that Mendel studied, it was necessary to assume not only that the unit, the gene, existed in two forms (e.g., tall or short) but also that one form was dominant, the other recessive. When the pair of genes for a particular organism consisted of one of each allele, the dominant form was expressed in the physical appearance of the organism, while the recessive remained hidden. Only when both genes were of the recessive form would that character be expressed.

Mendel began with pure strains for the two alleles of each gene. For these to have been pure (i.e., to have bred true for many generations), all the individuals in each strain must have had both pairs of genes corresponding to a particular allele. If we represent the tall allele by the letter T and the one for short by S, the genetic form of the two pure strains must be TT and SS. When these are crossbred, the hybrid must derive one gene from each parent and will have the form TS. Since in this case T is dominant over S, the short character has no effect on the hybrid offspring (see fig. 28). When the first hybrid generation is self-fertilized, there is an independent redistribution of the units which gives approximately equal numbers of all four possible combinations of T and S. Allowing again for the dominant-recessive relationship, one-quarter of the second hybrid generation will show the recessive character (see fig. 29). A twin-unit theory of heredity, coupled with the dominant-recessive relationship, thus explains the effects Mendel observed. His work also showed that each of the character differences was inherited independently of the others—although this is not always the case, because Mendel failed to discover the effect now called linkage.

Why was Mendel's paper ignored for thirty-five years if it contained insights essential for modern genetics? The journal of the Brno society was not prestigious, but it circulated even in Britain, and there were a few references to Mendel's paper in later literature (Olby and Gautry 1968; Weinstein 1977). Mendel was also in contact with Carl von Nägeli, but the latter failed to see the significance of the results and persuaded Mendel to work on another experimental subject (hawkweed), one quite unsuited to his techniques. The scientific community of the time was simply not willing to take this approach to heredity seriously (Gasking 1959; Posner and Skutil 1968). The paper may have been interpreted—perhaps as Mendel intended—as a contribution to the debate on the origin of species by hybridization. No one was willing to believe in unit characters or nonblending inheritance. Mendel's results were seen as a rare exception to the normal rule. To be fair to the scientists of this era, most cases of heredity are more complex than the clear-cut example that Mendel picked out. Most visible characters are influenced by more than one gene, so the simple ratios he studied cannot be observed and the parental characters do appear to blend. To realize that Mendel's laws contained the key to a whole new theory of heredity, biologists would have to make a major theoretical leap, and they were not willing to do this until other factors convinced them of the need to abandon their old ideas.

The Rediscovery

By the end of the century, some biologists were convinced that the time was ripe for a new initiative. Weismann's germ plasm had established the concept of hard heredity in which characters are predetermined by material in the chromosomes of the cell nucleus. This new version of preformationism

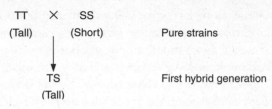

28. First hybrid generation.

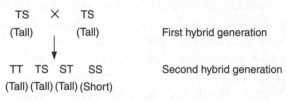

29. Second hybrid generation.

was highly controversial, since many still believed that the environment influenced both the development of the organism and its inheritance (Gilbert 1978; Maienschein 1978, 1984). But the new theory offered advantages to those biologists who were determined to put their field on a more experimental basis, throwing off the old traditions of natural history and morphology. Other approaches certainly were tried: there was an attempt to create a science of "developmental mechanics" which would explain the development of the embryo in materialistic terms. This turned out to be beyond the capacity of the techniques available at the time, so emphasis gradually switched to the study of how characters were actually transmitted. To study this process, biologists looked for discrete characters that easily could be traced through successive generations. Their views also were influenced by the growing interest in the idea that new characters in evolution were created by sudden leaps, something that even T. H. Huxley and Galton had taken seriously. If characters were created as units, it would seem likely that they might be inherited as units. A situation thus emerged in which the phenomena that Mendel had revealed might be seen as the basis for a new approach to heredity.

Several biologists were led to study effects similar to those reported by Mendel, and then realized that they had been anticipated by his 1865 paper. From this rediscovery of Mendel's laws came a new science, at first often called Mendelism but increasingly known as genetics. This new discipline marked a significant break with the old developmental tradition within which both Darwinism and more especially Lamarckism had flourished. Heredity and individual development no longer were seen as aspects of an integrated process that would allow modifications of development to be transmitted (Horder, Witkowski, and Wylie 1986). Heredity was now to be studied without worrying about how the characters were produced in the developing offspring. Lamarckism became unacceptable because it blurred the boundary between heredity and development, and the recapitulation theory was now irrelevant because there was no obvious reason why development should be a model for evolution. The only process that could produce evolution was the production of new genetic characters by saltation (mutation), something that would be irrelevant to the needs of the developing organism. Genetics thus succeeded where Darwinism had failed: it eliminated the teleological approach to evolution implicit in the analogy with individual development.

Yet this transition was not instantaneous and, indeed, was resisted for decades in some areas of biology. In Britain and America, at least, genetics was a new discipline founded quite self-consciously by younger scientists anxious to create a professional niche for themselves in the new world of experimental biology. The emergence of genetics can be seen as a product of the move away from natural history toward a more manipulative, experimental form of science. Its conceptual break with the past was stimulated by a new attitude toward biology which sought to modernize it even beyond the standards of T. H. Huxley's generation. The aim was not to describe dead organisms but to control living ones, allowing genetics to be taken up eagerly by the animal breeders and horticulturalists. The old approaches to evolutionism, both Darwinian and Lamarckian, were challenged by scientists no longer interested in the complex problems of how organisms existed in the wild.

Geneticists led the campaign to discredit Lamarckism, but, to begin with, they were equally hostile to the selection theory. In principle, genetics solved some of the problems that the original form of Darwin's theory had encountered, most obviously Fleeming Jenkin's claim that blending heredity would dilute a favorable variation too quickly for it to be of value to the species (de Beer 1964; Vorzimmer 1968). If a new character were transmitted as an unchanged unit, it could not be diluted, and its proportion in the population could be increased only when the organisms carrying it bred more than the others, that is, by selection. But the geneticists were not interested in gradual changes within populations produced in response to adaptive advantage. They at first were determined to explore the logic of the theory of saltations, in which individuals with the new character establish a separate breeding population—that is, a new species—immediately, without any need for selection. The concept of genetic mutation was introduced in this context, as a means of explaining evolution without recourse to adaptation or natural selection. On this model, the continuous range of normal variation studied by the biometricians was trivial and could have no relevance for the appearance of new genetic characters and hence for evolution.

By 1900 the search for a discontinuous model of heredity had led to a duplication of Mendel's work and the rediscovery of his laws (Henig 2000; Olby 1966; H. Roberts 1929; Wilkie 1962). Two biologists, Carl Correns and Hugo De Vries, published accounts of the laws and acknowledged Mendel's priority. De Vries probably did not understand the results he had obtained until after reading Mendel (Campbell 1980; Kottler 1979; Meijer 1985; Zirkle 1968). The claims of a third rediscoverer, Erich von Tschermak, are no longer accepted because he clearly did not understand the laws (Tschermak's paper is not reprinted along with the others in Stern and Sherwood 1966). Perhaps in part to head off the possibility of a priority dispute between Correns and De Vries, Mendel was now hailed posthumously as the founder of the new approach to heredity. De Vries subsequently lost interest in Mendelism, but in the meantime other biologists began to establish the science of genetics.

The leading British Mendelian was William Bateson (see B. Bateson 1928). Originally an evolutionary morphologist working on the origin of the vertebrates, Bateson became frustrated by this approach and looked for a way of studying the actual process of evolution. The work of Galton and the American biologist W. K. Brooks convinced him that evolution took place by sudden leaps, so that the continuous range of variation highlighted by the Darwinians was irrelevant. Bateson's *Materials for the Study of Variation* (1894, reprint 1992) showed that discontinuous characters were far more numerous than Darwinians admitted and could be explained only by saltations. When flowers on the same species of plant showed different numbers of petals, for instance, it was most unlikely that a petal was lost or added by small increments. Almost certainly the petal was lost or added as a whole unit. Bateson also launched into an attack on the whole concept of evolution driven by adaptation:

> We knew all along that Species are approximately adapted to their circumstances; but the difficulty is that whereas the differences in adaptation seem to us to be approximate, the differences between the structure of species are frequently precise. In the early days of the Theory of Natural Selection it was hoped that with searching the direct utility of such small differences would be found, but time has been running now and the hope is unfulfilled. (1894: 11)

To Bateson, it was obvious that the saltative changes he postulated would occur and be inherited whether or not they were of any advantage, although he conceded that, in the long run, truly maladaptive structures would be eliminated. This position provoked the clash with the biometricians described above. Bateson now began breeding experiments to trace the inheritance of discontinuous characters, and when he read of the rediscovery of Mendel's work, he realized that here was a theory that would account for his results. He published the first English translation of Mendel's paper (in W. Bateson 1902), and went on to champion the creation of a new science of heredity for which he himself coined the name *genetics* (Cock 1973; Coleman 1970; Darden 1977; Olby 1989; Provine 1971).

Most of the early Mendelians assumed that what soon were being called genetic mutations were responsible for creating new characters—and hence new species—by saltation. But Bateson, who was suspicious of the concept of the gene as a material unit on the chromosome, became increasingly unhappy with this assumption. He now suspected that all mutations were really degenerative, that they destroyed genetic characters rather than creating them. He argued that the appearance of "new" characters was actually the result of the destruction of genes which had originally masked the characters, preventing their expression. At one point, he even suggested that this process might explain the whole of evolution (W. Bateson 1914), as though all the characters later expressed in the history of life on earth were contained in the genes of the first living things, masked by a series of genes that gradually were destroyed by mutation. Later on he expressed doubts as to whether scientists had any real idea about how evolution occurred (Bowler 1983: chap. 8).

The Mutation Theory

Few of the other Mendelians took these concerns seriously. To most of them, it seemed obvious that mutations change the material structure of the gene and, hence, change the character it expresses in the developing organism. The concept of mutation was in fact pioneered in Hugo De Vries's "mutation theory," which became the most popular theory of evolution in the early decades of the twentieth century. De Vries had published a theory of "intracellular pangenesis" in 1889, which postulated discrete units of inheritance (translation 1910a; Darden 1976). To test this theory, he began breeding experiments and eventually was led to Mendel's paper, thus becoming one of the rediscoverers. He soon lost interest in Mendel's laws, however, and developed his theory of mutations, published in 1901–1903 (translation 1910b) and introduced in a series of lectures given at the University of California (1904; see also G. Allen 1969b; Bowler 1983: chap. 8).

De Vries did not use the term *mutation* in the modern sense, because he thought that the mutated forms bred only among themselves and formed varieties or even species distinct from the parents. Instead of being formed gradually by natural selection of continuous variation, varieties were created instantaneously by saltation. The complex debates of the naturalists over isolating mechanisms thus were rendered superfluous. Unlike Bateson, De Vries claimed that there were positive mutations which produced new characters, and he assumed that all species underwent occasional bouts of rapid mutation when they threw off numerous new varieties in this way. This rapid production of new forms explained the gaps in the fossil record. The overall process of evolution would also work faster than Darwin supposed, fitting in with Kelvin's reduced estimates of the age of the earth.

De Vries supported his theory with studies of the evening primrose, *Oenothera lamarckiana.* He had found this species growing wild in Holland, apparently producing significantly mutated forms before his very eyes. He believed that each mutated form bred true and thus counted as a distinct variety. De Vries even gave the most drastically changed forms new specific names, implying that a single mutation had established an entirely new species. Only in 1910 was it first suggested that there might be something wrong with this evidence, and in the 1920s it was shown that *Oenothera* is a complex hybrid species, so the apparently new forms were not due to genetic mutations at all—they were merely new combinations of existing characters.

In the meantime De Vries's theory gained wide popularity as a rival to Darwinism that was more compatible with the new experimental approach to biology. De Vries himself did not present the theory as a complete alternative to the Darwinian worldview. Although the actual production of new varieties was not controlled by natural selection, he believed that, in the long run, only those mutated forms that conferred some advantage would survive. In effect there was competition among the mutated forms—they could all survive for a short period, but eventually the maladaptive ones would be eliminated. Sooner or later mutation would create a new form even better adapted to the prevailing conditions than the parent species, which would then become extinct. For De Vries, this element of selection was essential to ensure that the new theory was not tainted with the old tendencies toward mysticism and teleology.

Many of De Vries's supporters thought his efforts to retain a role for selection were misguided. The American biologist Thomas Hunt Morgan used the mutation theory as the basis for an attack on Darwinism even more vitriolic than Bateson's. Since Morgan subsequently became one of the founders of classical genetics, his early enthusiasm for an anti-Darwinian saltationism is a remarkable illustration of the transformation needed for the new approach to heredity and variation to become compatible with the selection theory. Morgan's *Evolution and Adaptation* (1903) rejected both the selection mechanism and the idea that evolution could be driven by the demands of adaptation (G. Allen 1968, 1978; Bowler 1983: chap. 8). He seems to have been disturbed by the moral implications of a theory in which struggle and suffering were integral parts of nature. Morgan picked up De Vries's point that mutations are not produced for any adaptive purpose and went on to argue that there was no need to invoke selection at any level. He believed that any mutated form not grossly incompatible with the environment would be able to survive and reproduce as a new species. Thus the course of evolution would be determined solely by the kinds of mutations that appeared, presumably as the result of spontaneous rearrangements within the germ plasm (although at first Morgan rejected the concept of the unit gene).

Another argument against the selection theory came from the Danish biologist Wilhelm Johannsen and his work on the breeding of "pure lines" in beans (translation 1955). Johannsen was not concerned with the wider problem of evolution, and his studies involved only selection within a population. The species he chose reproduced by self-fertilization, and by a pure line he meant all the descendants of a single individual, which in these circumstances would be genetically identical. In this species, there was a continuous range of variation of the kind studied by the biometricians, but Johannsen showed that the variation could be resolved into a number of discrete pure lines which overlapped with one another to give the impression of a continuous range. When applied to the population, all selection could do was to eliminate most of the pure lines, leaving one at the favored end of the range. Any remaining variation in this pure line was purely somatic: it had no genetic foundation, and selection had no further effect. There was thus a limit to the power of selection, just as the critics of Darwinism had maintained. Johannsen believed that the only way selection could move further was if a mutation altered the character of a pure line in the favored direction.

The biometricians challenged Johannsen's analysis, but his work was widely seen as evidence that mutation, not selection, was the real source of new characters. Johannsen went on to argue that the same principle would apply in populations where there was normal sexual reproduction. The recombination of characters would complicate matters, but selection would still only be able to isolate genes that corresponded to one extreme of the normal range of variation. Mutation would be necessary to produce genuinely new characters. Yet Johannsen's work did shift the emphasis of the mutation theory, because it forced biologists to abandon the artificial distinction between continuous and discontinuous variation. Discrete genetic units were relevant not only in cases of discontinuous variation but also in normally varying populations where a number of genetic factors contributed to determining a particular character and thus overlapped one another. This meant that mutations (which presumably produced new pure lines) did not establish separate breeding populations, as De Vries and Morgan supposed. They merely added to the range of variation in the existing population. This paved the way for a reconciliation between the biometrical and Mendelian concepts of variation. At the same time, Johannsen introduced the distinction between genotype and phenotype (Churchill 1974). The genotype is the genetic constitution of the organism, the phenotype its physical appearance. Because of dominance, these might not be the same: in Mendel's original experiment, the tall parent and the first hybrid generation had the same phenotype (both were tall) but different genotypes (TT and TS).

The best evidence that mutations feed new genetic characters into the existing population came from experiments on the fruit fly *Drosophila melanogaster* by T. H. Morgan and his colleagues (G. Allen 1978; Kohler 1994; Shine and Wrobel 1976). Morgan had converted to Mendelism by this time (Lederman 1989) and accepted that mutations were changes in the structure of the gene producing new somatic characters. He and his coworkers identified a number of different mutations appearing within their experimental populations, showing that the mutations were increasing the species' range of variability. Because the fruit fly has large and easily visible chromosomes, they were also able to show that the inheritance of genetic characters according to Mendel's laws could be explained in terms of the way the chromosomes function in reproduction. In the production of the ovum or sperm by meiosis, the reproductive cell, or gamete, was given only one of the homologous pairs of chromosomes present in normal cells. The fertilized ovum derived one of each pair of its chromosomes from the sperm and one from the ovum, allowing both parents to contribute an equal amount of genetic material to the offspring. It was even possible to work out the relative positions of the genes on the chromosomes. These results were summed up in *The Mechanism of Mendelian Heredity* (Morgan, Sturtevant, Muller, and Bridges 1915), which established the foundations of classical genetics. In the English-speaking world, only Bateson continued to resist the chromosome theory of the gene (Coleman 1965, 1970).

It would now be possible to heal the split between the biometricians and the Mendelians. Genetics did not depend on the idea that new species are produced by saltation, nor was a Mendelian analysis of inheritance incompatible with the belief that the continuous range of variation shown by most characters is significant for evolution. In principle, at least, it would be possible to create a theory of natural selection based on the differential reproduction of the genes according to their adaptive advantage. In fact, Morgan himself moved toward a theory of natural selection in the later part of his career (1916). Even so, much work remained to be done before the modern genetical theory of natural selection was formulated, and even more to render possible a complete reconciliation between the geneticists and the field naturalists. These developments would lead to the creation of the modern Darwinian synthesis (see chap. 9).

The eventual success of the new form of Darwinism was ensured by the fact that genetics had eliminated all of the alternative mechanisms of evolution. The Mendelian concept of the gene was firmly based on Weismann's vision of a germ plasm completely isolated from the body it created. Geneticists such as Bateson led the way in the campaign to discredit the inheritance of acquired characters, since in their view there was no way in which minor changes to the body could influence the character of the unit gene. They rejected the possibility that the cytoplasm, the extranuclear material of the cell, might play a role in heredity to supplement the chromosomes (Sapp 1987). Genetics also eliminated the possibility of orthogenesis. A few geneticists, including Morgan at one point, had toyed with the possibility that mutations might occur consistently in a single direction, thus producing an orthogenetic trend by "mutation pressure." But the work on the fruit fly indicated no such trends; a whole range of apparently random mutations was observed, suggesting that Darwin had been right to suppose that variation by itself imposed no direction on evolution. Once the geneticists' initial enthusiasm for saltationism had waned, natural selection became the only plausible mechanism of evolution.

In conclusion, however, it is important to note that the geneticists' growing consensus in favor of nuclear preformationism and natural selection was confined to Britain and America. In continental Europe, genetics did not develop such a hard-line stance against Lamarckism and other non-Darwinian mechanisms. In France, genetics failed to gain an independent place in the highly centralized academic system (Burian, Gayon, and Zallen 1988). There was a strong preference for nonmechanistic theories, so the idea of nuclear preformationism was not accepted, and Lamarckism survived well into the twentieth century. In Germany, too, genetics did not become rigidly institutionalized, and cytoplasmic inheritance remained plausible, along with various non-Darwinian models of evolution (Harwood 1985, 1993; Reif 1983, 1986; Rinard 1988). Here, the study of heredity did not become alienated from field studies and paleontology, and the developmental model of evolution, heredity, and embryology survived. Genetics did not eliminate the non-Darwinian approach to evolution, so there was much less emphasis on the eventual synthesis of genetics and the selection theory. When the German biologist Richard Goldschmidt went to America as a refugee in the 1930s and tried to promote the idea of saltationist evolution as an alternative to Darwinism, he was received with hostility (Goldschmidt 1940; see also G. Allen 1974; Dietrich 1995). The emergence of the modern Darwinian synthesis was very much a product of the way science was institutionalized in the English-speaking world, and we should be careful not to assume that it was typical of how genetics or evolution theory developed across the whole scientific community.

8   Evolution, Society, and Culture, 1875–1925

By 1875 the majority of educated people in Europe and America had accepted evolution. Even religious thinkers were now trying to come to terms with the prospect of a natural origin for humanity. Opponents of religion openly rejoiced at the prospect of replacing ancient superstition with a philosophy based on a scientific understanding of human nature. Across the spectrum of theological, philosophical, and political thought, the idea of evolution exerted a powerful effect on the imagination. This was an age where everything would be transformed by the idea that humanity was the product of a natural process that had operated long before we actually appeared on the earth. Themes of evolution, progress, and struggle permeated even the literature of the period, although most writers had only the vaguest understanding of the Darwinian theory (Beer 1983; Bender 1996; Carroll 1995; Henkin 1963; G. Levine 1988; Morton 1984).

This lack of any precise Darwinian focus should not surprise us. The late nineteenth century was the era of the eclipse of Darwinism, a time in which scientists questioned the adequacy of the Darwinian selection theory even more fiercely than in the 1860s. Since scientists were suspicious of the selection theory, it would be surprising if thinkers in other areas blindly accepted it. Our perception of this period is distorted by popular images of a rampant social Darwinism in which, it is alleged, all moral standards were abandoned. People were taught that only material success mattered. But if the most materialistic form of evolutionism was not accepted by the scientists, why should we assume that the whole age was mesmerized by the implications attributed to that theory by its opponents? In fact, the story of evolutionism's cultural impact is a good deal more complicated than implied by the simple model of a transition from natural theology to ruthless social Darwinism.

Our vision of the whole Darwinian revolution has been skewed by the fact that the evolutionary movement became known as Darwinism even though the theory of natural selection was not widely accepted. What we call Darwinism today—a reliance on natural selection as the sole mechanism of change—corresponds to the neo-Darwinism of late-nineteenth-century biology, which was highly controversial at the time. It is all too easy for a historian unfamiliar with the range of evolution theories then available to be misled into thinking that the Darwinism of the late nineteenth century is the Darwinism of today. Such a projection of the present onto the past misses the significance of the selection theory's long eclipse. The age of evolutionism may have been called an age of Darwinism, but the ideas discussed were Darwinian in only the loosest sense and would not be counted as such by modern standards.

Another source of misunderstanding is the ease with which anyone unfamiliar with the detailed theory of natural selection may confuse it with some of the alternatives. Darwin has been so closely associated with the idea of the struggle for existence that any theory evoking struggle is automatically assumed to be Darwinian. The possibility that competition might be seen as the spur to self-improvement in a Lamarckian theory is forgotten— and to be fair, some nineteenth-century thinkers were unclear about the distinction between selection and Lamarckism. Many who rejected individualistic natural selection accepted that there would be competition between the rival species produced by some other mechanism, and assumed that this was a form of Darwinism. Darwin did believe that species were driven to extinction by the appearance of better-adapted rivals, but those who invoked this mechanism of group selection may not have been Darwinians in the modern sense. The fact that Darwin became identified with any mechanism of change involving struggle is obviously important in the attempt to survey the interaction of science and society at the time. It tells us that a scientific theory is as likely as anything else to serve as the label for a set of cultural values. But to gain a properly nuanced understanding of the interaction, we must allow for the existence of non-Darwinian alternatives. When dealing with the eclipse of Darwinism, it is important not to use the label *Darwinian* too casually.

The dominant theme of late-nineteenth-century thought was progress, and evolutionism became popular because it was perceived as a scientific expression of this broader principle. But the concept of progress itself could exist in different, and to some extent contradictory, forms. Western culture was profoundly affected by the technological progress associated with industrialization, but there were various groups within society which reacted to social change in different ways. Traditionalists wanted to turn the clock back to a hypothetical age of stability, while industrialists and entrepreneurs demanded political power consistent with their newly earned wealth. The latter were likely to endorse a philosophy of progress which made their demands seem no more than acceptance of the next step in an inevitable advance.

Other groups were less sure of the benefits of progress and less willing to endorse a philosophy in which economic success was the only measure of development. The appeal of evolutionism to scientific experts such as T. H. Huxley was that it offered the hope of heading off working-class discontent by promising slow but sure progress in the future—while leaving the experts to gain ever more control over modern society. As the industrial revolution gave way to the age of imperialism, new expressions of the idea of progress manifested themselves as white people sought means of justifying their domination of the "less advanced" branches of the human stock. And the possibility that progress was not uniform but included occasional episodes of degeneration also began to loom larger in the minds of those who faced the cultural fragmentation of the fin de siècle, the end of the century of progress. (On nineteenth-century thought, see Copleston 1963, 1966; Mandelbaum 1971; Willey 1949, 1956. On the idea of progress, see Bowler 1989a; Bury 1932; Pollard 1968; Van Doren 1967.)

The simplest model of progress is based on a "ladder": the advance is defined in terms of a linear hierarchy. This linear system played a major role in the life sciences and was linked strongly to the model of embryological development and the recapitulation theory. A parallel to this emerged among sociologists, anthropologists, and archaeologists who were convinced that history revealed stages of social development culminating in Western culture. Such a model presupposes that development moves in a certain direction, and there is little incentive to inquire about the mechanism generating the advance. The linear model is almost teleological in its willingness to depict the advance as preordained. Radical thinkers such as Herbert Spencer concealed the goal-directed nature of progress by implying that there was no fixed line of advance, only a more general tendency for things to become more complex. In practice, however, even Darwin and Spencer invoked a linear hierarchy of progress with reference to the human species. They still saw the universe as a system designed to produce—in the long run—states morally preferable to the earlier, more primitive states. By portraying progress as the outcome of natural law, however, they were able to formulate a system in which the advance might be irregular or might even be temporarily reversed.

By the early twentieth century, the progressionist consensus which had dominated late-nineteenth-century thought was becoming fragmented. However complex the process of biological evolution, psychologists and anthropologists now began to doubt that the human mind and its achievements could be explained as consequences of a universal evolutionary process which had begun with the origin of life itself. A more relativistic view of the different human races and cultures emerged, one no longer based on the assumption that the Western pattern was superior. Some now doubted the idea of progress altogether, especially when the Great War revealed the depths to which supposedly civilized peoples could descend. The legacy of evolutionism continued to play a role, however, even if turned on its head. The analytical psychology of Sigmund Freud still rested on the assumption that the human mind has advanced through animal stages of development—only now those stages lurked within the subconscious, evading all the efforts of the rational mind to control them.

This chapter surveys the impact of evolutionism in three broad areas. First comes the subject of human origins: what does biology tell us about our ancestry, and how can the idea of evolution be used to help us understand the development of the mind, of society, and of culture? Here the study of hominid fossils interacted with prehistoric archaeology and anthropology. While anthropologists looked for remnants of the earliest forms of human society, evolutionary psychologists tried to define the hierarchy by which new mental functions were added in the rise from animal to human. These issues lead naturally into the second broad area: the impact of evolutionism on social thought. If social Darwinism is a problematic concept, we must explore the rival ways in which models of social evolution were used to defend new ideologies. Race became an important biological concept, with evolutionism being called in to justify existing prejudices about the hierarchy of human types. By the early twentieth century, the claim that human nature is rigidly fixed by heredity was becoming an important part of social policy as the eugenics movement called for limits on the breeding of the "unfit." Finally, we look at the influence of evolutionism on philosophy and religion. The debates of the 1860s on the theological implications of evolution died down in the later part of the century because so many modes of thought took it for granted that the world was the product of a progressive and purposeful development. But the gulf separating the traditional and the modern viewpoints was to some extent papered over rather than confronted, and the emergence of a fundamentalist opposition to evolution in the 1920s symbolized the underlying tensions.

THE MISSING LINK

The question of human origins was hotly debated in the 1860s even though Darwin had excluded it from the *Origin*. The implications of an apelike ancestry for the human race were immense, but the whole project depended on convincing everyone that the evolutionary link was real. Huxley made the case for humanity's close relationship to the apes but had been forced to concede that there was no good fossil evidence for the missing link. When he accepted that the famous Neanderthal remains were fully human, despite their superficially apelike character, he abandoned the hope of using them as the link between modern humans and their hypothetical ancestors. More of the Neanderthal-like specimens subsequently were discovered, and some authorities began to challenge Huxley's position, arguing that the Neanderthal race or species was indeed a key step in the ascent from ape to modern human. In the 1890s the discovery of "Java man," or *Pithecanthropus*, began a new era of intensive research. By the early decades of the twentieth century, several important new discoveries had been made, although one of the most important subsequently turned out to be a hoax (the now notorious Piltdown remains). Paleoanthropology—the study of fossil hominids—only now became a recognizably distinct area of science. In one respect, these discoveries provided belated confirmation of the evolutionists' case by revealing creatures which could not be classified as either truly ape or human. But far from providing a clear picture of human origins, they served only to confuse the issue. There were several different forms of early hominid, none of which corresponded exactly to what the evolutionists had expected. They had to accept that the evolution of the human race was a complex process, as indeed were most of the major steps in evolution (see fig. 30). Instead of ascending a simple ladder defined by increasing brain size, nature seemed to have experimented with several forms of humanity and driven all but one to extinction (Bowler 1986; Lewin 1987; Reader 1981).

The discovery of additional Neanderthal-like remains, especially at Spy in Belgium in 1886, confirmed that this was a genuinely ancient human race or species. The Neanderthals were heavily built and had skulls with an apelike brow ridge, although as Huxley had pointed out, their cranial capacity was equal to that of a modern human. They used fire and tools, and the Mousterian industry of stone toolmaking was now identified as theirs. Some authorities began to suggest that they were indeed an intermediate stage in the evolution of modern humanity from the apes. Physical anthropologists anxious to stress the allegedly primitive character of the living

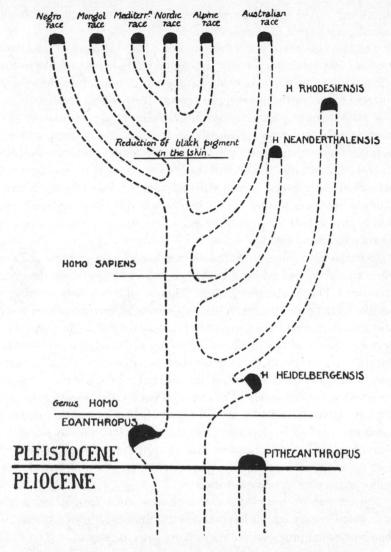

Negro race   Mongol race   Mediterr. race   Nordic race   Alpine race   Australian race

H RHODESIENSIS

H NEANDERTHALENSIS

Reduction of black pigment in the skin.

HOMO SAPIENS

H HEIDELBERGENSIS

Genus HOMO

EOANTHROPUS

PITHECANTHROPUS

PLEISTOCENE

PLIOCENE

30.   Elliot Smith's tree of human evolution. This tree of evolution from
Grafton Elliot Smith's *Evolution of Man* (1924: 2) shows a typical early-
twentieth-century view of human origins. Note that the Nordic race forms
the endpoint of the main trunk, while the other races have diverged away
from the main line at different points in its development. The Neanderthals
have branched much earlier than any of the living races and are now extinct
(*Homo heidelbergensis* was thought to be an even earlier branch). On this
diagram, *Pithecanthropus* is so far removed from the main line that it
appears only as the endpoint of a parallel branch at the bottom right.
Piltdown man *(Eoanthropus)*, however, is very close to the main stem.

black races of Africa and Australia sometimes depicted them as little more than slightly improved Neanderthals.

The idea that there was a Neanderthal phase in human evolution remained in play (e.g., Brace 1964), although it has been undermined by modern genetic evidence. This evidence supports a rival hypothesis which emerged following the discovery of another Neanderthal skeleton in 1908. The paleontologist who described it, Marcellin Boule (translation 1923), went out of his way to exaggerate the apelike character of the specimen, ignoring evidence that its bent posture was the result of arthritis. For Boule and his followers, the Neanderthals were so primitive that they could never have had time to evolve into modern humans. They were a degenerate or primitive side branch of the human family tree which had survived in isolation in Europe before disappearing as modern humans invaded from elsewhere (Hammond 1982).

In the preceding decade, much attention had focused on a new and much more primitive hominid fossil unearthed in Java in 1891–1892. Its discoverer was a Dutch scientist, Eugene Dubois, who had been inspired by Haeckel's prediction that the closest link between humans and apes was the orangutan rather than the African ape (Theunissen 1989). His discovery revealed a creature with a thighbone suggesting a completely upright posture, but with a brain capacity only halfway between that of an ape and a modern human. To Dubois, this was the real missing link, and he borrowed Haeckel's term to name it *Pithecanthropus erectus*. Haeckel welcomed the discovery (translation 1898), as did his disciple Gustav Schwalbe, and they depicted a linear sequence of development from the ape to *Pithecanthropus* and the Neanderthals to modern humans. (See fig. 31.) But most paleoanthropologists remained doubtful, finding the combination of a primitive skull and fully upright posture unconvincing.

Darwin had suggested that our ancestors stood upright before their brains grew bigger, but his hypothesis went unnoticed in an age convinced that intellectual progress was the driving force of evolution. The British cerebral anatomist Grafton Elliot Smith (1924) was particularly influential in promoting the theory that the expansion of the brain had come first, with the upright posture being acquired only after our ancestors had become intelligent enough to see the advantages of moving out of the trees onto the open plains. On this model, Dubois's Java man simply did not fit in and had to be dismissed as a less successful side branch of the human family tree. This in turn encouraged the rejection of the Neanderthals as human ancestors. Already the pattern of human origins was looking more complex than a simple ladder. The new theory treated human evolution more like a tree

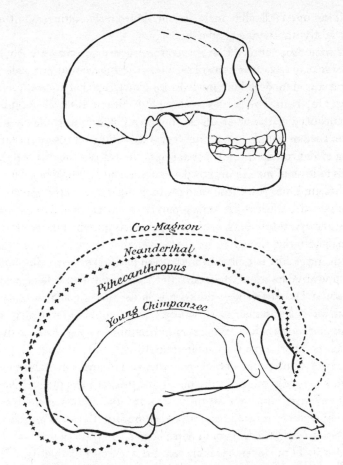

31. The skull of *Pithecanthropus*. This diagram from
Haeckel's *The Last Link* (1898: 25) shows a restoration of the
skull of *Pithecanthropus* (top). Beneath is a comparison of the
skulls of a Cro-Magnon (almost modern human), a
Neanderthal, *Pithecanthropus*, and a young chimpanzee. In direct contradiction to the branching model provided by fig. 30,
this model gives the impression of a steady expansion of skull
size indicating a continuous evolutionary development. Note,
however, the large lower rear section of the Neanderthal skull,
which provides it with a cranial capacity as large as that of the
Cro-Magnon skull.

but still assumed that all branches stemmed from the same tendency to improvement, with varying degrees of success.

It is sometimes argued that this episode was inspired by a virtual rejection of evolution, because none of the known fossils were admitted as the direct ancestor of modern humanity (Brace 1964). But this suggestion ignores the fact that early-twentieth-century phylogenetic research had revealed the complexity of the whole evolutionary process. It was widely accepted that, because of parallel evolution, it was difficult to reconstruct the exact origins of any modern form. Landau (1990) stresses that all the theories seeking to explain human origins use the same elements (bipedalism, brain enlargement, emergence from the trees) combined in different ways. She argues that the theories all have a narrative structure and thus resemble folktales or creation myths, a suggestion which horrified modern paleoanthropologists who thought that it implied that they too were still only "telling stories." In fact, all explanations of particular events in phylogeny which invoke adaptation have a narrative structure (often called an adaptive scenario), so Landau's suggestion is not as threatening as it sounds. More problematic is her tendency to ignore the role of progressionist ideas in these early theories, thus exaggerating their similarity to modern explanations conceived within a Darwinian framework.

The fascination of the "brain first" theory of human evolution explains several subsequent events in the history of paleoanthropology. It accounts for the ease with which one of the most notorious frauds in the history of science was accepted by the scientific community. This was a group of remains discovered by an amateur archaeologist, Charles Dawson, at Piltdown in the south of England in 1912. There were fragments of a skull, fairly thick but otherwise relatively modern looking, and a jaw that was apelike apart from the pattern of wear on the teeth. Arthur Smith Woodward of the Natural History Museum in London described the remains as those of a new species, *Eoanthropus dawsoni*, Dawson's dawn man. For a while, most authorities took the remains to be genuine because they fitted in so well with the hypothetical line of human evolution then in favor. The fact that the cranium was large but the jaw still apelike fitted Elliot Smith's brain-first theory, while the existence of a non-Neanderthal line of human evolution confirmed the prediction that the Neanderthals themselves were on a side branch of the human family tree.

Arthur Keith, an anatomist who had come out in support of the theory of Neanderthal extinction, was a leading advocate of the significance of Piltdown man (Keith 1915). Keith also argued that, if the main line of human evolution was paralleled for vast periods of time by other side branches, then it was reasonable to suppose that the racial types within modern humanity were also of great antiquity. If the "lower" races could no longer be dismissed as surviving Neanderthals, they could at least be branded as ancient types with no close relationship to the whites, who had advanced farther up the scale of mental development. Keith had little doubt about what happened whenever higher and lower forms came into contact: "What happened at the end of the Mousterian period we can only guess, but those who observe the fate of the aboriginal races of America and Australia will have no difficulty in accounting for the disappearance of *Homo neanderthalensis*. A more virile form extinguished him" (1915: 144). Significantly, Keith went on to champion racial conflict as a major factor in human progress (1949).

The Piltdown remains were too convenient for the paleoanthropologists of this period to ignore. As more discoveries were made, the Piltdown remains gradually came to be seen as anomalous, but not until 1953 were they exposed as a fraud (J. S. Weiner 1955). An ancient human skull had been planted alongside an ape jaw, which had been stained to make it look ancient and filed to produce a human pattern of wear on the teeth. Dawson was probably one of the culprits, but a minor literary industry has grown up around the effort to identify the scientific brains behind the scheme (Blinderman 1986; Millar 1972; F. Spencer 1990). Even Elliot Smith and Keith have been implicated, somewhat implausibly considering the amount of time they both wasted on describing the remains. The most likely candidate is Martin Hinton, a jealous subordinate of Smith Woodward's, who may have intended to embarrass his boss by revealing the hoax but realized too late that the whole thing had got out of hand. Few of the detective-style efforts to uncover the culprit acknowledge the theoretical preconceptions that made the Piltdown combination seem so plausible when it was first reported.

One of the discoveries that made Piltdown seem anomalous even before the fraud was exposed was a report by Davidson Black of more *Pithecanthropus*-like remains in China in the late 1920s. "Pekin man" (as it was popularly known) was reported as a new species but is now regarded as belonging to the same species as *Pithecanthropus*, renamed *Homo erectus*. Few at first accepted what has become the modern view, that *Homo erectus* was probably ancestral to *Homo sapiens*. Instead, the discovery seemed to confirm a long-standing belief that Asia was the cradle of humankind, not Africa as Darwin had supposed. Black had in fact gone to China in the hope of proving that the harsh climate of central Asia had stimulated the advance of humans from their ape ancestors, not the lush tropical environment of

Africa. His fossil discoveries disappeared in the chaos surrounding the Japanese invasion of China preceding World War II, and only plaster casts now remain.

The popularity of the Asian theory of human origins helps to explain the initially negative reaction to what was subsequently accepted as a far more significant find. Raymond Dart discovered a juvenile hominid skull at Taungs, South Africa, in 1924, which he named *Australopithecus africanus*. Dart saw evidence from the skull that this creature had walked fully upright, yet the brain was scarcely larger than an ape's. His claim that it was the true ancestor of humankind was rejected because all eyes were focused on Asia, and no one expected the line of human ancestry to have achieved bipedalism so early. In fact, Dart had confirmed Darwin's hypothesis that the key breakthrough defining the human family was the acquisition of an upright posture, with the enlargement of the brain coming later. But at the time, no one was prepared to admit the possibility that an adaptive transformation could have played so vital a role, because everyone was obsessed with the belief that the drive to gain greater intelligence was the motivating force. Dart's views were treated with skepticism until further discoveries of australopithecines by Robert Broom in the late 1930s confirmed that this early hominid type was indeed bipedal. The australopithecines' significance as the founders of the human family was not admitted until the 1950s, when the modern Darwinian synthesis highlighted the crucial role of adaptation and exposed the hidden teleology of the assumption that an increase in brain size was the central driving force of evolution.

THE ORIGINS OF CULTURE AND SOCIETY

The late nineteenth century's vision of progress was based on the model of a ladder of developmental stages. This model was based not on a study of fossils—because there were hardly any available—but on indirect arguments provided by archaeology and anthropology. The 1860s had seen a revolution in ideas about the antiquity of the human species which paralleled, but was to some extent independent of, the Darwinian revolution in biology. While Huxley lamented the missing link in the fossil record of humanity, geologists and archaeologists uncovered stone tools dating back to the ice ages, confirming that humans had been on the earth for a vast period of time. These tools suggested a primitive level of technology widely assumed to imply a primitive level of culture. Perhaps the early humans who made the tools had had a level of mental ability inferior to that evolved by modern humans. Anthropologists looking at "savage" cultures in Africa and Australia saw their level of technology as parallel with that of Stone Age cultures, and thus used these peoples as models for the early stages of human evolution. Since the explorers of the Victorian era routinely pictured the peoples they encountered as mentally and morally less advanced than themselves, an image of humanity evolving from a level of savagery equivalent to that of the most despised living peoples was established. The resulting developmental model of mental and cultural evolution postulated a linear hierarchy of stages leading up to the level of modern Europeans. The parallel between this model and the linear sequence of evolutionary stages from the apes to the Neanderthals to modern humans was obvious to many prehistorians, although some doubted that the racial diversity of modern humanity could be fitted into so neat a sequence.

Archaeology and Anthropology

A key element in the emergence of the developmental form of evolutionism was the study of prehistory (Bowler 1989a; Daniel 1975; Grayson 1983; Trigger 1989; Van Riper 1993). Archaeologists such as John Lubbock (1865) distinguished between the Old and the New Stone Ages and saw these as stages in the development of technology preceding the discovery of bronze and then iron. By the 1870s it had become possible to recognize a number of different toolmaking cultures in the Paleolithic (the Old Stone Age), and most archaeologists assumed that these could be ranked in an evolutionary hierarchy of increasing sophistication. The French archaeologist Gabriel de Mortillet proposed four levels of Paleolithic culture, each named after a characteristic site. Thus, the overall sequence of technological progress was as shown in fig. 32.

The number of Paleolithic cultures was subsequently expanded even further (de Mortillet 1883). Although some archaeologists thought that the same cultures sometimes coexisted at a single period in time, de Mortillet was convinced that they formed a universally valid evolutionary sequence. As a prominent socialist, he saw this model of linear progress as a symbol of the inevitability of further social progress in the modern world (Hammond 1980). Since the Mousterian culture seemed to be linked to the Neanderthal race or species, de Mortillet was eager to link the cultural progress he saw in the archaeological record with the sequence of evolutionary stages being proposed by some biologists to cover the emergence of modern humans from the apes. Thus, the makers of the most primitive stone tools were pictured as primitive ape-men, allowing both the Neanderthals and those mod-

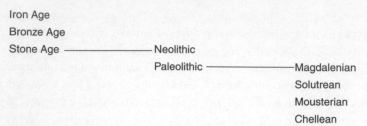

32.   De Mortillet's chronological sequence of cultures in the archaeological record.

ern peoples who still used stone tools to be dismissed as mentally inferior to the white race.

Lubbock played an important role in linking the emerging discipline of anthropology to this evolutionary synthesis (1870). He described the more "savage" modern tribes in the harshest terms, depicting them as both congenitally stupid and immoral, and assumed that their behavior was characteristic of the primitive humans of the Old Stone Age. The leading British anthropologist, Edward B. Tylor, arranged all living cultures into a single developmental hierarchy, with Europeans at the top (Burrow 1966; Kuper 1988; Stocking 1968, 1987). He, too, assumed that the sequence corresponded to a historical development—his *Researches into the Early History of Mankind* (1865) was a study of living, not ancient, cultures. Modern savages became, in effect, living fossils left behind by the march of progress, relics of the Paleolithic still lingering on into the present. Tylor envisioned only one Culture, with different levels of development and a clear goal toward which it was progressing. The only problem was to explain why some peoples lagged behind others in the ascent. This apparent teleology was no relic of the old theological viewpoint: Tylor went out of his way to describe religion as a product of the primitive level of mental development which had to be exposed so that it could be eliminated from modern life. But the radicals' faith in progress was so secure that they built their own values into evolution by assuming that their goals were also those of nature. Tylor's disciple J. G. Frazer achieved international popularity at the end of the century with his *Golden Bough* (new edition 1924), a depiction of classical Greek and Roman beliefs as stages in the development of religion from primitive times to the modern world. It was assumed that rational analysis would reveal how our ancestors had been misled into believing in the supernatural.

The American Lewis Henry Morgan arrived at a similar system of cultural evolutionism from a study of languages and kinship systems (Kuper

1985; Trautmann 1987). Morgan's *Ancient Society* (1877) proposed stages of development from savagery to barbarism to civilization. He was more aware of the role played by material factors in defining the level to which a particular people could rise, but the sequence of stages he defined was, like Tylor's, predetermined by the inherent logic of how the human mind seeks to understand its environment. Morgan followed Lubbock's lead in accepting that the humans who still lived in a state of savagery remained intellectually inferior to Europeans, relics of the past mentally as well as culturally. The evolutionary scheme remained in use throughout the rest of the century by American anthropologists, such as Daniel Brinton, who were charged with studying the indigenous peoples of the West. Significantly, these studies were carried out mostly by geologists and biologists who saw the evolutionary scheme as integral to the natural sciences.

The decision to treat savage peoples as though they were mentally and morally inferior to Europeans was also inspired by the evolutionism of Herbert Spencer (Duncan 1911; J. Greene 1959b; Kennedy 1978; Peel 1971). Spencer's philosophy was widely admired from the 1860s onward, and although it was eclipsed in Europe by the end of the century, it remained influential in America. His was a vision of progress driven by universal law, in which successive phases of equilibrium were achieved as the whole moved from a state of homogeneity to heterogeneity—that is, from simple to complex. Spencer invoked both natural selection and the inheritance of acquired characters to explain biological, mental, and social evolution, but his main concern at the social level was to understand the history of the human race in progressionist terms.

Although in principle Spencer's philosophy promoted the vision of evolution as a branching tree rather than as a ladder, he saw a main line in social progress leading toward modern industrial civilization. Peoples who had not achieved a modern form of civilization by themselves were being left behind because a more complex culture stimulated mental improvement, just as greater mental powers made it possible to advance to higher levels of culture. Sociology thus concurred with anthropology in recognizing an evolutionary hierarchy of social and cultural levels. Both used biology to explain why races which advanced to higher levels of social organization did so by increasing their mental powers. Liberal thinkers who insisted that all humans had the same level of mentality, whatever their culture, were now on the defensive against the prevailing image of a racial hierarchy created by the combination of biological and cultural evolutionism.

The linear progressionism of this social evolutionism was related more to a developmental, Lamarckian view of biology than to Darwinism. The linear model broke down among paleoanthropologists in the early twentieth century, as noted above, yet the scientists who founded this discipline remained wedded to a more complex model of racial origins which still allowed biology to be seen as a determinant of human character. In anthropology and the social sciences, there was a far more decisive break with the old evolutionism. In order to free themselves from the yoke of evolutionary biology, the anthropologists and sociologists repudiated the evolutionary paradigm altogether. They rejected the assumption that biological evolution had anything to say about how human societies and cultures developed, and in so doing freed themselves from the shackles of the linear model in which modern Western values were seen as the natural goal of evolution. Evolution might have shaped the human mind, but in so doing it had created something capable of transcending all the dictates of its biological origin (Cravens 1978; Greenwood 1984; Harris 1968; Hatch 1973; Ingold 1987).

In Europe, scholars such as Max Weber and Émile Durkheim began to treat each society or culture as a functioning whole which cannot be evaluated by the standards of any other. The assumption that the rational structure of the mind drives cultural change in a fixed direction was rejected, and along with it went the need to rank all societies into a linear hierarchy with Europeans at the top. The British psychologist W. H. R. Rivers returned from an expedition to the Torres Straits, between Australia and New Guinea, convinced that the cultures he had encountered there were so diverse that they could not be arranged along a linear scale. Rivers at least remained interested in the history of cultures, but saw each as having its own course of development unrelated to any other. Later British anthropologists such as the Polish émigré Bronislaw Malinowski adopted a functionalist approach similar to that of the Continental sociologists, who rejected history as irrelevant to the understanding of how each society actually works to satisfy human psychological needs (Kuklick 1991; Kuper 1972; Stocking 1996).

In America, Franz Boas and his students introduced a system of cultural relativism which, like Rivers's, allowed each culture to be seen as a product of its own unique history and repudiated the attempt to measure other cultures by Western values (Cravens 1978: chap. 3). Since they no longer saw cultures as forming a hierarchy with the West at the top, they could reject the racism of the evolutionists who had labeled cultures as "inferior" and perceived the people engaged in them as being endowed with lesser mental abilities. Cultural factors alone accounted for the differences, and as A. L. Kroeber (1917) proclaimed in a paper on the "superorganic," those factors could not be reduced to biology and had nothing to do with the biological origins of the human mind. When Boas's student Margaret Mead returned from Samoa to proclaim that the "adolescent trauma" which plagued Western teenagers was unknown in the sexually relaxed atmosphere of the South Seas, her message that biology placed no restrictions on behavior was popularized throughout the English-speaking world (for a critique of her work, see Freeman 1983). Boas was convinced that he and his followers had thrown off the shackles of Darwinism, little realizing that the linear paradigm of the cultural evolutionists had owed little to Darwin's theory in biology.

Psychology

The evolutionary model of culture and society was intimately connected with a developmental account of how the faculties of the mind had been produced (R. Richards 1987). In the early nineteenth century, many liberal thinkers had retained the classic notion of the mind as a tabula rasa, or blank slate, whose structure—that is, the individual personality—was constructed through interaction with the natural and social environment. Opposed to this was the view that the faculties of the mind were innate, and to some extent this had been reinforced by materialist principles arising from phrenology. If the mind was a product of the physical operations of the brain, then any preexisting structure in the brain must predetermine the faculties of the mind (Young 1970a). The scientific naturalists of the 1860s were committed to the view that the mind was governed by natural law, Huxley in particular seeing the mental world as a powerless epiphenomenon generated by the brain.

Such a position would be reinforced by showing that behavior was governed by inherited instincts as well as learned habits. Darwin himself accepted an important role for instinct and believed that instincts could be altered by natural selection. But he also accepted another possibility which was central to the philosophy of Herbert Spencer, the belief that learned habits could be turned into inherited instincts by the Lamarckian process of the inheritance of acquired characteristics. These evolutionary processes offered an explanation of how the faculties of the human mind have been created by evolution. New levels of activity and new instincts were created by our ancestors' interaction with their environment, including the increasingly complex social environment.

Darwin hinted at a darker side of human evolution in his *Expression of the Emotions in Man and the Animals* (1872), which sought to explain much of our emotional behavior in terms of instincts inherited from our animal ancestors. But there was no follow-up to this initiative because most evolutionists were anxious to distance themselves from the idea that we still carry with us the legacy of our ancestry among the brutes. The emphasis was on how the higher faculties were created, especially the intellectual and social faculties, which were, in Darwin's view, the foundation of our moral values. It was assumed that, as the mind became more complex, its reasoning powers would increase and gradually gain a greater influence over behavior. Modern humans had acquired rational powers great enough to allow science to emerge. Only in the area of social behavior were instincts created by evolution still active in determining our lives.

Evolutionary psychology rested on an attempt to draw up a phylogeny of the mind, a reconstruction of the steps by which the ladder of mental ability had been ascended through the animal kingdom up to the modern human level. Like many of his contemporaries, Darwin had accepted anecdotal evidence suggesting that the rudiments of most of the higher mental faculties could be seen at work in animals. This anthropomorphic view of animal behavior made it easier to argue that the human mind was only an extension of the animal mind, not a totally new spiritual faculty as religious opponents claimed. Darwin's leading disciple in this area was G. J. Romanes, who expounded a developmental model of mental evolution in which social activity promoted the emergence of language and hence the development of higher mental faculties (Romanes 1888; on Darwin and Romanes, see Schwartz 1995). This was a linear model of mental development which made considerable use of the recapitulation theory. Romanes and his contemporaries identified the stages in mental evolution though the animal kingdom with the steps visible in the mental development of a human child. They often assumed that modern "primitives" still exhibit a childlike way of thinking (Gould 1977b: chap. 5; Morss 1990). One of the last developments in evolutionary psychology was an attempt by William McDougall and others to investigate the extent to which human social behavior was conditioned by instinct.

This developmental approach to psychology began to break down in the last years of the nineteenth century. Although initially inspired by evolutionism, Conwy Lloyd Morgan proposed his famous "canon" which required the psychologist to attribute to animals only the minimum level of mental ability required to perform their observed behavior. Morgan rejected much of the anecdotal evidence for the high level of animals' mental powers. He observed that when his dog was asked to bring a stick through a narrow gap in a fence, it was unable to solve the problem rationally—it simply kept running at the fence until by accident it lined the stick up parallel to the gap and got through. It could learn from this experience, but it could not think the problem out in advance the way a human could. Morgan's canon (commonly known as Lloyd Morgan's canon) undermined the evolutionists' attempt to project the higher mental functions back down the scale of animal organization. In particular it reinforced the claim that animals do not have true language (Radick 2000). This buttressed the objections of those traditionalists who maintained that there was something unique about the human mind. Its reasoning powers, its language abilities, and its ability to recognize moral values were all features which raised it above anything observable in the animal kingdom. In Britain there was a revival of interest in the old-fashioned kind of psychology which treated the mind as a free agent capable of transcending natural law. Morgan himself went on to propose his theory of "emergent evolution" (1923), in which mind and spirit were new categories somehow added to the material world at certain stages in the advance of life.

In Germany, however, psychology was being established as a laboratory-based discipline which studied behavior without concerning itself with the evolutionary origin of mental functions. This new approach came to the English-speaking world in the form of behaviorism, which swept through American psychology in the early decades of the twentieth century. Under the leadership of J. B. Watson, behaviorism helped to establish psychology as an independent academic discipline, just as Boas's rejection of the evolutionary model helped to create modern anthropology (Cravens 1978: chap. 2). And, like the anthropologists, the behaviorists rejected all claims that the animal or the human mind was constrained by instincts generated in the course of evolution. Once again, the mind became a pure learning machine, a blank slate upon which experience (or the experimenter) could impose any form of behavior. For Watson it was illegitimate even to talk of the mind, since all that can be observed is behavior.

Academic psychology thus emancipated itself from the evolutionary paradigm. But outside the universities, the legacy of the developmental approach to the mind was anything but dead—although one of its major products would have horrified the previous generation. In France, the educational psychologist Jean Piaget continued to treat the mind of the growing child in terms of the recapitulation theory, stressing that the learning process must be adapted to the stage of mental development the child has reached at the time (Messerly 1996). Far more pervasive was the influence of Sigmund Freud's analytical school of psychology, which sought to treat mental dysfunctions by postulating an unconscious level to the mind. For Freud and his followers, the conscious mind often has difficulties controlling the biological (mostly sexual) desires programmed into the unconscious. Freud's theories had their roots in many areas of nineteenth-century thought about the mind, but a number of historians have noted the role of developmental evolutionism (Morss 1990; Ritvo 1990; Sulloway 1979b). His image of the unconscious was of a deep reservoir of animal instincts overlaid with a superficial layer of rationality. The animal past was still buried within us, but far from transcending it as we mature, our conscious mind has to struggle to contain its influence within socially acceptable bounds. For both Freud and Piaget, it was not so much Darwinism as the recapitulationist-developmental model of evolution which was crucial, and both continued to believe that the Lamarckian theory must be true in order to explain how new levels of behavior are added in the course of evolution. Given the immense impact of Freud on twentieth-century thought—he himself compared his revolution to that of Darwin—this link to nineteenth-century evolutionism is of great significance. But the optimistic progressionists of Darwin's era would have shuddered to see their model of increasing rationality undermined by this revelation of the power of the animal urges still buried within our minds.

EVOLUTION AND RACE

The nineteenth-century anthropologists who created the linear hierarchy of cultural evolution were exploiting a model which owed little to Darwin's theory of natural selection. Nor indeed did this model necessarily commit them to the belief that humans had evolved from apes, although the assumption that humanity had risen from a primitive state certainly resonated with the theory of biological evolution. The idea of progress was central to the anthropologists' vision, as it was to many nineteenth-century philosophies, and the assumption that social and cultural progress were inevitable would form the basis for many attempts to see a parallel between biological and social evolution. The theory of evolution thus influenced social thought in many different ways. One obvious extension of the belief that humans have evolved from apes was to map the anthropologists' cultural hierarchy onto a parallel hierarchy of mental evolution. Those races which had not developed a sophisticated culture and technology were branded as evolutionary failures, living fossils with primitive mental powers who survived only because they were isolated from competition with more advanced races. Evolution also challenged those who insisted that the races were distinct biological entities, perhaps even separate species: how had the human family tree become divided into these different branches?

Europeans were aware of the physical differences between themselves and other peoples, and as they began to exert control over other regions, they were increasingly self-conscious about those differences. There was a tendency to view other peoples as distinct racial types with their own physical and mental characters. As Europeans began to conquer, enslave, and even exterminate other races, there was a tendency to exaggerate these racial differences to justify the exploitation. If the nonwhite races were less than fully human, it was easier for whites to feel comfortable with a situation in which the superior race determined the fate of the inferior. Even within Europe, there was a sense that the various nations had different racial origins, and archaeology encouraged the feeling that whether a nation was descended from Celts or from Teutons (Anglo-Saxons) was a crucial part of its identity. The old Enlightenment ideal of a single, unified human nature, in which everyone started with the same mental and moral equipment, was coming under threat.

Writing originally in 1850, the anatomist Robert Knox proclaimed, "With me[,] race, or hereditary descent, is everything; it stamps the man" (1862: 6). Knox regarded Africans as a totally alien people, but his condemnation was not confined to the non-Europeans. This is how he characterized the Celtic race: "furious fanaticism, a love of war and disorder, a hatred for order and patient industry; no accumulative habits, restless, treacherous, uncertain: look at Ireland" (25). Knox's sense of the inferiority of many racial groups was taken up by anthropological societies founded in London and Paris during the 1860s. Although seen at first as an extremist position, this kind of thinking began to make serious headway as Western countries became more self-consciously imperialist toward the end of the nineteenth century (Banton 1987; Barker 1998; Bolt 1971; Frederickson 1971; Gould 1981; J. Haller 1975; Lorimer 1988, 1997; Snyder 1962; Stepan 1982).

Evolutionism helped to justify the belief that the nonwhite races were inferior by offering a new explanation of how the hierarchy of races had formed. Even before evolutionism became popular, the blacks of Africa and Australia had been portrayed as more apelike in their physical appearance than the whites. The intolerances of the age of imperialism encouraged the belief that these races had lower levels of intelligence and a weaker moral sense. The hierarchy of cultural stages erected by anthropologists such as Tylor and Louis Henry Morgan was increasingly identified with a hierarchy of mental development produced by biological evolution.

Not all anthropologists moved in this direction. Tylor himself began from the assumption that all humans have the same mental capacity. But later even he came to doubt this, and those more open to racist assumptions took it for granted that peoples with inferior levels of culture were stuck at that level because they were more primitive mentally. For evolutionists such as John Lubbock and Herbert Spencer, it was obvious that the non-white races were equivalent to the Stone Age ancestors of the whites, preserving an apelike appearance and a lower level of mental and moral powers.

The cultural hierarchy had been erected without reference to Darwinism, and the non-Darwinian ideas of evolution which flourished in the late nineteenth century played a major role in sustaining the racial hierarchy. The neo-Lamarckian model of evolution as the addition of stages paralleled by the development of the embryo was especially powerful in this respect (Gould 1977b: chap. 5). Because ontogeny and phylogeny were supposed to be equivalent, the supposedly inferior mental capacities of the savage could be equated with those of a white child (Muschinske 1977). American neo-Lamarckians such as E. D. Cope (1887) listed the features that were supposed to indicate the blacks' "retardation of growth" (J. Haller 1975). Ernst Haeckel expressed similar views in Germany, and his philosophy has been identified controversially as a key influence on the later development of fascism and Nazism (Gasman 1971, 1998). The Italian criminal anthropologist Cesare Lombroso identified the "criminal type" within the white race as a throwback to an earlier stage of evolution (Nye 1976).

The developmental model of evolution was thus a powerful addition to the arsenal of those seeking to attack the character of other races. As late as the 1920s, the Lamarckian embryologist E. W. MacBride dismissed the Irish Celts as a lower race which should be eliminated from the British population (1924; Bowler 1984). To explain why some races had advanced farther up the scale than others, it was assumed that they had been exposed to a more stimulating environment. The whites, who evolved in the harsher climate of the north, were more advanced than the blacks of tropical Africa. For MacBride, the Anglo-Saxons were more advanced that the Celts because the latter originally had evolved in a softer Mediterranean environment. All these developmental models tended to ignore the element of divergence characteristic of Darwinism (see fig. 33).

The hierarchical scale of mental development could be imposed on a more complex model of racial origins in which multiple branches evolved in parallel. Traditionally, the whole human race was supposed to have descended from Adam and Eve, according to the hypothesis of "monogenism." But some scholars had long challenged this hypothesis on the grounds that the few thousand years of history accounted for in Genesis was not enough to allow the differentiation of a single human species into such diverse racial types. The alternative was "polygenism," the claim that the various races were separate creations, with only the whites having descended from Adam

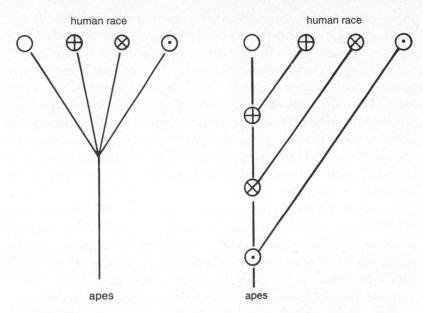

33. Darwinian and developmental views of race. The left-hand diagram shows how the various human races might be related on a Darwinian model of branching evolution. They share a common ancestor but are divergent branches which cannot be ranked into a hierarchy. In the developmental model (right), which is consistent with the Lamarckian and recapitulationist theories, the living races form a hierarchy because the "lower" races have simply preserved earlier stages in the development of the highest. The assumption is that the side branches have not changed since they split off from the main line, perhaps because they were no longer stimulated by a challenging environment. The lower races are equivalent to the links missing in the fossil record.

and Eve. This view was congenial to Knox and the racial anthropologists of the mid–nineteenth century. There were even efforts, contrary to all experience, to claim that whites and blacks could not interbreed successfully. The Swiss-American naturalist Louis Agassiz found it difficult to accept that the blacks he encountered in America were of the same species as himself—and his support for polygenism was seen by Southern slave owners as a useful endorsement of white superiority. On this view, the blacks were primitive in a more fundamental sense—they were a separate species whose very essence was fixed at a lower point on the chain of being (Priest 1843).

The archaeological revolution of the 1860s opened up the vast timescale of human prehistory and provided enough time for significant racial differentiation to have occurred. But evolutionism also affected thinking on this issue. Darwin was not immune to the growing conviction that whites were more highly developed than the other races, but he realized that all were still members of the same species precisely because they were interfertile. Some subbranches of the human family tree had developed further beyond the ancestral ape than others, but all shared a comparatively recent common ancestry. A. R. Wallace suggested a different view which inclined rather more toward the polygenist position (1870, 1891: chap. 8). He argued that the various racial types had become differentiated before the final stages in the emergence of humanity from the apes. Each racial group had completed the last phases of the development independently. Wallace, unlike many of his contemporaries, was prepared to allow that all branches had attained the same level of mental development.

For exponents of parallel evolution such as Cope and Haeckel, the races were distinct species, even though they could interbreed, because they had been separated from one another for so long. The most extreme form of this model was Carl Vogt's theory in which the split went all the way back to the apes, each human species having its origin in a different ape. The whites came from the chimpanzee, blacks from the gorilla, and the oriental races from the orangutan. This was always a minority view, although one occasionally expressed even in the early twentieth century (Bowler 1986). But its influence was felt among those paleoanthropologists who began to argue that there were multiple lines of human evolution, with some of those lines, including the Neanderthals, being driven to extinction at a later stage in the process. The theories in which many lines independently evolved toward the goal of becoming human were the last expression of the developmentalist viewpoint. Only within this framework of vast antiquity for even the living races did the more extreme form of white supremacy maintain a tenuous link with science. The Nazis' claim that the fair-haired Aryan was a distinct and superior form of humanity destined to rule the "sub-men" of other types had its origin in a cultural tradition which drew on many other factors besides the sciences of physical anthropology and evolutionism (Mosse 1978; Poliakov 1970; Tennenbaum 1956).

In America, the race question was focused on the problems created by slavery and the difficulties encountered by emancipated slaves after the Civil War (J. Haller 1975; Smedley 1993). We have noted the efforts of neo-Lamarckians such as E. D. Cope to define the blacks as relics of the evolutionary past. In the twentieth century, biologists and paleontologists such as H. F. Osborn continued to claim that the races were distinct species which had evolved in isolation. They exploited the prevailing view that many apparently closely related forms were independent products of parallel evolution. There were close links between American race scientists and those working in Nazi Germany (Kuehl 1994). The race question was also important in the American eugenics movement (see below).

Opposition to race science began among anthropologists and social scientists in the early years of the twentieth century (Cravens 1978). For some time, social scientists and biologists battled for supremacy over this issue, with Franz Boas himself suffering considerable persecution, both professional and social, from the powerful race lobby within American biology. Only gradually in the 1930s did some more liberal biologists, including Julian Huxley, begin to throw off the legacy of racism (Barkan 1992). The emergence of the Darwinian synthesis did much to discredit the theories of parallel evolution on which the more extreme theories of racial differences were based. In the end, however, the challenge to the race science of the early twentieth century was created by the growing sense of horror at the extremes to which the Nazis pushed their drive to eliminate the Jews and other racial types deemed undesirable by their ideology.

SOCIAL EVOLUTIONISM

The classic image of social evolutionism focuses on the mechanism which was supposed to drive the escalator of progress. It was easy to assume that a more challenging environment was the spur which developed improved mental powers, and the image of a people struggling against the limitations imposed by nature easily took on Darwinian connotations. A more direct link with the Darwinian theory could be made by suggesting that competition within the species was crucial for progress, either at the individual or the group level. The classic and much-maligned concept of social Darwinism is the assumption that struggle is the motor of progress, spurring on development and weeding out those who do not keep up. But social Darwinism could be expressed in many forms: beginning as a link between free-enterprise capitalism and the Darwinian theory of individual competition, the ideology of progress through struggle was increasingly transferred to the level of national or even racial competition. Nor was Darwin's theory the only way of forging a link between biology and social thought. The idea that struggle spurred individual self-improvement had distinctly Lamarckian overtones, and many who welcomed the concept that the elimination of "unfit" races occurred as a part of evolution did not believe that the original racial differences were created by natural selection. Selection was only one among many models of evolution, but use of the term *social Darwinism* blinds us to the role of non-Darwinian ideas in promoting harsh social policies.

Lamarckism required one to believe that individuals are not totally constrained by their biological inheritance. For new characters to be acquired and transmitted, inheritance had to be "soft" enough to allow for some modification. But natural selection included no such requirement: heredity could be "hard" in the sense that it allowed no room for individual modifications, and still evolution would occur because only the fit individuals would transmit their rigidly defined characters. Even Darwin himself did not believe that heredity was as hard as this, but in the later nineteenth century there was a growing conviction that heredity defined the individual's capacities and temperament so rigidly that no modification was possible. Nurture (environment and upbringing) was incapable of altering what nature (biological heredity) had predetermined. Although the selection theory played a role in fostering this belief, it was not the only source, and the Mendelian genetics of the early twentieth century was equally hereditarian in outlook without being sympathetic to the selection theory.

Social Darwinism

The area where evolutionism impinged most obviously on social thought was the implication that there might be a struggle for existence among individuals, nations, or races. This points us toward the controversial topic of social Darwinism, the use of the Darwinian notion of struggle to justify social policies in which there was little sympathy toward those who could not support themselves. But the possibility of a social Darwinism focused on race conflict (see below) highlights the complex nature of the analogy between natural selection and human struggle. The most prominent form of social Darwinism stressed not racial or national struggle but the individual competition which flourished in free-enterprise capitalism. The two levels of struggle—between individuals and between groups—are not necessarily compatible, because the late-nineteenth-century imperialists who worried about their nation's status in the world wanted a strong central government which would limit commercial rivalries. There is no single form of social Darwinism, only a complex of often contradictory ideologies exploiting the model of the survival of the fittest in different ways (Paul 1988). To make the situation even more complex, Darwinian natural selection was not the only means of articulating struggle as either a scientific theory or a social philosophy. The expectation that struggle promoted individual self-improvement generated an ideology based on Lamarckian rather than

Darwinian evolutionism, if those improvements were supposed to contribute to the progress of the race. It is often assumed that any social philosophy advocating struggle must be Darwinian, but for this to be logical, the term *Darwinian* must be used in a broad sense which goes far beyond the detailed theory of natural selection.

The whole subject generates endless controversy, the arguments being all the more heated because they bear upon issues still relevant today. The term *social Darwinism* was introduced only in the late nineteenth century and was used from the start in a pejorative context (Bellomy 1984). To call someone a social Darwinist was to insult them by implying that they had abandoned all moral standards to make success the only criterion for what is good. Those who portray modern interpretations of human nature as direct continuations of the harsh Darwinian viewpoint of the late nineteenth century use history to label those interpretations as morally unacceptable. The ideological debate is polarized between the right and the left: socialists see any attempt to reduce humans to the level of animals struggling with each other as an illegitimate use of science to bolster right-wing ideologies. The debate also bears on the question of scientific objectivity, because opponents of social Darwinism often regard Darwinism itself as an illegitimate projection onto nature of values derived from the ideology they distrust (Young 1985). Scientists seeking to defend the independence of their theorizing from social constraint are branded as social Darwinists because they will not admit that Darwin derived any inspiration from his social environment. For the anti-Darwinians, it is obvious that the importation of right-wing ideology into science was necessary so that science could be presented as objective evidence for the inevitability of humans behaving in a selfish manner.

The classic expression of the view that late-nineteenth-century social thought was dominated by the Darwinian metaphor of the struggle for existence is Richard Hofstadter's *Social Darwinism in American Thought* (1959); for a more recent exposition see Mike Hawkins's *Social Darwinism in European and American Thought* (1997). But Hofstadter's account acknowledges that there were many different forms of social Darwinism besides the classic justification of capitalism. Some historians, especially Robert C. Bannister (1979), argue that the level of support for an ideology of struggle has been overestimated, and that the direct input from Darwinian biology is obscure (see also Bowler 1993; Fichman 1997; Halliday 1971; Heyer 1982, G. Jones 1980; Rogers 1972). Bannister has been attacked for giving credence to those who seek to portray science as purely objective. His approach certainly implies that we should be careful not to assume that every casual reference to struggle as the driving force of progress reflects a considered evaluation of Darwin's theory. But the critics have a point when they argue that the influence of Darwinism may be pervasive at a level which reflects rhetoric and metaphor rather than detailed scientific analysis. Bannister's critique is better understood, however, as an attempt to suggest that there were also non-Darwinian evolutionary ideas that could be used as social models. The aim is not to whitewash Darwinism but to show that the whole range of evolution theories available at the time was incorporated in the effort to portray biology as a foundation for social thought.

The disagreements among historians are reflected in the debate over Darwin's own views on society (J. Greene 1977). Opinion ranges all the way from accusations that he openly promoted aggressive individualism (Harris 1968) to denials that he had any sympathy for such views (Freeman 1974). Unfortunately, Darwin's writings contain passages that can be interpreted in favor of both positions. He was aware of the role played by the model of economic individualism in his thinking, especially as expressed in Malthus's population principle. He saw both individual and tribal struggle as important in human evolution, and feared that the relaxation of selection within a civilized society (where charity helps the unfortunate to survive) would harm the race by allowing the unfit to breed. Yet he was surprised when a newspaper article accused him of justifying the actions of Napoleon and of tradesmen who cheat. For Darwin, at least, "fitness" in the human context did not include the kind of immorality which would justify any action by the motto "Might is right." The fit were the able and energetic, not those who cheated or forced their way to success.

Darwin's liberalism was a long way short of the ruthless individualism worshiped by some successful industrialists (G. Jones 1980). He certainly wanted to restrict the powers of the landed aristocracy, who, he believed, had no hereditary right to rule. But he accepted that there was a natural aristocracy of talent which should have the freedom to rise to the top in every generation. The debate between Darwinism and conservative religious thought can be seen in part as an element in the bid by the new commercial and professional elite to gain power in society (Desmond 1982; Desmond and Moore 1991; Turner 1978). Darwin himself came to play the role of a country squire at Down House, and James R. Moore (1982) argues that his burial in Westminster Abbey was used by his followers to establish their position as the new ruling class. But selfishness had no role in the ideology of Huxley and the other professionals, many of whom worked endless hours in education and public service for only limited reward. The problem would come when critics began to argue that the unfit poor were proliferating in the slums precisely because of the efforts made by the new rulers to protect them from disease and starvation. Then it might become necessary, according to this argument, for the state to play a role in regulating human reproduction.

The more ruthless form of individualism is usually associated with the name of Herbert Spencer, although as R. J. Richards (1987) and other have argued, Spencer—like Darwin—would have been horrified at the thought that his philosophy was being used to undermine moral values. His Synthetic Philosophy became the most broadly articulated version of the progressionist philosophy into which Darwinism was received. Spencer accepted natural selection as an important mechanism of biological evolution, and he coined the phrase "survival of the fittest." He was also an exponent of an extreme laissez-faire individualism, seeing the struggle between individuals jockeying for position as the driving force of social progress (Kennedy 1978; Peel 1971; Taylor 1992). His ideas were welcomed with enthusiasm by the robber barons who masterminded the development of American industry in the late nineteenth century. It is easy, then, to see why Hofstadter should take Spencer as the archetypical social Darwinist, responsible for transmitting this harsh philosophy of progress through struggle across the Atlantic. Spencer's opposition to socialism was based on the assumption that state support for the poor would encourage them to be idle. In his later life he came to place increasing emphasis on the fear that a state-funded welfare system would permit an ever greater number of "unfit" people to survive and breed, thereby undermining social progress (see his 1884 book *The Man versus the State*, reprinted in 1969). But Spencer was a biological Lamarckian, and an equally important aspect of his support for free enterprise was his belief that competition would stimulate individuals to improve themselves. The aim was not so much to eliminate the unfit as to make everyone fitter, and in this sense Spencer's ideology endorsed the Victorian sense of the need for personal development articulated in Samuel Smiles's classic book, *Self-Help* (1859; see also Jarvis 1997). Since the acquired improvements were supposed to be transmitted to future generations, this was more social Lamarckism than social Darwinism. But because struggle was seen as the spur to progress, the Lamarckian element has gone largely unnoticed by later writers. The emphasis on the virtues of thrift, industry, and initiative was an attempt to revise the old Protestant work ethic, and for this reason Spencer's apparently agnostic philosophy was welcomed by some liberal religious thinkers (Moore 1985a).

At the same time, however, Spencer's gospel of progress through struggle was seized upon by those who advocated a more ruthless social policy which really did seem to abrogate all traditional moral values in favor of the worship of success at any price. This was especially the case in America (Hofstadter 1959; for a collection of primary sources, see Ryan 2001). The Yale economist William Graham Sumner endorsed the motto "Root, hog, or die" and challenged anyone to displace him from his position by displaying superior ability. Idleness and inefficiency would be punished by miseries inflicted by nature itself. Yet Sumner seemed more concerned about the struggle of the human race as a whole against the limitations imposed on it by the natural environment. This concern may have been expressed in Darwinian language, but it deflected attention away from the competition between individuals. Leading American industrialists also claimed that Spencer's philosophy justified their own enthusiasm for unrestricted competition. Andrew Carnegie became his avowed disciple, while John D. Rockefeller and the railway magnate James J. Hill used the phrase "survival of the fittest" to endorse the capitalist system. In their view, the fact that the most successful firm drove its competitors into bankruptcy simply allowed the most efficient producer to dominate the market, thereby ensuring economic progress.

Hofstadter's claim that these endorsements illustrate a widespread enthusiasm for free-enterprise social Darwinism has been disputed (Bannister 1979; Russett 1976; Wyllie 1959). Sumner did not, in any case, hold out much hope for social progress in the future. Many small businessmen, fearful of being gobbled up by their more powerful competitors, openly called for the state to restrict competition. The capitalist form of social Darwinism was self-defeating—the goal of every industrialist was to achieve a monopoly, thereby eliminating all rivals and hence all further competition. Nor did the concentration of wealth in the hands of a few able individuals guarantee that the resources would be well-used by their children—the families of the rich often became drones who lived a life of luxury while employing managers to look after their affairs. Carnegie realized that he should use his wealth for the public good by founding libraries and other institutions that would help ordinary people improve themselves. The analogy between commercial competition and natural selection is so vague as to be virtually meaningless, because the inheritance of wealth does not correspond to the inheritance of biological qualities.

Racial and National Conflict

By the end of the century, Spencer's influence was already waning in Britain, although he seems to have kept a stronger hold on the American imagination. This was in part because the enthusiasm for unrestrained free enterprise was being overtaken by an ideology of imperialism, which focused attention on the white nations' efforts to conquer or colonize the rest of the world and on the resulting national rivalries. In this era, another form of social Darwinism seemed more appropriate: perhaps the main focus of the struggle for existence was between the races of humankind or between nations within the dominant white race. Even those who distrusted natural selection could accept that the struggle for existence would eliminate the less successful of nature's productions. Many now began to claim that the domination of one race or nation over another was a natural part of the process by which the human species had advanced.

Darwinism implied that species and races must compete for territory. Not everyone accepted this struggle as inevitable, because some thought it was impossible for a race to adapt properly to a territory different from the home in which it evolved. But the success of whites in conquering and settling territory in America and Australia made it difficult to sustain this view. There was an increasing assumption that populations would try to expand into new territory, and thus would compete with the indigenous races. Darwin himself seemed to endorse this racial struggle, as in the subtitle of the *Origin of Species:* "The Preservation of Favoured Races in the Struggle for Life." This was not the main driving force of natural selection, but competition between species and varieties was an important subsidiary theme within Darwinism, and the image of racial competition is thus a genuine form of social Darwinism. Some even thought that blacks in America would decline to extinction once freed from the protection offered by slavery. The Darwinist Karl Pearson welcomed whites' conquest of the world: "It is a false view of human solidarity . . . which regrets that a capable and stalwart race of white men should replace a dark-skinned tribe which can neither utilize its land for the full benefit of mankind, nor contribute its quota to the common stock of human knowledge" (Pearson 1900: 369). Pearson did at least accept the theory of natural selection working within populations, but many of those who promoted race conflict did not, attributing the actual origin of races to Lamarckian or orthogenetic factors. Paleoanthropologists such as Arthur Keith compared the extinction of the Neanderthals with what was happening to the natives of America and Australia—and saw these replacements as essential to human progress. In this respect, at least, the Nazi theory of Aryan supremacy was not significantly different from the form of racial Darwinism prevalent among many scientists in the early twentieth century.

Pearson and Keith's confidence in the white race's ability to expand its territory was not universally shared. Pearson himself became concerned about the extent to which Jews from eastern Europe were multiplying in the East End of London. In America—where there was a much greater flow of immigrants from both Europe and the Far East—the threat that the biological character of the white race might be undermined by faster-breeding but intellectually inferior races became a major concern. Race thus played a role in the emergence of the eugenics movement (see below). The assumption that a race's superior mental abilities guaranteed its success had seemed natural in the age of imperialist expansion, but the more pessimistic worldview of the early twentieth century acknowledged that the rate of reproduction might be far more important in the long run.

Europeans were convinced of their superiority over other races, but there were major rivalries among the European nations themselves. In the late nineteenth century, a wave of nationalist fervor swept through many countries, fueled by the West's increasing power to dominate the world. Everyone wanted their share of the economic benefits of conquest and colonization. National hostilities were sharpened, and the result was a return of militarism—denounced by Spencer as a return to a more primitive social organization. The intensification of national rivalries offered another level at which the Darwinian model could be applied, since each nation could feel that it was struggling to demonstrate its superior "fitness." But only a strong central government could guarantee success in this national struggle for existence—the free enterprise advocated by Spencer would undermine solidarity because individuals or firms would seek private profit at the expense of the nation's long-term interests.

This transition from an individual to a national struggle for existence had already been made in Walter Bagehot's *Physics and Politics* (1872). Setting out deliberately to apply the principle of natural selection to society, Bagehot argued that, throughout history, the strongest nations have always dominated their neighbors, and that the strongest have always been the fittest, in the sense that they have contributed more to the development of civilization. Inferior nations might not be eliminated in the struggle, but they would be subjugated and taught the advantages developed by their conquerors. Bagehot's alienation from the spirit of the individualistic selection theory is evident in the fact that he saw the increasing power of governments as essential for progress. Anything that helped the state to control unruly individuals was a benefit, including religion. Church and state should unite to suppress freedom of thought in the name of national unity.

Bagehot's views were anathema to the British enthusiasts for free enterprise, but as the century progressed toward its end, Britain became caught up, along with the other European countries, in international rivalry (Crook

1994). In France, too, competition between nations was assumed to be the most obvious social extension of Darwinism (L. Clarke 1984). In Germany, however, the nationalist form of social Darwinism became more explicit. In the years preceding World War I, the German general Friedrich von Bernhardi wrote openly of his newly united nation's desire for conquest and claimed that this would be justified because it allowed the superior German culture to displace inferior rivals in a struggle for existence. Ernst Haeckel, originally a liberal, threw his weight behind Germany's assertion of its role as Europe's leading culture. He endorsed Germany's position when war finally broke out in 1914, and was bitterly disappointed when Germany was defeated. Haeckel had developed a philosophy of "monism," in which spirit and matter were different aspects of the same underlying substance (Di Gregorio 1992; Holt 1971; Weindling 1989a). His Monist League promoted this philosophy throughout Germany and linked it with an ideology of German supremacy. The league's subsequent influence on the rise of Nazism has been stressed by Daniel Gasman (1971; see also Zmarzlik 1972; on Haeckel's influence on fascism in other countries, see Gasman 1998). Gasman's account has been challenged by other scholars, who see Darwin's influence as more pervasive and hence as affecting a variety of different political philosophies (e.g., A. Kelly 1981). The views expressed by Haeckel were typical of many contemporary right-wing intellectuals, not all of whom were influenced by Darwinism, although they may have exploited the metaphors provided by the selection theory. Significantly, the philosopher Friedrich Nietzsche—who gained notoriety for his proclamation of a new morality based on "might is right"—repudiated any link with the Darwinian theory in biology (Bannister 1979; Call 1998). Nevertheless, many British and American biologists were worried that Darwinism's name had been tainted by its association with German militarism. The American biologist Vernon Kellogg visited the German army occupying Belgium in the early part of World War I and reported that the officer corps was pervaded by the ideology of nationalist social Darwinism. For this reason, some biologists were encouraged to persevere with the non-Darwinian theories promoted during the "eclipse of Darwinism" (Mitman 1990).

Alternatives to Social Darwinism

Darwin's theory has acquired a reputation for promoting ruthless social attitudes. But the use of Darwinian catchphrases such as "the survival of the fittest" by a variety of right-wing thinkers creates an exaggerated sense of the theory's influence and conceals the fact that many had only the vaguest understanding of science. Modern scholars have noted a wide range of social applications of the theory which went against the spirit of ruthless social Darwinism (G. Jones 1980; A. Kelly 1981). Even socialists could gain some comfort from the theory while repudiating the idea of struggle at both the individual and the national level (Pittenger 1993). The codiscoverer of natural selection, Alfred Russel Wallace, wrote actively in support of socialism and found a justification for the ideology in the Darwinian theory (Durant 1979). He argued that the inheritance of wealth distorted people's choice of a marriage partner: a biologically fit individual might be tempted to marry someone of inferior character because the latter had inherited a fortune, thereby weakening the constitution of the race as a whole. If equality of wealth were imposed by a socialist government, the fittest individuals would naturally tend to partner one another, to the benefit of the race. Wallace's vision of a biological rationale for social equality was by no means unique (Weikart 1999). The socialist politician J. Keir Hardy referred to group selection to argue that progress was achieved through the success of those groups in which the individuals felt sympathy for one another. In the end, though, this line of argument was developed more by those who favored the Lamarckian theory.

One possible link between socialist ideology and Darwinism has been exaggerated, however. It is sometimes implied that Darwinism was associated with the philosophy of Karl Marx (e.g., Barzun 1958), and at one time it was believed that Marx had offered to dedicate a volume of his *Capital* to Darwin. Marx and Engels certainly welcomed evolution theory because of its support for a materialist view of human nature. But they also realized from the start that there was an analogy between natural selection and the capitalist system of economic competition, and so were suspicious of Darwin's theory. Marx's concept of class struggle has different roots lying in Hegel's idealism—in the clash and synthesis of coherent social entities reflecting the stages of social evolution rather than competition between individuals or tribes (Heyer 1982). We now know that the claim that Marx offered to dedicate his book to Darwin was based on a misunderstanding of the relevant correspondence (Colp 1974, 1982; Fay 1978; Feuer 1975; on Marx and Darwin, see Pancaldi 1994). In the twentieth century, Soviet communism was always hostile to the Darwinian theory, and Lamarckian theories such as Lysenko's flourished in Soviet Russia.

By the end of the nineteenth century, confidence in the assumption that struggle is the driving force of progress was undermined by a growing fear that human civilization was no longer, in fact, permitting the struggle to take place. It might even be weakening the constitution of the human race by creating artificial environments and stresses that people could no longer cope with. Some now feared that progress would be replaced by degeneration as the human race undermined the very forces which had created it (Bowler 1989c; Chamberlin and Gilman 1985; Pick 1989). The French social writer Max Nordau achieved wide notoriety after publication of his *Degeneration* (1895) proclaiming the effects of nervous stress created by modern living. On a very different tack, the evolutionary biologist E. Ray Lankester called attention to the possibility that a civilized lifestyle might be so lacking in challenges that the external stimulus necessary for the species to progress was diminished. Whatever the mechanism of adaptive evolution, species only progressed when they faced a challenge from the environment. Just as some animal groups had degenerated when they took up a sessile life on the seabed, so humanity might slip back from the standards achieved by the ancient Greeks and Romans once all material wants were met. As Lankester put it, "Possibly we are all drifting, tending to the condition of intellectual Barnacles or Ascidians [sea squirts]. It is possible for us—just as the Ascidian throws away its tail and eye and sinks into a quiescent state of inferiority—to reject the good gift of reason with which every child is born, and to degenerate into a contented life of material enjoyment accompanied by ignorance and superstition" (1880: 61).

It is significant that when H. G. Wells came to write his science fiction story *The Time Machine*, depicting a future human race reduced to bestiality and futility, he had been reading Lankester's warnings. Faith in the idea of progress, central to any form of social Darwinism, was now beginning to crumble as late-nineteenth-century thinkers realized that industrial civilization was a mixed blessing.

BIOLOGICAL DETERMINISM

One response to this more pessimistic view of social evolution was the eugenics movement, which called for governments to impose a breeding program on the human race. In an age of increasing reliance on management by experts, why not let the experts on heredity determine who should have children and who should not? The problem with free-enterprise social Darwinism was that it was ineffective if one thought in terms of the human race's biological character. The "fittest" people, those with enough intelligence and initiative to gain a dominant place in society, might gain more wealth and power, but they were notorious for limiting the number of children in their families. Meanwhile, the least able and energetic drifted into the slums of the great cities where, in an age of better public health, they managed not only to survive but also to breed prolifically. Far from natural selection weeding out the unfit, in a civilized society it was the unfit who produced most of the next generation. Darwin himself had worried about this, and by the end of the century the eugenics movement was openly pleading for a policy of artificial selection applied to the human race.

Eugenics typified the increasingly popular ideology of genetic determinism, the claim that a person's character and abilities were predetermined at birth by the power of inheritance. In a sense, the racist policies discussed above express a form of this ideology, since they assume that an individual's character is fixed by the race to which the person belongs. For this reason, race could also play a role in eugenics policies. But the fully developed eugenics of the early twentieth century insisted that even within the race, individual character was biologically predetermined. All too often, it turned out that when the expected differences were mapped onto the social classes, the poor class contained the larger proportion of unfit individuals.

There were other forms of genetic determinism, however, one obvious version being the widely held opinion of Victorian men that women were intellectually inferior. There are remarkable parallels between the applications of the determinist ideology in the areas of race and gender. Yet the late nineteenth and early twentieth centuries by no means were dominated by the determinist model of human nature. There had always been those who had insisted that the individual is shaped not by heredity but by environment and upbringing. In the twentieth century, the social scientists proclaimed their independence from biology by stressing this aspect of human nature. But there were biologists, too, who emphasized the role of the environment, and this model could be linked to evolutionism by exploiting the Lamarckian theory of the inheritance of acquired characters. If people could be improved by education, perhaps the benefits could be passed on by heredity to become part of the character of the whole race.

Eugenics

The Darwinian selection theory played a role in the emergence of the eugenics program. Thus, eugenics may be counted as a form of social Darwinism—although it is a form which violates every principle of the free-enterprise version popularized by Spencer. Eugenics had a closer link with the nationalist form of social Darwinism, since one argument for improving the biological fitness of the population was to resist the threat posed by rival powers. But the movement also drew on another biologically inspired view of human nature which acquired a momentum of its own, independent of evolution theory. The logic of eugenics rested on the ideology of hereditary or genetic determinism. No amount of education or improved conditions could modify the characters imposed by the genes. In the great debate over whether nature or nurture determined a person's abilities, eugenics was firmly on the side of nature (Pastore 1949). Once this point was accepted, the policy of selective breeding could be supported without reference to evolution theory. It did not matter where the bad genes came from, if they existed, they should be prevented from reproducing. Many of the early geneticists were not Darwinians, and much of the modern controversy over the role of the genes in determining human characters takes place without reference to evolution.

Whatever these later developments, the link between eugenics and Darwinism was initially very strong. Darwin's cousin, Francis Galton, first proposed methods of testing the strength of heredity in predetermining character and called for policies to manipulate the race's biological constitution. He coined the term *eugenics* in 1883. Although he himself favored the theory of evolution by saltation, he accepted that selection was the best way of maximizing the fitness of each population, while his disciple, Karl Pearson, founded the biometrical school of Darwinism. Ronald Aylmer Fisher, one of the founders of the modern genetical theory of natural selection, was also a convinced eugenicist. With the emergence of genetics, however, the logic of hereditary determinism took on a life of its own, supported by various social groups with an interest in arguing that expensive reforms benefiting the poor were a waste of money (for general surveys, see G. Allen 1975b, 1976; Bajema 1977; Blacker 1952; Bowler 1989b; Farrall 1979; Kevles 1985; Roll-Hansen 1988). Some have argued that eugenics was essentially a movement of the professional middle classes, who were convinced of their own high qualities and worried that state-sponsored reform programs would require high taxation. It was popular among those who wanted a well-managed society with themselves as the managers (Semmel 1960). For this reason, eugenics was occasionally taken up by left-wing thinkers, although it is normally associated with the political right (Paul 1984). Even Pearson called himself a socialist (1894), although by this he meant an enthusiast for a strong central government.

Galton began to advocate the importance of heredity in human character in the 1860s as a means of gaining recognition among the Darwinian circle (Waller 2001a). He had made an expedition to Africa during which he had become convinced of the white race's superiority over the black (Fancher 1983; on Galton's life and work, see Buss 1976; Cowan 1977; Fancher 1983; Forrest 1974; Gilham 2001; Pearson 1914–30). He now looked for a similar form of biological determinism within the white population, and used careful pedigree studies to argue that high levels of intelligence ("genius") tended to run in families. His *Hereditary Genius* of 1869 (reprint 1892) developed his case and explored the implications of this hereditarian viewpoint for the future of humanity. Hereditarianism was the ideal foundation for Galton's pioneering efforts to apply a statistical method to biological problems (Waller 2002). Galton was already convinced that the poorest individuals were breeding faster than those with high levels of intelligence, the latter forming mostly the professional and commercial elites. The drive to halt the resulting decline in the quality of the race became a moral crusade for Galton. He wanted a "positive eugenics," in which the most able individuals would be encouraged to have more children, although later he came to place much more emphasis on "negative eugenics," in which the state would compel the unfit poor to breed less.

In some respects Galton's fears about the transmission of harmful characters only echoed a common Victorian theme about the possibility of a "hereditary taint" such as insanity in a family bloodline (Waller 2001b). But his assumption that no amount of improved conditions or education could raise the intellectual standard of someone with bad heredity flew in the face of the liberal ideology of self-help. His warnings were ignored at first, and Galton's subsequent efforts to develop statistical techniques in biology were intended to back up his position (Cowan 1972b). Eventually he was joined by Pearson, who refined Galton's statistical methods and used them to found the biometrical school of Darwinism. Pearson pointed to the poor quality of the recruits coming into the British army for the Boer War to illustrate the decline of the nation's biological fitness and the dangers this posed for the empire. In the early years of the new century, the eugenics movement finally expanded to become a serious lobby group seeking to change government policy. The Mental Deficiency Act of 1913 was eventually passed by Parliament, ensuring in theory that those diagnosed as having low intelligence would be institutionalized and prevented from bearing children. Galton was the figurehead for the movement, although he had played a negligible role in building up popular support—serious political action was never his strong point. He did, however, found a National Eugenics Laboratory in 1904 and soon afterwards headed the Eugenics Education Society (on British eugenics, see Barker 1989; G. Jones 1986; Mackenzie 1976, 1982; Mazumdar 1992; Searle 1976, 1979).

In America too, eugenics flourished in the early twentieth century (M. Haller 1963; Ludmerer 1972; Pickens 1968). The American Breeders Association, a Mendelian group, set up a Eugenics Committee in 1906, and in 1910 the Eugenics Records Office was founded (G. Allen 1986). Efforts were made to trace the alleged hereditary taints of insanity, feebleminded-ness, and immoral behavior through generations of poor families. Thanks to constant lobbying, a number of states set up programs requiring the com-pulsory sterilization of the mentally handicapped. One side effect of the movement was the support it provided for the development of intelligence tests that would simplify the identification of the feebleminded (Evans and Waites 1981; Gould 1981).

Eugenics became popular in many other countries (Adams 1990). The first International Eugenics Congress was held in 1912. A well-established eugenics program existed in Germany long before the Nazis came to power in 1933 (Weindling 1989b; Weingart 1989; Weiss 1986, 1988), although the Nazis certainly applied far more extreme methods to sterilize and ulti-mately to liquidate the "unfit" (Harmsen 1955).

The influence of eugenics on the biology of the time has been widely de-bated by historians. Donald Mackenzie (1982) argues that the statistical methods developed by Galton, Pearson, and Fisher were shaped by their de-sire to provide support for hereditarian social policies. This would imply that biometrical Darwinism and population genetics both were contaminated deeply by ideology. Eileen Magnello (1999) has shown, however, that Pearson's research on human heredity used techniques different from his biometrical Darwinism and was kept institutionally separate. Fisher is a more complex case. He was certainly a convinced eugenist: his wife eventu-ally left him because he insisted on having a large family to perpetuate his own superior genes. Yet his own work demonstrated that eugenic policies would be ineffective except on an enormous timescale, even if the basic hereditarian principle were valid (Bennett 1983; Depew and Weber 1995; Norton 1983). It must also be remembered that Pearson rejected Mendel's laws as the basis for a reformulation of Darwinism, while many early ge-neticists in turn rejected the selection theory. In America, genetics rather than Darwinism provided the central biological support for eugenics, with early enthusiasts such as C. B. Davenport insisting that there was a single gene for each identifiable character, including feeblemindedness. The Mendelian eugenists claimed it would be easy to identify those carrying the harmful genes and prevent them breeding, thus purifying the race within a few generations. It was soon pointed out that if the relevant genes were re-cessive (i.e., they did not manifest themselves when combined with a "nor-mal" gene), it would be much more difficult to identify the carriers. This was why Fisher's work showed that selection—while effective on an evolu-tionary timescale—was too slow to be an efficient social policy.

The enthusiasm of some evolutionists and geneticists for eugenics cannot be doubted, and although scientific support began to wane as the difficulties became apparent, many biologists refused to speak openly against it (Gould 1974a; Ludmerer 1972; Provine 1973). But the suggestion that theoretical developments were distorted by this ideology is less easy to sustain, given the very different theories of heredity proposed at the time. It was even possible for a Lamarckian to be a eugenist, as demonstrated by E. W. MacBride, who wanted to see the Irish component of the British population reduced (Bowler 1984).

MacBride's shifting of the focus from social class to race was not typical of the British eugenics movement. In America, however, race became a central theme of eugenic concern. Given the large number of freed black slaves and the ever-increasing flood of immigrants from eastern Europe and Asia, white Americans began to fear that the biological quality of their race would be contaminated by fast-breeding but inferior types. A host of writers harped on this theme and called for the influx of immigrants to be stopped (Burr 1922; Fairchild 1926; Grant 1918; Ross 1927). Their campaign was crowned by the passing of an Immigration Restriction Act in 1924. American eugenics thus concentrated on both the purification of the white race and the effort to prevent it from being contaminated by blending with genetically inferior types.

By the late 1930s, all strands of biological determinism were coming under suspicion as the excesses of the previous decades came to a head in Nazi Germany. Biologists now recognized that few characters are controlled by single genes, while the difficulties created by the existence of recessive genes made the hope of purifying the human race seem an illusion in practical terms. More important, even those biologists who accepted the importance of genetics now conceded that environment was crucial too. We cannot fairly assess the genetic difference between two individuals if one has been raised in middle-class comfort and the other has been raised in a slum. If R. A. Fisher remained an enthusiast for eugenics, the other British founder of the genetical theory of natural selection, J. B. S. Haldane, became a Marxist and campaigned against it on the grounds that all people had to be given a decent standard of living before one could even begin to think of assessing their genetic potential (Haldane 1938; see also Werskey 1978). Long before this, social scientists and psychologists had turned against the hereditarian viewpoint. For them, the human ability to learn and acquire a culture was far more significant in determining how people behave (Cravens 1978).

The subsequent history of the debates over genetic determinism lies outside the scope of this book, since much of the enthusiasm for a revival of the hereditarian position has been driven by hopes raised by the improvements in genetics. The opponents of hereditarianism have promoted many programs for social improvement, and these have often failed to produce the expected benefits—but is this because those who were offered help were genetically incapable of benefiting from it, or because the reforms were superficial (Gould 1974a)? More recently, the expectations raised by the Human Genome Project have encouraged people to think once again in terms of single genes determining single characters. The use of genetic counseling to discourage those with hereditary defects from breeding has raised the specter of a new eugenics based not on state control but on social pressure to conform (Paul 1998).

Biology and Gender

A parallel form of determinism focused on the question of gender; indeed, some of the arguments used to define women's place in society were remarkably similar to those use by the race theorists. Late-nineteenth- and early-twentieth-century biologists and social scientists, almost all of them men, were anxious to preserve a social order in which women occupied an inferior position. Invoking a biological foundation for the alleged inferiority of female intelligence, or for the assumption that women's role as mothers made them unfit for an active life outside the home, was a natural tactic for these men to use. Darwin's theory of sexual selection seemed to reflect Victorian stereotypes about what was "natural" behavior for males and females and thus formed another source of prejudice. It has even been suggested that the theory of natural selection itself reflects an essentially masculine view of nature, because of its emphasis on the role of struggle (Easlea 1981). When carried to this extreme, the analysis betrays a flawed understanding of history because it ignores the fact that Lamarckism also emphasized the stimulating effect of struggle. But feminist writers have become increasingly sophisticated at detecting ways in which male prejudice seems to have influenced biological thinking, and there can be little doubt that most scientists in the post-Darwinian decades were certain that women were fated to play a subordinate role (Alaya 1977; Conway 1973; Duffin 1978; Haraway 1990; Russett 1989).

To some extent, evolutionism merely reinforced value judgments attributed to other scientific foundations. T. H. Huxley was not the only anatomist to insist that women's brains were smaller or less convoluted than men's, which was considered a sure sign of intellectual inferiority. Although professing himself a liberal, he campaigned actively to keep women out of the scientific and medical professions (E. Richards 1989b). But his distrust was based on other grounds too: like many of his contemporaries, Huxley believed that women were temperamentally unsuited to the rigors of intellectual, professional, and political life. This sense that the female temperament was predetermined by biology found many opportunities to express itself, not all within the context of Darwinism. Herbert Spencer maintained that the female sex had to devote most of its vital energies to reproduction, leaving less for intellectual development (significantly, he conserved his own reproductive energies and did not marry). Like other Lamarckians, he believed that one consequence of this would be the adaptation of the female moral sense to cherish family values rather than the sterner virtues needed to face the world outside the home. The sociologist Patrick Geddes and the biologist J. Arthur Thomson wrote their *Evolution of Sex* (1889) to argue that this fundamental difference of temperament was expressed at the cellular level and dated back to the very origins of sexual reproduction. On this model, if women insisted on trying to get an education or a career, they would undermine their femininity and lose the capacity to reproduce.

Darwin's theory of sexual selection seems to reflect the tendency for male assumptions to become embedded in scientific thinking (Bender 1996; E. Richards 1983). Darwin assumed that males were always self-assertive through either combat or display, while females were passive and coy, waiting to see which male offered the better prospects. It would not be surprising, then, for psychological evolution to have entrenched these instinctive behavior patterns in the two sexes. It would be easy to assume that Darwinism is thus nothing more than an expression of Victorian values masquerading as science, here as in the case of free-enterprise capitalism. That the situation is not quite so simple is evident from the fact that, during the eclipse of Darwinism, sexual selection was even less popular among biologists than natural selection was. The theory may have been a perfect articulation of contemporary male prejudices, but that did not ensure it would be taken up as science. Indeed, it was another hundred years before sexual selection became a key part of modern evolutionism, with the emergence of sociobiology—and that was after the feminist movement had emerged to alert everyone to the issues involved. Feminists also draw attention to the extent to which paleoanthropologists' theories of human origins focused on the role of hunting as the key to our ancestor's success in transforming themselves from apes. Since hunting is assumed to be a male activity, the woman's role as the gatherer and child minder is thus relegated to a subordinate role in the process which made us human (Haraway 1990; see also Cartmill 1993).

The emergence of genetics required a reassessment of the factors which predetermined the female character, because both sexes were required to transmit the same basic genes (although the recognition that sex is determined by the chromosomes leaves plenty of room for speculation about inbuilt behavioral differences). The eugenics movement concentrated more on persuading middle-class women that to fulfill their duty to the race they had to give up the hope of a career in order to produce children. Modern feminism has attacked such stereotypic images of women's place in society, but the possibility that there might be some biological differences between the male and the female brain, and hence between the sexes' mental faculties, continues to stir debate.

NEO-LAMARCKISM AND SOCIETY

The psychologists and social scientists who opposed the hereditarian position revived the old liberal tradition in which all people were supposed to share a common human nature. Nineteenth-century liberals such as John Stuart Mill had long argued that women and nonwhite races should not be discriminated against on the grounds of any alleged biological inferiority. In the debate over the relative powers of nature and nurture, liberals (and social scientists) were on the side of nurture, because this held out the hope of producing a better society through improved conditions, especially education. But one biological theory blurred the distinction between nature and nurture. The Lamarckian theory of the inheritance of acquired characters implied that a learned habit might eventually become an inherited instinct. Thus improvements made by better education and social conditioning might eventually become part of the race's biological inheritance. Here was the potential for a social evolutionism which did not depend on the Darwinian struggle for existence.

Because Lamarckism is seen as the natural alternative to Darwinism, there has been a tendency for later liberals to regard it as a morally preferable theory (e.g., Easlea 1981; Koestler 1971). But apart from the fact that Lamarckism has turned out to be scientifically unsound, this position ignores the many cases from the historical record where Lamarckians enthusiastically promoted racism, sexism, and even an ideology of struggle. There *is* a reformist interpretation of Lamarckism, and it played an important role by providing an alternative to social Darwinism. But it was by no means the only social interpretation of the theory, and in the rush to identify Darwinism as the principal source of harsh social values, we all too easily overlook the extent to which the rival theory could be adapted to similar political ends. We have noted the involvement of Lamarckians in the establishment of race science. Herbert Spencer's social Darwinism also had a strong Lamarckian component based on the assumption that struggle was the best spur to self-improvement—Spencer thought that any attempt by the state to impose reforms would only interfere with this process. The rival view of Lamarckism's human implications extended the traditional reformers' belief in the state's ability to generate improved behavior patterns in its citizens via the educational system. Lamarckism held out the hope that the improved behavior would not serve only this generation—eventually it would become an inherited instinct, and human nature itself would have been changed for the better.

The boundaries between the Darwinian and Lamarckian positions were not always clear. In Russia, the prevailing view of Darwinism paralleled that favored by Western socialists: the main struggle for existence was that of the species as a whole against the environment, and it promoted cooperation, not competition, among individuals (Todes 1989). In the 1890s the émigré Russian prince Peter Kropotkin published a series of articles later collected as his *Mutual Aid* (1902). He claimed to have observed animals cooperating with others of their species to survive in the harsh Russian winters. Evolution's main driving force was the development of the cooperative instincts, exactly the reverse of the Darwinian prediction. Kropotkin was an anarchist who believed that the human race eventually would evolve cooperative instincts, making government unnecessary. He later wrote explicitly in favor of Lamarckism, recognizing that, without the inheritance of learned habits, group selection would be required to explain how the cooperative instincts were formed.

In America, too, there was considerable enthusiasm for the Lamarckian view of human nature (Stocking 1962). F. J. Turner's "frontier hypothesis" presented the West as a stimulating environment which worked directly on the constitution of immigrants to produce a superior form of humanity (Coleman 1966). Some members of the American neo-Lamarckian school in biology stressed similar optimistic prospects, especially Joseph Le Conte (1899; see also Stephens 1982). The psychologist G. Stanley Hall saw the development of the child's mind as a recapitulation of mental evolution but also stressed the role of education in shaping further developments (1904; see also Gould 1977b: chap. 5). Hofstadter (1959) notes the writings of Lester Frank Ward as an influential source of opposition to the Darwinian view of society (see also Scott 1976). Ward insisted that our efforts to promote human progress would be wasted if Lamarckism was not valid. It was not enough to improve only the next generation; the whole nature of humanity would have to be transformed, and that could happen only if the improvements society encouraged in people's behavior became inherited as instincts.

In the early decades of the twentieth century, the Lamarckian theory was gradually eliminated from biology by the rise of Mendelian genetics, although the playwright George Bernard Shaw continued to defend it under the name "creative evolution." Shaw's moral indignation against the selection theory was both profound and effective—in the preface to his *Back to Methuselah* (1921: liv), he declared, "If it could be proved that the whole universe had been produced by such selection, only fools and rascals could bear to live." Shaw's vision of the life force struggling to dominate its material environment reflected a brief flurry of interest in nonmechanistic biology at the turn of the century, but it was out of date in terms of the science of the 1920s. One of the theory's last great supporters in science, the Austrian biologist Paul Kammerer, also cited its human implications when trying to defend his work (1924), and he generated newspaper headlines about the breeding of a race of super beings. Arthur Koestler subsequently tried to revive interest in Kammerer's work in his *The Case of the Midwife Toad* (1971). Unfortunately, he praised Kammerer's leading British defender, E. W. MacBride, as "the Irishman with a heart of gold," oblivious to MacBride's explicitly anti-Irish racism (he was an Ulster Protestant; see Bowler 1984). Such slips illustrate how easily the image of social Darwinism can blind us to the harsher implications that can be derived from other evolutionary theories. No one now cites Lamarckism in the hope of improving human nature, and the reformers' plans are limited to changing culture through habits which would have to be relearned in every generation. This does not stop them from invoking the harsh image of social Darwinism in their efforts to brand all forms of biological determinism as morally suspect. It may be worth remembering that one of the most pessimistic predictions about the future of humanity produced in the last century, Aldous Huxley's *Brave New World* (1932, reprint 1955), foresaw social conditioning through learning and environmental manipulation as the means by which our masters might enslave us—all the time claiming that it was for our own good.

EVOLUTION AND PHILOSOPHY

Evolutionary ideas were absorbed into late-nineteenth-century thought in a variety of ways. The worldview of the period was shot through with images of progress and often linked to the idea that some form of struggle or effort was needed to advance toward higher things. Spencer's was by no means the only philosophy to take this idea as its foundation. But increasingly, people recognized that the trend toward progress might not be absolutely predetermined; evolution could advance in many possible ways, and our actions might decide which of the possibilities would be realized. Many liberal religious thinkers became comfortable with this view of things, content to assume that the human race had retained its dignity as a key step forward in achieving the divine purpose.

The scientific naturalism advocated by Huxley and John Tyndall remained influential into the later decades of the century. Spencer, too, still had a wide audience for his evolutionary philosophy, especially in America. But scientific naturalism had never been popular among academic philosophers; and in Britain, the last decades of the century saw the emergence of an influential idealist movement drawing inspiration from German thought (Copleston 1966). At a more popular level, there was increasing confidence among those who sought to retain a role for mind and purpose in the world—although evolutionism itself remained unchallenged. Idealism was, in any case, well adapted to a developmental view of things, since it saw everything as an expression of mental power. The conservative politician and philosopher Arthur Balfour attacked Huxley's position in 1895, defending an intuitive sense that there was a divine purpose in nature. Huxley died while still composing his response (Lightman 1997). Many other thinkers now were expressing their dissatisfaction with scientific naturalism, often in a form compatible with some type of religious belief (Turner 1974; on the complexity of Victorian science's impact, see Lightman, ed. 1997).

Tensions had emerged even within the scientific naturalists' camp. Huxley at first had been confident in the power of science to dominate nature and had enshrined his lack of religious faith by coining the term *agnosticism* (Lightman 1987). In the later years of his life Huxley himself became disillusioned with the optimistic progressionism of Spencer's philosophy. He saw little evidence that natural evolution was progressive, and became suspicious of Spencer's efforts to found an evolutionary ethics on the basis that whatever succeeded must be defined as good. To be fair to Spencer, he had never held that progress justified mere ruthlessness: his philosophy was meant to show that the virtues of thrift, industry, and initiative triumph in the end. But he had little sympathy for those who could not make the grade, and became increasingly worried that these failures no longer were being eliminated by natural processes. In this sense, Spencer became more of a social Darwinist, while Huxley seems to have become more aware of the unpleasant implications of a truly Darwinian worldview. Huxley's campaign culminated in his Romanes memorial lecture, "Evolution and Ethics," of 1893 (Huxley 1894; see also Desmond 1997; Helfand 1977; Paradis 1978). Here, he depicted nature as being without purpose or pity, a scene of unrelenting struggle with no apparent tendency for the "fittest" (in any moral sense of the term) to succeed. All efforts to impose order and purpose on this ceaseless activity were illusions created out of vain hope and anthropomorphism. But if evolution was not progressive, why should we accept its harsh values as a guide to our lives? Huxley now presented moral values as something developed in defiance of nature's laws: the human race had somehow transcended the system which gave it birth.

At the same time, Huxley saw no prospect that we could impose our will on the universe to give it a moral purpose. By a cosmic accident, we had been given the power to recognize the meaningless character of the world, and it was precisely that capacity which made us human. But in the end, the world would reclaim its own and civilization would perish. We struggle to maintain our values in a hostile world, not in the hope of ensuring progress but because to do so makes us human. In the twentieth century, such a sense of cosmic pessimism would engender a sense of existential insecurity verging on moral paralysis. But Huxley felt that we had to fight to improve the lot of our fellow humans, hoping to stave off at least for a while the encroachments of a blind and mechanical nature. By opting for unrestrained individualism, Spencer was giving in to nature; Huxley now wanted to take on the world in defense of something like traditional moral values.

This cosmic pessimism, so different from the confident progressionism of the earlier Victorian era, now began to gain a hold in the intellectual world. The early decades of the twentieth century saw thinkers from diverse backgrounds turning their backs on the hope of gaining certain knowledge. Science, the arts, and philosophy all seemed to reflect a concern that the human race had to make its own way in a basically incomprehensible world. This was the movement known as modernism (Everdell 1997). The mood of pessimism it engendered was fueled by the sense of cultural degeneration which swept through Europe at the turn of the century (Chamberlin and Gilman 1985). We have seen how this was reflected in science in the degenerationism of E. Ray Lankester and H. G. Wells (who also studied under Huxley). Huxley's scientific naturalism was repudiated by the new generation of analytical philosophers, but his cosmic pessimism seemed to resonate with the mood of the time. In his essay "A Free Man's Worship" of 1903, Bertrand Russell summed up the image of humanity's place in the world thus:

That Man is the product of causes which had no prevision of the end they were achieving; that his origin, his growth, his hopes and his fears, his loves and his beliefs, are but the outcome of accidental collocations of atoms; that no fire, no heroism, no intensity of thought and feeling, can preserve individual life beyond the grave, that all the labours of the ages, all the devotion, all the inspiration, all the noonday brightness of human genius, are destined to extinction in the vast death of the solar system, and that the whole temple of Man's achievement must inevitably be buried beneath the debris of the universe in ruins—all these things, if not quite beyond dispute, are yet so nearly certain that no philosophy which rejects them can hope to stand. (1961: 67)

This was exactly what the previous generation had feared would be the consequence of Darwinism's triumph, although many other factors had conspired to drive the point home.

Yet as Russell himself admitted, not everyone shared the intellectual elite's sense of the purposelessness of nature, and some even welcomed a world in which an element of uncertainty seemed to unlock the straightjacket of deterministic materialism. Perhaps the idea of progress could be retained in a less structured form, with the human race as only one possible outcome of the upward strivings of nature. The developmental version of progressionism had imposed a new form of determinism—the goal of evolution was assumed to be inevitable. But the real logic of Darwin's open-ended branching model of evolution was that no such single goal could be identified. In America, John Dewey (1910) argued that Darwinism undermined the hierarchical view of nature and showed us that we have the freedom to shape our own destiny. The concept of freedom was also important to pragmatists such as Charles Peirce and William James (Wiener 1949). They, too, saw that the lesson of Darwinism was its destruction of determinism. Nature was inherently creative, and the lack of constraints on evolution guaranteed the freedom of the individual will. Peirce saw evolution as the growth of "cosmic reasonableness," retaining the idea of progress in a less structured form.

This position was now widely accepted, although paradoxically it was often seen as a form of opposition to Darwinism, so strongly was the selection theory identified with materialism. The clearest illustration of this is the wave of enthusiasm for the French philosopher Henri Bergson's *Creative Evolution* (translation 1911; Gallagher 1970; Grogin 1988). Bergson insisted that there was no harmonious plan of creation, nor any sign of intelligent design in the structure of each species. The history of life was progressive but in an irregular way. This could be explained if we postulated a creative life force, the élan vital, struggling against the limitations of matter. Evolution strove to progress but was fragmented into a host of separate branches by the need to cope with the material world. Intelligence was one facet of the life force, which had become intensified in the branch leading toward humanity. Our consciousness thus symbolized the creative heart of nature. Bergson's philosophy was welcomed by liberal religious thinkers looking for a way to accommodate the idea of evolution, but it also fascinated a number of scientists, including Julian Huxley.

Similar implications were seen in the "philosophy of organism" proposed by Alfred North Whitehead. Originally an analytical philosopher who worked with Russell on the foundations of mathematics, Whitehead moved toward a vision of cosmic teleology eventually summed up in his *Process and Reality* (1929; see also Emmet 1932). He maintained that the world should be seen not as a collection of discrete objects but as a complex of ongoing processes in which nothing was isolated from the whole. Atoms themselves were quasi-organic entities capable of interacting with their surroundings. Unlike Bergson, Whitehead believed that the processes of nature were meaningful, harmonious, and orderly, with humankind being their highest product. Life and mind were not in conflict with matter but were essential components of a universe in which nothing was completely inorganic or lacking in awareness. It was still possible to recognize a kind of Platonic order in the way processes unfolded, evidence of the God who stood as the ideal toward which the whole universe aspired. There might be no single plan of creation, but the world created its own order in each epoch of history.

The psychologist Conwy Lloyd Morgan proposed his philosophy of "emergent evolution" (1923) to avoid having to postulate that life and mind are present even in so-called inert matter. According to emergent evolution, mental properties began to manifest themselves as a new level of reality only once evolution had reached a certain stage of complexity. Life, mind, and spirit "emerged" at key points in the development of nature and, once formed, began to play an active role in directing further progress. The theme was taken up by Samuel Alexander (1920) and Roy Wood Sellars (1922), Alexander implying that God would be the final emergent reality. The philosophy of emergence has continued to play a role for philosophers and biologists who stress the ability of complex interactive systems to display new properties (Blitz 1992). But for Morgan himself, and for many of his readers, it was yet another way of trying to read a spiritual purpose into evolution. Mind was not the underlying driving force of nature, but it was a level of reality designed to emerge as soon as evolution reached a certain level of complexity. Once again, progress was retained, but in an unstructured way that implied the possibility of nonhuman forms of mentality.

EVOLUTION AND RELIGION

By the later decades of the nineteenth century, most liberal religious thinkers had accepted the idea that evolution was the unfolding of the divine plan of creation. But tensions remained as liberal theologians pushed for further reassessment of traditional doctrines in order to accommodate the implications of science. Conservatives became increasingly afraid that the basic foundations of Christianity were being undermined rather than modernized. The problem ultimately centered on the doctrines of sin and atonement. Every effort to adapt the faith to evolutionism seemed to center on the assumption that the development of life was progressive and purposeful, thereby displaying God's intelligence and purpose. The human species retained a key role as the final product of organic evolution and the agent by which God's further purpose would be achieved by conscious control of nature. But there was no room in this scheme for the traditional belief that humans were fallen creatures alienated from God and in need of salvation. In the early decades of the twentieth century, these misgivings were articulated in a number of ways. Historians' attention has been mesmerized by the outburst of fundamentalist opposition to evolutionism in America. But this was by no means the only expression of Christian discomfort with evolutionism, nor should that discomfort be allowed to distract us from recognizing the efforts being made to transform the faith in a way that would allow an evolutionary natural theology to emerge (Moore 1979; Numbers and Stenhouse 1999). Other faiths also had to engage with the theory, and Islamic thinkers in particular sought means of confronting Darwinism along with other potentially disturbing Western influences (Ziadat 1986).

In Britain, the Anglican Church endorsed the idea of teleological evolutionism; Charles Gore's edited volume of 1889, *Lux Mundi*, was particularly influential (Elder 1996). Liberal evangelicals in the Free Churches were able to make a similar accommodation (Livingstone 1987; Livingstone, Hart, and Knoll 1999). Henry Drummond's widely read *Ascent of Man* (1894) drew an explicitly religious message from the argument that evolution promoted the development of altruism, the willingness to sacrifice one's own interests for those of others (Moore 1985b). Thus, God could be seen as the Creator of a process designed to promote moral values, even if the early stages relied on harsher instincts which ultimately had to be transcended. By the early years of the new century, a concerted effort had emerged to create a new natural theology based on evolutionism (Durant 1985).

In the Anglican Church, the movement known as Modernism promoted this effort to forge a reconciliation with science. This was not, of course, the modernism of the artistic and philosophical avant-garde—it was based firmly on the progressionist viewpoint established in previous decades, and increasingly it depended on the older generation of biologists for its scientific credibility (Bowler 2001). Conwy Lloyd Morgan's emergent evolutionism was typical of the kind of scientific writing which appealed to liberal religious thinkers. Some if its supporters, including the Anglican Modernist Charles Raven, still openly supported non-Darwinian theories such as Lamarckism (Raven's 1942 biography of the seventeenth-century naturalist John Ray was a product of his campaign to restore natural theology). Others, such as Bishop E. W. Barnes, were aware that biology was moving steadily against Lamarckism, but still were anxious to preserve the progressionist viewpoint. In the 1920s Barnes was notorious for preaching "gorilla sermons" in which he insisted that the evolutionary viewpoint required a rejection of the old notion of original sin (Bowler 1998). As a mathematics don at Cambridge, Barnes had taught R. A. Fisher, and he could see how Fisher was creating a new Darwinism based on the genetical theory of natural selection. Fisher himself was an Anglican, but as yet few were willing to follow his claim that natural selection was the kind of creative force a Christian could endorse.

The problem with the Modernist position was that it presented humankind as the agent of progressive evolution, ignoring the traditional Christian belief that we are fallen creatures. Original sin was no more than the awakening of the moral sense in our apelike ancestors. The Modernist position was, in fact, very close to that presented by some explicitly non-Christian thinkers such as George Bernard Shaw and Julian Huxley (J. Greene 1990). Conservative Christians, both evangelical and Catholic, were concerned that the liberals, in their efforts to create a religion that would be credible to evolutionists, had abandoned the fundamental teachings of their religion. The conservatives realized that evolution was widely perceived as a component of the rationalist campaign against organized religion. It was thus necessary to challenge evolutionism, at least as an account of the origin of the human soul. In Britain, popular Roman Catholic writers such as Hilaire Belloc and G. K. Chesterton were the most effective opponents of evolutionism, Belloc being particularly active in challenging rationalists such as H. G. Wells (Bowler 2001). The Catholic Church had hardened its attitude against evolutionism at the turn of the century, thanks in part to the influence of a conservative group of Jesuits in Rome (Brundell 2001). This position soon softened on the question of the evolution of the physical body, as writers such as Henri de Dorlodot (1925) and Ernest Messenger (1931) pointed out—as Mivart had noted earlier—that the church fathers did not interpret Genesis literally on this issue. The Church has, however, remained opposed to the idea that the soul could have emerged from an animal mentality.

In late-nineteenth-century America, there were also many liberal Protestants prepared to endorse evolutionism as the unfolding of God's plan (Livingstone 1987; Moore 1979, 1985a; J. Roberts 1988; for a collection of primary sources, see Ryan 2002). Spencer's influential philosophy showed that evolution promoted the traditional values of the Protestant work ethic. This tradition was presumably carried into the twentieth century, although historians have been remarkably slow to explore its influence. Instead, all attention turned to the form of Protestant fundamentalism which emerged as the flagship for opposition to evolutionism. The most visible symbol of this campaign was the "monkey trial" of John Thomas Scopes in Dayton, Tennessee, in 1925. In response to mounting concerns that evolutionism was undermining traditional Christianity, Tennessee had passed the Butler Act forbidding the teaching of the subject in its public schools. Scopes deliberately violated the act and was prosecuted in a high-profile trial. The fundamentalist politician William Jennings Bryan led the prosecution, and the agnostic lawyer Clarence Darrow the defense. The trial has gone down in history as a watershed in the modern relationship between science and religion, resulting in a plethora of books (e.g., De Camp 1968; Ginger 1958; Scopes 1967; Settle 1972) and a play (subsequently twice filmed), *Inherit the Wind.* Yet modern historians have challenged most of the myths surrounding this celebrated trial. Not all fundamentalists were opponents of evolution, and the conservative theological position concealed a wide range of differing positions (E. Larson 1998; Livingstone 1987; Numbers 1992, 1998; for primary sources, see Numbers 1994–95). Not all southern states followed Tennessee in banning the teaching of evolution, nor was the Scopes trial the high point of the movement's influence. And although the popular myth has the opponents triumphing over the discredited fundamentalists, the latter's campaign was remarkable effective in keeping evolutionism out of high school textbooks over the next several decades (Grabiner and Miller 1974). A later, and still active, wave of opposition to evolutionism emerged in reaction to a renewed onslaught by the evolutionists in the 1950s and 1960s, following the consolidation of the modern Darwinian synthesis.

9   The Evolutionary Synthesis

Scientists had received Darwin's theory of natural selection with many reservations, and in the early years of the twentieth century, the level of hostility increased. Teleological theories such as Lamarckism were still accepted in some traditional areas of biology, but now the new experimentally based study of heredity had created a model of variation by mutation which also made selection seem superfluous. Even the basic idea that evolution was shaped by the demands of adaptation came under fire from geneticists convinced that mutated forms would breed even if they offered no additional fitness. But by the 1920s, the situation had begun to change in a direction few would have anticipated. Genetics steadily undermined the plausibility of Lamarckism and orthogenesis by showing that there was no process by which acquired characters could be incorporated into the fixed units of heredity. It also showed that mutations were not directed along any predetermined path. Variation was effectively random, as Darwin had supposed, and studies of the genetics of whole populations increasingly suggested that the extreme saltationism advocated in the first flush of enthusiasm for the new science was untenable. The variability of large populations was not controlled by small numbers of character differences, because in most natural (as opposed to artificially bred) populations, there are many small genetic differences affecting the same character. Mutations simply provide more random variation, not new species created by a single character difference. Even the geneticists began to concede that selection by the environment might affect the degree to which a particular gene could spread within a population. Meanwhile, field naturalists accepted that the Lamarckian effects they once had favored could be explained by a genetical model of natural selection. From these developments came a revived form of Darwinism which became known as the evolutionary, or modern, synthesis.

No one doubts the significance of the evolutionary synthesis, but there has been much debate over its origins and its true nature. At one time it was assumed that the *synthesis* referred to the integration of the Darwinian selection theory and Mendelian genetics after the period of initial hostility. William B. Provine's *Origins of Theoretical Population Genetics* (1971) encapsulated this view. But one of the founders of the synthesis, Ernst Mayr (1959c), had long argued that the true story was more complex. While acknowledging the conceptual initiative of the genetical theory of natural selection, he insisted that field naturalists, too, had played a role by focusing on traditional Darwinian concerns such as geographical isolation, ignored by some of the founders of population genetics. For Mayr, the synthesis was more a unification of the various branches of biology which had ceased to communicate with one another during the eclipse of Darwinism. Provine (1978) defended his position but, subsequently, collaborated with Mayr to edit a more comprehensive overview of what had happened (Mayr and Provine 1980). Another survey (Grene 1983) included an article by Stephen Jay Gould (1983) pointing out that the earliest version of the synthesis allowed room for some non-Darwinian effects which subsequently were ruled out as the Darwinian element "hardened." Others have been more critical of the synthesis's achievements (e.g., Eldredge 1985), while Provine (1988) stressed the variety of theoretical perspectives adopted by different research traditions claiming to operate under the umbrella of the synthesis. On this model, there was no genuine integration of ideas, only a "constriction" ruling out the more extreme anti-Darwinian mechanisms debated at the start of the twentieth century. Precisely because of a growing awareness of the tensions involved, some recent historians have defined the synthesis more in terms of the professional structures created to establish evolutionary biology as a recognized field within the scientific community (Smocovitis 1996). This has created its own problems, however, by inflaming fears already expressed by biologists from outside the English-speaking world who complain that the contributions of their national traditions have been neglected (Reif, Junker, and Hossfeld 2000).

Whatever the true nature of the synthesis, it gave evolutionary biologists new self-confidence in an age increasingly dominated by the demands of biomedical science. The scientists who sought to popularize the synthesis could draw on parallel developments in other fields to offer the hope of constructing a general evolutionary worldview. The question of the origin of life itself was at last opening up to scientific investigation. At the opposite end of the scale, the discovery of new hominid fossils backed up efforts to reinforce the Darwinian explanation of human origins, exploding in the process the racist assumptions of an earlier generation. Eventually it would be argued that even human behavior could be explained, at least in part, in terms of Darwinian sociobiology (see chap. 10). The early proponents of the synthesis adopted a less deterministic stance, however, and were still willing to see evolution as having developed new mental and moral powers in humans. But were their beliefs the philosophical spin-offs of genuine scientific advances, or was the philosophy already there, buried in the preconceptions of the scientists themselves? Michael Ruse (1996) has stressed the extent to which the founders of the synthesis still clung to the progressionist philosophy so popular in the previous century. He emphasizes, however, that the actual science has taken on a momentum of its own because only concepts which can be expressed without explicit reference to nonscientific values may be openly debated. Later opponents of Darwinism have no doubts that the revival of the selection theory had dire consequences for wider issues. Sociobiology has joined other expressions of genetic determinism on the list of ideologies to which socialists and liberal social scientists object.

POPULATION GENETICS

The initial hostility between the biometricians and the Mendelians was a product of the very different research programs they had initiated. Where Pearson and the biometricians looked at large populations exposed to the pressures of the natural environment, Mendelians such as Bateson had crossbred artificial varieties of plants and animals in the laboratory. The biometricians could not believe that the distinct characters traced by the Mendelians were typical of the normal range of variation in a species, while the Mendelians tended to dismiss the bell curve of normal variation as an environmental distortion of underlying genetic differences. By the 1920s, however, experimental studies of mutations by T. H. Morgan and his school had undermined the plausibility of De Vries's original mutation theory. Mutations were modifications of existing genetic characters, not all of which involved drastic changes, and it became clear that the mutated individuals bred with other members of their species, so that the mutation became, in effect, an addition to the previously existing range of genetic variation. Even Morgan began to concede that selection by the environment might play a role in determining which mutated characters would spread into the population.

At the same time, the founders of modern population genetics were beginning to explore the processes that would govern the genetic structure of a large, freely interbreeding population, including natural selection. Three men are usually credited as the founders of population genetics: R. A. Fisher and J. B. S. Haldane in Britain and Sewell Wright in America (Depew and Weber 1995: chaps. 10–12; Gayon 1998; Provine 1971; Sarker 1992). The key was statistics, the mathematician's ability to see how a vast array of minute changes, each too small to study in itself, could be summarized to give a measurable overall effect. But as Mayr (1959c) claimed, both Fisher and Haldane worked with what has been called a "beanbag" approach to genetics. Each gene was seen as a fixed unit, and selection worked by adding and subtracting units from the gene pool of the whole population. Wright, who approached the problem with the interests of an animal breeder, recognized that genes were not discrete units because they could interact with each other to create a greater degree of variability, especially within smaller subpopulations where inbreeding was common. This approach turned out to have greater relevance for the field naturalists' experience with populations divided by geographical barriers.

An important first step toward the reconciliation of genetics and Darwinism was taken by those geneticists who reformulated the concept of mutation. For De Vries, a mutation was a new character so distinct it could form a new subspecies or even species. Thomas Hunt Morgan and his group created the modern concept of mutation as the spontaneous transformation of a gene which produced a new character (G. Allen 1968, 1978; Bowler 1983: chap. 8, 1989b; Shine and Wrobel 1976). His work on the chromosomes of the fruit fly *Drosophila* had converted Morgan to Mendelism and allowed him to demonstrate the chromosomal mechanism by which the genes are transmitted (Morgan, Sturtevant, Muller, and Bridges 1915). Noting the appearance of apparently new characters in his flies, Morgan realized that these were true mutations, caused by the (then unknown) constitution of the gene being restructured so that it now coded for a new character. The changes seemed to occur at random, in the sense that a wide range of apparently useless and sometimes deleterious new characters was produced. Many of the changes were quite small, and the individuals carrying the new genes simply reproduced with other members of the population, thereby spreading the mutated character into the species. By 1916, Morgan—originally an opponent of the selection theory—had begun to recognize that external factors might affect the reproductive success of an organism carrying a mutated gene. Harmful genes would be eliminated or kept to a very low level within the population. Conversely, if a new gene were created which conferred adaptive benefits, the affected individuals would breed more rapidly and the proportion of the new gene in the population would increase. Thus Morgan came to accept a form of natural selection, although he would never admit that a large number of maladaptive variations were eliminated. New genes were evaluated more or less immediately after they were produced by mutation, and the harmful ones eliminated.

Morgan's very simple model of selection ignored an important point: if the mutated gene were recessive, it could not be eliminated completely, because many carriers would not exhibit the phenotypic character of the gene. The true genetical theory of natural selection arose from recognition that the gene pool of the species contains a wide fund of genetic variability, all originally created by mutation, but with many useless or even slightly harmful genes retained in the population at a very low level. Morgan's simpler view of variability was preserved in the "classical" version of population genetics defended by his student H. J. Muller (1949), in which variability was seen as very limited. Most organisms possessed the "wild type" of each gene, corresponding to a character with maximum adaptive fitness, with abnormal or harmful genes being rapidly eliminated by selection. The population genetics which became fundamental to the modern synthesis adopted the rival, or "balance," view of variation, which assumed that genetic variation was extensive, with many different forms of each gene circulating in the gene pool of the species. On this model, genetic variation made up the range of normally distributed variation for each character studied by the biometricians. In effect, the fund of variability built up by mutation and retained by sexual reproduction provided the "random" variation which Darwin and the neo-Darwinians had assumed to be the raw material of selection (Hodge 1989).

As early as 1902, G. Udney Yule had pointed out that Mendel's laws were not incompatible with the biometrical approach to the study of variation. The Mendelians had focused on simple examples of variation in which a character was determined by a single gene with only a small number of alleles (two in Mendel's original experiments). But many characters in wild populations are influenced by a number of different genes, each with its own alleles. Even if the genetic characters are themselves discrete units, the constant circulation of so many differences within a large population will generate a continuous range of variation for each measurable feature. This suggestion had important consequences: it eliminated Fleeming Jenkin's old "swamping" argument against natural selection—because individual genes cannot be diluted—while allowing overall variation to be seen as a continuous range whose extremes could be influenced by their adaptive fitness.

In 1909 the Swedish biologist H. Nilsson-Ehle undertook a series of breeding experiments with grain which confirmed Yule's insight. He showed that some characters must be accounted for in terms of three or four genetic differences, each segregating independently according to Mendel's laws. Calculating what would happen if there were ten such factors, he showed there would be nearly sixty thousand different phenotypes. Because the differences between them would be very slight, they would appear as a continuous range of variation. The same point was made in America by Edward East (1910). Over the next decade, more geneticists came to accept the multiple-factor explanation of continuous variation, and some, including Nilsson-Ehle, began to argue that selection might act on this wide range of genetic differences. In 1915 the geneticist R. C. Punnett published a study of mimicry in butterflies, which included a table prepared by the mathematician H. T. J. Norton showing how selection would allow a favorable gene to spread into a population. Punnett himself believed in discontinuous evolution and explained mimicry in terms of nonadaptive resemblances caused by a single gene. But Norton's table had the effect of undermining Punnett's own position, forcing the geneticists to reconsider the possibility of selection.

The way was now open for the creation of a genetical theory of natural selection based on an assumed wide range of genetic variation within wild populations. One of the leading figures in this movement was the statistician Ronald Aylmer Fisher, who was trained in mathematics at Cambridge and who became interested in Pearson's techniques (Bennett 1983; Box 1978; Depew and Weber 1995: chaps. 10–12; Hodge 1992; Mackenzie 1982; Norton 1975b, 1983). Fisher built on the new insights of the geneticists to show that their work now reinforced the biometrical model of Darwinism. The hostility between biometry and Mendelism was still so great that his first paper on the topic was rejected by the Royal Society of London and appeared instead in the transactions of the Royal Society of Edinburgh (Fisher 1918; Norton and Pearson 1976). Over the next decade, Fisher refined and extended his work, which culminated in the appearance of his *Genetical Theory of Natural Selection* (1930).

To create a workable mathematical model, Fisher assumed that selection acted uniformly on a large population in which the individuals interbred freely. His analogy was the kinetic theory of gases, where the effects of innumerable molecular motions and collisions could be interpreted as pressure and temperature. The difference was that, in the case of population genetics, natural selection could actually create new characters by accumulating the effects of otherwise random mutations. Fisher showed that if a particular gene conferred an advantage resulting in a faster breeding rate, its frequency in the population would be increased. Harmful genes would be held at a very small level, but could not be eliminated if recessive. Most genetic differences would be very small, so mutation merely enhanced the range of variation in the population. Although selection worked by changing the frequency of discrete genetic units, its visible effect was gradually to shift the continuous range of variation. Fisher demonstrated that selection could maintain a balance between two alleles when the heterozygote was fitter than either homozygote. He knew that most mutations were deleterious, but that their tendency to occur at a fixed rate balanced the effect of selection in reducing their frequency. In a small population, however, a rare gene could be eliminated by chance, and for this reason Fisher believed that evolution occurred most rapidly in a large population which maintained wide variability.

For Fisher, selection was a deterministic process taking place by the summing up of a vast number of individual events grinding away, slowly but surely, to increase the frequency of favored genes. Even so, he conceived of it as a truly creative process, a view which fitted his own liberal Christian beliefs because it allowed him to argue that selection was a divinely established mechanism of progress. Progress was certainly important to Fisher, as to most of the founders of the evolutionary synthesis (Ruse 1996). Inspired by Pearson, he was an ardent enthusiast of eugenics, which he saw as humanity's own contribution to fulfilling the divine purpose by improving the race's moral and intellectual character. He devoted a significant proportion of *The Genetical Theory of Natural Selection* to the human implications of the theory on the assumption that most human characters were rigidly determined by heredity and thus were subject to selection. Curiously, however, the implication of Fisher's own theory was that the effect of artificial selection via a eugenic policy would be so slow that its results would become significant only after many generations.

The other British founder of population genetics, J. B. S. Haldane, was a larger-than-life figure who contributed to many areas of science and followed a philosophy very different from Fisher's (Adams 2000; R. Clark 1969). From an early age, he favored a Darwinian view which negated any idea of divine purpose in the universe, but which allowed him to see humanity as engaged in a struggle to dominate an intractable universe. He was inspired to work on natural selection by the assertions of religious writers such as Hilaire Belloc that "Darwinism is dead" (McOuat and Winsor 1995). His first paper in the area was published in 1924, and his survey, *The Causes of Evolution*, in 1932. Like Fisher, he assumed that selection acted on a large, freely interbreeding population, but he used practical examples to show that selection could act much more rapidly than Fisher supposed. Best known was the case of industrial melanism in the peppered moth *Amphidasys* (now *Biston*) *betularia*. A dark or melanic form of this moth first had been noticed in 1848 and had begun to dominate the population in industrial parts of Britain, where its color offered protection from predators by concealing it against the soot-covered background. By 1900 the melanic form had replaced the normal one almost completely in these areas. Haldane showed that so rapid a spread must indicate a 50 percent greater production of offspring by the melanic form, a far more intense selective effect than anything imagined by Fisher. Haldane also became engaged in the debate over the political implications of genetics and selection. He did not dispute the eugenists' assumption that human characters were predetermined by heredity, but he was far more conscious of the fact that the environment also had an effect. Turning to Marxism, he campaigned vigorously against eugenics (Werskey 1978).

Fisher and Haldane both treated genes as entirely independent units, and thus were working with what Mayr disparagingly called beanbag genetics. But by this time, developments in genetics were already beginning to undermine the plausibility of this assumption in ways that had important implications for Darwinism. The British researchers also assumed that populations are normally large and uniform, which did not square very well with the work of field naturalists, who often dealt with species split into small subpopulations by geographical barriers. The collapse of beanbag genetics began when William E. Castle (1911) used breeding experiments with hooded rats to show that additional variation sometimes seemed to emerge within a small population without the need for mutation. Continued selection in a small inbreeding population produced unusual genetic combinations which exhibited characters lying outside the original range of variation. This was enhanced by the existence of "modifier genes," which could influence the effectiveness of those genes responsible for producing phenotypic characters. The breeding population of a whole species now had to be seen as a complex system capable of generating an enormous amount of variation when stimulated by selection and inbreeding.

Castle's student, Sewall Wright, incorporated these points into a new form of population genetics (Crow 1990; Provine 1986). Experiments with coat color in guinea pigs had convinced Wright that interaction systems between genes were important, and his participation in the hooded rats experiments had shown him how inbreeding within small populations could stimulate variation. By 1920 he had developed a powerful mathematical technique to analyze the effects of inbreeding, enabling him to show that gene-interaction systems could be fixed by inbreeding in small populations and then acted on by selection. He began to apply these insights to natural evolution, which he assumed would take place most readily in small populations where inbreeding was intense enough to create new interactions systems through a random effect he called genetic drift. The most effective situation for evolution would be a large population imperfectly broken up into partially isolated local populations. Natural selection would pick out the best-adapted systems, which gradually would spread through the whole species. The random effects of inbreeding would allow a local population to move away from the adaptive peak for its species so that it could cross a relatively nonadaptive intermediate zone to reach a new adaptive peak. Depicting the interaction between the species and its environment as an adaptive "landscape" with peaks and troughs was one of Wright's most effective metaphors. He wrote a critical review of Fisher's theory (1930) and then expounded his own theory of evolution at greater length (1931). By the mid-1930s, the potential for a new Darwinian evolutionism began to emerge, and Wright's approach would become particularly useful in addressing the concerns of field naturalists.

THE MODERN SYNTHESIS

In the early decades of the twentieth century, most paleontologists and field naturalists expressed little interest in the newfangled science of genetics. Many of them continued to support Lamarckism and orthogenesis. By the 1920s the situation was beginning to change. The geneticists, especially in Britain and America, had rejected Lamarckism as completely without experimental foundation. In Continental Europe, the situation was more fluid. Here, genetics had not achieved the disciplinary independence it enjoyed in the English-speaking world, and the less specialized biologists of France and Germany still hoped that mechanisms of heredity outside the cell nucleus might allow for Lamarckian effects (Burian, Gayon, and Zallen 1988; Harwood 1993; Sapp 1983, 1987). But anyone moving into the study of evolution in Britain or America had to confront a unified body of geneticists, most of whom had rejected the developmental viewpoint within which anti-Darwinian ideas had flourished. Harwood (1994) argues that the synthesis did, however, draw on antireductionist philosophies popular in Germany and introduced into the English-speaking world by scientists such as Ernst Mayr who fled to avoid the Nazi regime.

Now that the theory of natural selection was being revived on a genetic basis, scientists in the older disciplines within the natural history tradition had to rethink their priorities. Few field naturalists could follow the complex mathematics employed by the founders of population genetics. The mathematical conclusions would have to be translated into a descriptive language they could use in their studies of local populations, if the new selection theory was to be employed in their research. But Ernst Mayr argues that the field naturalists were already making moves which anticipated the revival of Darwinism. Bowler (1996) and Steven James Waisbren (1988) make the same point for the paleontologists and morphologists. The old typological and developmental viewpoints were giving way to a perspective in which adaptation, migration, isolation, and geographical distribution played more important roles. Field naturalists now realized that each species could not be treated as a fixed type with only minor local variations. Many species were broken up into a collection of subpopulations by geographical barriers, and each population was adapting to its own local environment as best it could. There was no way of telling what the "true" form of the species was. At the same time, paleontologists such as A. S. Romer (1933) were presenting the history of life as a series of almost unpredictable developments, each depending on interactions between species and their environment. No one doubted that evolution was progressive, but the progress was not a uniform ascent of the scale. This growing emphasis on the adaptive nature of evolution provided fertile ground in which the synthesis could take root.

The field naturalists were increasingly certain that geographical barriers were crucial for speciation, the splitting apart of a once homogeneous population into a group of divergent subpopulations that might eventually become true species. They also focused on the adaptations by which each separate population accommodated itself to its local environment. The fusion of this kind of research with the insights of the population geneticists provided the impetus for the synthesis (Mayr and Provine 1980). Techniques used by Fisher and Haldane were less easily applied to these biogeographical insights, but their work helped to convince many biologists that Lamarckism now had to give way to natural selection. Both Bernhard Rensch (1983) and Mayr note they were forced to abandon their initial reliance on Lamarckism. But Wright's work most easily fitted the field naturalists' requirements, and Theodosius Dobzhansky's adaptation of Wright's techniques in *Genetics and the Origin of Species* (1937) was a key step forward.

Pioneering work was done in Russia under the direction of Sergei S. Chetverikov (Adams 1968, 1970; Mayr and Provine 1980). Russian naturalists were not influenced by the anti-Darwinism so prevalent in the West around 1900, so Chetverikov was in a better position to appreciate the importance of genetics. The Russian school started from a conviction that natural populations contain much unseen variation in the form of recessive genes, a view confirmed by crossbreeding wild populations of the fruit fly *Drosophila* with genetically pure strains brought from America (Chetverikov 1961 and 1997 are translations of his paper of 1926). Chetverikov's students developed techniques for studying statistical effects in various population sizes, confirming his belief that the underlying variability would be revealed most readily in small populations. The concept of the gene pool was introduced to signify that the population should be seen as a reservoir of genetic variation, although the combinations actually realized would depend on the accidents of inbreeding (Adams 1979). Chetverikov's school was eliminated during the Lysenko affair (see chap. 7), but his work influenced N. W. Tomofféef-Ressovsky, who moved to Germany in 1925 and promoted a more sophisticated view of the role mutations played in maintaining variability. Dobzhansky was also influenced by Chetverikov, although he was never a member of his school.

In Britain, E. B. Ford explored the results of Fisher's work in his *Mendelism and Evolution* (1931). Ford was interested in ecological questions, and his subsequent work confirmed Haldane's suggestion that selective pressures were often much greater than Fisher supposed. Gavin de Beer's *Embryology and Evolution* (1930) undermined the evidence for the recapitulation theory on which the Lamarckians had relied so heavily (Rasmussen 1991; Waisbren 1988). Because genetic mutations work by changing the developmental pathway, not by adding new stages to the existing course of ontogeny, there is no reason why the development of the individual should pass through the adult stages of its evolutionary ancestors.

Perhaps the most influential British advocate of the new Darwinism, certainly outside the scientific community, was Julian Huxley (1970; Waters and Van Helden 1992). A grandson of T. H. Huxley, Julian was an early enthusiast for Darwinism, although he preferred to interpret the theory as the basis for a philosophy of cosmic progress which the human race now had a duty to continue. He collaborated with the writer H. G. Wells to produce *The Science of Life* (Wells, J. Huxley, and Wells 1930), a popular account of biology which already stressed the selection mechanism as the driving force of evolution. He worked on animal behavior, explaining this in terms of inherited instincts implanted by evolution, but was also interested in embryology and the latest developments in genetics. His survey *Evolution: The Modern Synthesis* (1942) pulled many of the latest strands of thought together and gave the synthesis its name. Huxley went on to become an advocate for a humanist philosophy in which the idea of evolutionary progress played an important role. To many working scientists, though, he was a dilettante and a popularizer, not a serious contributor to the research on which the synthesis was based.

In America, Francis B. Sumner's work on the geographical distribution of deer mice began to establish the genetic basis of geographical variations within the species (Provine 1979). Although he began from a Lamarckian position, Sumner soon began to argue (1924) that the morphological differences between different populations had a genetic basis. The work of Theodosius Dobzhansky, however, did most to convert the field naturalist tradition to a theoretical position compatible with population genetics (Adams 1994). When he joined T. H. Morgan's team in 1927, Dobzhansky brought with him the experience of the Russian school's populational approach. He was thus in a position to appreciate what the field naturalists needed from the new genetics, with the result that his *Genetics and the Origin of Species* (1937) served to bridge the gap between them. Dobzhansky summarized the evidence showing the true nature of mutations, emphasizing how small their effects could be and how they sustained the natural variability of populations. He also summarized the mathematicians' conclusions, paying special attention to Wright's techniques. He presented his own work on the genetic basis of geographical variation in insects along with other demonstrations of the same effect. From 1938 on, he collaborated with Wright in a series of investigations of the genetics of natural populations (collected in Dobzhansky 1981). One aim of this research was to show that selection was not only a mechanism of change but could also act to maintain stability by means of balancing mechanisms such as that based on superior heterozygote fitness. The widespread existence of such dynamic equilibria was important to the synthesis because it demonstrated that populations do contain substantial reservoirs of genetic variation.

In 1930 Ernst Mayr came to America from Germany after several years of fieldwork on the birds of New Guinea and the Solomon Islands (Ruse 1994). Mayr had been influenced by Bernhard Rensch, who had revived the belief that there was a strong correlation between geographical variation and climate. Rensch and Mayr began with a Lamarckian explanation of this phenomenon, but in the 1930s they both realized that the Darwinian mechanism was now preferable. Mayr had not read the work of the population geneticists, so it was his own experiences which formed the basis of his claim that the field naturalists were moving independently toward a more Darwinian perspective. For Mayr and many others, Dobzhansky's 1937 book pointed the way toward a complete synthesis by presenting the mathematicians' conclusions in a form they could understand and use. Mayr's own *Systematics and the Origin of Species* (1942), emphasizing the role of geographic factors in speciation, was one of the founding works of the modern synthesis.

Mayr and the other founders of the modern synthesis were convinced that speciation occurred only through an initial phase of geographical isolation breaking a once homogeneous population into subpopulations which could begin to evolve in different directions according to the demands of their local environment. They were actively hostile to the efforts of Richard Goldschmidt, another German immigrant into America, to promote a theory based on macromutations (Goldschmidt 1940; see also G. Allen 1974). Mayr introduced the term *allopatric speciation* for the process in which geographical rather than biological forces were responsible for dividing a population. If the barriers subsequently were removed before the populations had separated so far that they were genetically unable to interbreed, then in principle the process of speciation would cease as the populations blended together again. Dobzhansky introduced the term *isolating mechanisms* to denote the fact that there might be nongenetic factors besides geographical separation which would prevent two populations from merging. Differences in mating behavior might prevent interbreeding even if it were genetically possible. Thus, the process of speciation could proceed even when the geographical barriers were removed. The new synthesis did not require any special genetic mechanisms for speciation: as long as isolating mechanisms were in place, natural selection would continue to build up genetic differences until eventually they resulted in sterility between the two new species.

George Gaylord Simpson extended the synthesis into paleontology. His *Tempo and Mode in Evolution* (1944) demonstrated that the evidence for macroevolution revealed by the fossil record could be explained by the accumulated effects of the short-term microevolutionary process postulated by the naturalists and geneticists. Whatever the claims of previous generations of paleontologists, the record was compatible with Darwinism (Gould 1980a; Laporte 1990, 2000). Simpson used quantitative analysis to show that major evolutionary developments took place in the irregular and undirected manner predicted by the selection theory. The evidence used to promote linear trends supposedly due to Lamarckism or orthogenesis did not stand up to scrutiny. The evolution of the horse family, for instance, was not a direct trend toward specialization; it was an irregular branching tree with many branches leading to extinction. On the much debated question of the discontinuity of some sequences, Simpson suspected that something more than the imperfection of the record would have to be invoked to explain the apparent jumps. He postulated occasional episodes of nonadaptive "quantum evolution" driven by Wright's genetic drift. These would occur quite rapidly in small populations and thus would be most unlikely to leave fossil evidence. In later works (e.g., 1953b), he adopted a more rigorously Darwinian approach, stressing the adaptive character of all evolution.

The synthesis was cemented in the mid-1940s. In 1939, a short-lived Society for the Study of Speciation had been formed to promote the integration of the various disciplines involved (Cain 2000). In America, the most visible sign of the new confidence was the creation in 1946 of the Society for the Study of Evolution, in which Dobzhansky, Mayr, Simpson, and Wright played prominent roles. Many of the founders seem to have been explicitly motivated by the philosophical calls made earlier in the century for the unification of the sciences (Smocovitis 1996). Evolutionary biology was at last to become a recognizable scientific discipline which transcended the barriers that had divided so many areas of traditional natural history, to say nothing of the new experimental sciences.

What exactly had been achieved has been a subject of controversy ever since. The synthesis was certainly a revival of the selection theory and of a Darwinian perspective more generally. The major non-Darwinian mechanisms were eliminated, and the extent to which even a minor role for nonadaptive processes could be retained was disputed. Simpson's rejection of Wright's genetic drift is typical of what Gould (1983) called the hardening of the synthesis to a more rigid Darwinian approach. At the intersection of population genetics and field studies, there were active research programs, such as Dobzhansky's, in which the different mathematical models were tested by direct application in the field (Provine 1978). The role of local adaptation was stressed in renewed studies of industrial melanism by H. B. D. Kettlewell (1955), although more recently this work has become the subject of reassessment by historians (Hagen 1999a; Rudge 1999). Paleontology was transformed along the lines suggested by Simpson, despite the fact that detailed links between the long-range effects of macroevolution and the microevolutionary processes studied by the field naturalists were still a matter of inference. There was little collaborative research, but there was renewed confidence that the disciplines were now pulling together in a more general sense. Notably, the later consolidation of the synthesis was much more coherent in Britain and America than in Germany. German biologists had made important contributions in the 1920s and 1930s (Reif, Junker, and Hossfeld 2000), but their involvement during the 1940s was necessarily limited by World War II. And given the less rigid focus on chromosomal determinism in German genetics, there was less pressure on field naturalists and paleontologists to move into line with the new Darwinism. Otto Schindewolf's "typostrophe" theory, still explicitly based on orthogenesis and macromutations, remained influential in the 1950s (translation 1993; Reif 1983, 1986).

THE ORIGIN OF LIFE

The wider influence of the revival of Darwinism was enhanced by a parallel development in another field: the study of the origin of life on earth. The early Darwinians had avoided linking their theory to the far more controversial issue of what was called spontaneous generation (Strick 2000). After Pasteur had demonstrated that life cannot be formed from inorganic material in the laboratory, T. H. Huxley and E. Ray Lankester began to argue that the origin of life must have occurred under the very different conditions prevailing in the earth's early history. Others tried to popularize the view that spores which generated the first living things had arrived on the earth from outer space (Arrhenius 1908). In the 1920s a new assault on the question began as scientists became confident that their new understanding of organic processes at last enabled them to ask meaningful questions. By the 1950s it was becoming widely accepted that life must have had a natural origin, greatly extending the scope of the more materialistic thrust of the evolutionary synthesis.

The most important step forward was taken by the Russian biologist Alexander Oparin (Farley 1977). He realized that spontaneous generation should not be thought of as a single event; instead there must have been a cumulative process of chemical evolution by which the building blocks of living organisms were gradually assembled. In 1923 J. B. S. Haldane had suggested that the most important step would have been the formation of a self-replicating, viruslike structure from simpler organic chemicals. This was long before Haldane became a Marxist, but John Farley (1977) argues that, for Oparin himself, accepting the philosophy of dialectical materialism was an important stimulus. The Lysenko affair led to much criticism of the Soviets' heavy-handed attempt to apply political ideology to science. Oparin actually supported Lysenko against the geneticists, but in his case, enthusiasm for dialectical materialism led to positive theoretical inspiration. Because this was a nonreductionist form of materialism, it encouraged Oparin to think in terms of new qualities emerging from quantitative changes at a lower level. The dialectic also implied that the appearance of a new quality negates what existed before. Thus, Oparin was led to the idea that once life had appeared as a result of chemical evolution, its very existence would prevent the same process from ever being repeated. Living things would themselves destroy the earlier stages in the process as soon as they occurred, which was why we could not observe spontaneous generation taking place today.

Oparin's *Origin of Life* of 1936 (translation 1938) postulated that the earth's atmosphere originally contained no oxygen; that element had appeared in a free state only as a consequence of the later activity of living things. This was another reason why the process leading to life could never be repeated in the modern world. Chemical reactions in the original atmosphere of hydrocarbons and ammonia gave complex organic molecules, which formed the "primordial soup" from which the first stages in the formation of living things began. As structures capable of a primitive form of replication appeared, natural selection began to operate, favoring those which could reproduce most effectively. It was assumed that these "prebionts" went on to form the first true living things, although the later stages in the theory were far more speculative.

Oparin's theory was taken seriously in the West despite its connections with Marxism. In the period after World War II, the experimental opportunities to confirm the early steps in the theory began to look more promising. The most widely publicized step came with Stanley Miller's classic experiment (S. Miller 1953), which showed that an electric spark could produce amino acids when passed through the kind of reducing atmosphere Oparin had postulated. Since amino acids are the building blocks of proteins, the experiment confirmed that relatively simple physical processes could nevertheless create the potential for significant developments toward life. Much speculation continued over how those later developments might take place, but the origin of life no longer seemed beyond the scope of scientific observation. Evidence from the fossil record now suggests that the appearance of life occurred at a surprisingly early stage in the earth's history.

WIDER IMPLICATIONS OF THE SYNTHESIS

By the 1950s, evolutionists had achieved greater confidence in the power of science to explain the development of life without recourse to supernatural design or vital forces. The new Darwinism was as materialistic as the old, but its advocates were also equally certain that their theory had a positive message to offer. If the old belief in a divine purpose had to be abandoned, the insights of science offered a more realistic view of our position in the world and of our aspirations and responsibilities. There were associated developments in the understanding of human origins and of a number of biological issues with important moral implications. Far from simply undermining the old religious view of human origins, the modern synthesis might have a positive role to play in founding a new world order following the traumas of World War II. In the aftermath of the Scopes trial in 1925, American science teachers had seen the subject of evolution largely eliminated from high school science teaching (Grabiner and Miller 1974). Now a number of eminent biologists began to declare that a hundred years without Darwin was enough (Muller 1959; Simpson 1963) and to insist that the implications of the theory be confronted.

Darwinism could be applied to the origins of the human species, offering a very different perspective on how we emerged from apelike ancestors (Bowler 1986; Lewin 1987; Reader 1981). Early-twentieth-century theories had focused on the development of the human brain as the key factor separating us from the apes. A progressive trend toward increased intelligence was supposed to have affected a number of different branches of the primate family in a form of parallel evolution. There was little interest in the question of why the human and the ape families had diverged from their common ancestor in two such different directions. Darwin himself had addressed this topic and had even suggested that our ancestors' adoption of an upright posture as a means of locomotion on the open plains of Africa marked the key breakthrough. But few of his contemporaries had been willing to accept that an adaptive transformation could have had such far-reaching consequences, preferring to believe that progress toward higher intelligence had been stimulated by the more challenging climate of central Asia. The discovery of the first australopithecine fossil in South Africa in 1924 had been largely ignored because, although it fitted Darwin's predictions, it did not correspond to those of his later followers. By the 1940s, though, the situation had been transformed. Robert Broom had uncovered a whole series of australopithecine remains, making it almost impossible to avoid the implication that the first members of the human family had lived in Africa, and had stood upright even though they had had brains no larger than an ape's (Broom 1950; W. Le Gros Clark 1949, 1955).

It is no coincidence that the reformulation of anthropologists' thinking on the question of human origins took place at the same time as the emergence of the modern synthesis. There was no room in the new evolutionism for mechanisms based on orthogenetic or progressive trends. In a Darwinian perspective, if humans had separated from apes, there must have been an adaptive benefit for each branch driving it in a different direction. Darwin's old idea that standing upright was an adaptation to the new environment of the open plains exactly fitted the new perspective, and this forced the fossil hunters to take the australopithecines more seriously. Applying the new systematics derived from the modern synthesis to the hominid fossil record, Mayr and others argued that the designation of a vast number of different human types was unjustified. As a result, scientists unified the Java *Pithecanthropus* remains with the very similar fossils discovered in China to create a single species, *Homo erectus*. Almost coincidentally, the fraudulent nature of the Piltdown remains was at last confirmed (J. S. Weiner 1955). It was now assumed that the australopithecines were the first members of the hominid family, from which *Homo erectus*, and then *Homo sapiens*, had evolved. The Neanderthals were no longer regarded as an extinct and very primitive species; they were an early variant of *Homo sapiens*, perhaps adapted to the colder climate of the Ice Age (Trinkaus and Shipman 1993). Mayr at first insisted that only one species of *Australopithecus* could have existed, on the grounds that if there were two rival species occupying the same territory, one would drive the other to extinction (the principle of competitive exclusion). He eventually accepted that, in this case, the degree of morphological difference between the specimens was so great that there must have been two species, a robust form which had died out and the more lightly built *Australopithecus africanus*, which had evolved into *Homo erectus* (Delisle 2001). The human family tree still had side branches, but not so many as previously believed, and the differences between them were defined clearly by adaptive specializations.

Scientists could now focus their attention on the subsequent development of the enlarged brain in the genus *Homo*, for which a variety of explanations, often controversial, were offered. One that became popular immediately following the new fossil discoveries was Sherwood Washburn's thesis that the emergence of hunting as the principal means of gaining food provided a requirement for enhanced intelligence and communications skills, a view that attracted critical comment from a later generation of feminist scholars (Haraway 1990). Whatever the explanation, though, it was clear that the new Darwinism made it much more difficult to portray the emergence of the human race as a preordained goal of the evolutionary process.

Another important consequence of applying the perspectives of population genetics and Darwinism to human origins was the elimination of the old tradition of race science. Physical anthropologists often had described the various human races as distinct species, and the old non-Darwinian evolutionism had encouraged this view by allowing the creation of a model of human origins in which several different branches independently evolved in the same direction. Already in the 1930s, some of the more liberal biologists, including Julian Huxley, had begun to protest against the misuse of science to emphasize racial differences, pointing out the obvious fact that humans everywhere were interfertile and hence were, by definition, members of the same species (Barkan 1992). In a study of the new fossil evidence for human evolution, the British anatomist W. Le Gros Clark (1949) emphasized that the Darwinian theory undermined the plausibility of the old ideas of parallel evolution. The triumph of the modern synthesis drove home this point, and as more genetic research was done on the human population, it became evident that the differences between the so-called races was trivial by comparison to the differences within each local group. Recognition of this point was encouraged by widespread revulsion against the old race science following the role it had played in the Nazi's ideology of Aryan supremacy and the horrors of the Holocaust. The pioneers of the modern synthesis could point to the destruction of the biology of race as an important moral consequence of their new perspective on evolution (Dobzhansky 1962; Simpson 1963).

The new generation of Darwinians also had to address the question of how their theory, traditionally associated with materialism, could be used as the basis for a humanistic philosophy that would replace traditional religious beliefs. Most of them were convinced that science revealed no evidence of an underlying plan or purpose built into the universe. Simpson (1963) was perhaps the most active in pointing out that, in the open-ended Darwinian model of evolution, it was impossible to regard the human species as the predetermined goal of the history of life. Religious thinkers were still trying to sustain the older, teleological form of evolution, in which the human race played a key role in fulfilling a divine plan of spiritual progress. The evolutionary mysticism of Pierre Teilhard de Chardin's *Phenomenon of Man* (translation 1959) was especially popular. Among the biologists, however, only Julian Huxley seems to have taken Teilhard seriously, writing a positive introduction to the English translation. Most of the new generation of scientific evolutionists saw this as exactly the kind of vague teleology they were trying to escape (see the vitriolic review of Teilhard's book in Medawar 1961). Yet Huxley's more sympathetic response fits in with the claim made by Ruse (1996) that the evolutionists were still enthusiasts for the idea of progress (see also Gascoyne 1991). Huxley promoted a philosophy of humanism (1953, 1961) which denied the existence of a transcendental creator but still sought to provide a meaning for life by stressing the human race's responsibility for continuing the moral progress visible in the evolution of life on earth (J. Greene 1990; Waters and Van Helden 1992). Thus he could identify with some aspects of Teilhard's cosmic teleology, although like his fellow Darwinists he knew that the progress of life was far more episodic and haphazard than the evolutionary mystics acknowledged. As he wrote in the conclusion to *Evolution: The Modern Synthesis:*

> Let us not forget that it is possible for progress to be achieved. After the disillusionment of the early twentieth century it has become as fashionable to deny the existence of progress and to brand the idea of it as a human illusion, as it was fashionable in the optimism of the nineteenth century to proclaim not only its existence but its inevitability. Progress is a major fact of past evolution; but it is limited to a few selected stocks. It may continue in the future, but it is not inevitable; man, by now become the trustee of evolution, must work and plan if he is to achieve further progress for himself and so for life. (1942: 578)

Almost all the evolutionists shared Huxley's view that humans were more highly developed than any other animal species, although opinions differed as to how best to measure the ascent, and a few skeptics wondered if it were possible to establish a meaningful criterion by which to measure progress. Our intelligence, adaptability, and social qualities were still seen as indications of a higher level of development. Yet it was clear that most evolution did not lead toward these qualities, and that many branches of the tree of life had died out as a result of the specializations evolution forced on them. Only in a few rare episodes did life ascend to a new level of mental activity, the last of these being the emergence of humanity from its apelike ancestors. At this point, most of the evolutionists were prepared to accept that humans were indeed something special, possessing characters which lifted them above the other animals into a world where language, morality, and culture had become important. As George Gaylord Simpson wrote:

> Man is one of the millions of results of this material process [Darwinian evolution]. He is another species of animal, but not just another animal. He is unique in peculiar and extraordinarily significant ways. He is probably the most self-conscious of organisms, and quite surely the only one that is aware of his own origins, of his own biological nature. He has developed symbolization to a unique degree and is the only organism with true language. This makes him also the only animal who can store knowledge beyond individual capacity and pass it on beyond individual memory. He is by far the most adaptable of all organisms because he has developed culture as a biological adaptation. Now his culture evolves not distinct from and not in replacement of but in addition to biological evolution, which also continues. (1963: 24)

There was no effort here to minimize the gulf between humans and their animal ancestors. Paradoxically, by stressing the open-endedness and unpredictability of evolution, the new Darwinism left room for the emergence of new characters and for a sense of human freedom. Dobzhansky (1956) was particularly anxious to defend the freedom of the will against the determinists, perhaps because of his roots in the Russian Orthodox Church. For the time being, the evolutionists had no interest in challenging the social scientists who insisted that culture and society do not have a biological foundation.

If the new generation of Darwinian biologists was reluctant to upset the social sciences, the same could not be said for the ethologists (students of animal behavior in the wild), who now initiated a major debate over how best to illuminate the animal foundations of human nature. The so-called anthropology of aggression drew little inspiration from detailed models of natural selection but took it for granted that human nature must have evolved from animal patterns of behavior. But what lessons did the study of animals suggest we might learn about our deepest instincts? To those who took the Darwinian message on board, there was no point in trying to gloss over the harsher side of animal behavior. Driven by the perpetual shortage of resources, animals were in constant conflict with one another—and it was unlikely that our development of higher levels of intelligence would have removed the resulting aggressive instincts. Arthur Keith (1949) had long been arguing that group competition played a major role in promoting human evolution. The founder of the modern science of ethology, Konrad Lorenz, demonstrated the existence of aggressive behavior in many animals and warned that it would be folly to assume that humans were not endowed with similar instincts (1966). Robert Ardrey (1966) suggested that the instinct to defend a piece of territory was the most basic, insisting that the "territorial imperative" had shaped the behavior of most animals and was still visible in human selfishness and our propensity for warfare. Desmond Morris's best-selling *Naked Ape* (1967) stressed the animal foundations of human behavior and also warned that our social instincts were underpinned by selfishness and a tendency toward violence. We have already seen how paleoanthropologists such as Sherwood Washburn argued that hunting was the key behavioral innovation which made us human. Without even exploiting the concept of natural selection directly, a new generation of students of human nature used the presumed link with our animal ancestors to promote something which the critics could portray as a revived social Darwinism.

This image of human nature as fundamentally selfish and aggressive was very much at variance with the more optimistic model promoted by Huxley and Dobzhansky. The anthropology of aggression did not go unchallenged even in popular debates, because there was now an intense fascination with the great apes as our closest animal relatives, and new studies began to suggest that the apes were by no means the ferocious beasts depicted in so many cartoons and caricatures. The paleoanthropologist Louis Leakey encouraged two women who would become legends in their own lifetimes as a result of their intimate studies of apes in the wild: Jane Goodall and Dian Fossey (Haraway 1990). Fossey's studies of the mountain gorillas showed them to be gentle and unaggressive. Her message about the "gorillas in the mist" came to seem all the more poignant after her murder by poachers in 1985. Goodall's work on the chimpanzees of the Gombe Stream Reserve near Lake Tanganyika demonstrated that some long-standing claims about the differences between humans and animals had to be rejected. The chimpanzees did, for instance, make use of tools, including stones to crack nuts. Reaching a wide audience though her connections with *National Geographic* magazine, Goodall promoted a pacific image of the apes that contrasted with the anthropology of aggression. Her later work, however, showed that the apes did engage in hunting and even what looked like warfare between groups, thus suggesting that scientists on both sides of the debate had been working with limited models of animal behavior.

The popular fascination in the 1960s and 1970s with ape behavior was driven by a renewed willingness to take seriously the idea of an evolutionary ancestry for humankind. In this respect, the modern synthesis had consequences well outside the technical debates over the mechanism of evolution. But in the end, the study of animals had produced conflicting guidance for those who would use evolution to throw light on human nature. All this changed over the following decades as a more aggressive form of Darwinism began to emphasize the biological foundations of human nature, again raising fears of a new social Darwinism. At the same time, the creationists would see that the evolutionists were breaking the rule of silence that had kept the peace between the two groups since the 1920s. They began a renewed assault on what they perceived as a revival of the materialist program. Meanwhile, debates continued within biology over the best way to move forward from the position established in the 1940s. The evolutionary synthesis was by no means the last stage in the completion of a Darwinian revolution.

10 Modern Debates and Developments

With the consolidation of the evolutionary synthesis in the 1950s, the field had at last come of age. The broad outline of the history of life on earth was now known, and although more details might be revealed by the fossil record, few surprises were expected. The acrimonious debates of the early twentieth century between a range of mutually incompatible theories had been resolved by a broad consensus based on a revived Darwinism. Natural selection was the basic mechanism of evolution, and biologists had to work out the details of how that mechanism operated to produce the diversity of species we observe. Humans were part of the story, of course, but even some biologists conceded that, in this case, progressive evolution had generated an entirely new level of mental activity.

If the founders of the evolutionary synthesis expected the next few decades to show nothing but the filling in of details following the revolution—what T. S. Kuhn (1962) calls "normal science"—they were to be disappointed. The diversity of new theories has never matched that of the "eclipse of Darwinism," but it has been possible to articulate divergent perspectives within the narrower frame of reference established by the synthesis. The evolutionary synthesis was founded on the Darwinian assumption that evolution was a continuous process. New theories such as punctuated equilibrium reintroduced an element of discontinuity, and there was evidence that mass extinctions form genuine discontinuities in the history of life. The synthesis, especially in its "hardened" form, took it for granted that evolution was driven solely by the demands of adaptation. Refinements of the selection theory led to new applications suggesting that the focus of selection was not the individual organism but the gene itself. For the sociobiologists, fitness now had to be defined in terms of successful reproduction, not adaptive fitness. In response, a new generation of critics has emerged to suggest that this ultra-adaptationism goes too far by minimizing the extent to which embryological development shapes the way genes are expressed. Some bolder skeptics have reintroduced the claim that evolutionary explanations are little more than storytelling, because the fossil record simply does not permit us to test theories about large-scale events in the history of life (for surveys of modern evolutionism, see Mayr 2001; Strickberger 2000; and Zimmer 2001; on its wider implications, see Rose 1998).

These criticisms are driven partly by the success of rival areas of biology where the experimental method rules and where immense amounts of money are generated by the hope of medical applications. The explosion of interest in molecular biology in the late twentieth century offered both opportunities and threats to the evolutionists (Dietrich 1998; Hagen 1999b). The opportunities came from the ability to exploit molecular evidence to provide new forms of evidence about the relationships between species, based on degrees of genetic rather than morphological similarity, and even a molecular "clock" by which to determine rates of evolutionary change. The threats came from the suspicion that any form of inquiry not susceptible to experimental testing is an inferior form of science or perhaps not even science at all. To the critics, it all looks too much like the old-fashioned natural history that was becoming outdated even in Darwin's time. To the naturalists and paleontologists, however, any attempt to apply the standards of an experimental science to the study of historical events in the earth's past is misguided. If there are different kinds of questions to ask about the structure of the natural world, different methodologies may be needed to tackle them.

To the critics outside the scientific community, these disagreements are heaven-sent. For them, the failure of the synthesis to impose a lasting consensus reveals the weakness of its claim to offer a truly scientific account of the development of life. Most scientists see the debates as a sign of vitality indicating that they are still grappling with significant issues and trying to resolve their differences. But to anyone not actively engaged in scientific research, an admission that theories cannot be immediately verified looks like a sign of weakness. Religious thinkers, convinced that God created the world directly in its modern form, exploit any sign of dissent among scientists as evidence that the materialistic worldview has major flaws. In America, at least, creationism remains a major threat to research and teaching in evolution. Paradoxically, some members of the academic community who stress a relativist view of knowledge are willing to make common cause with the creationists against evolutionism, forgetting that the alternative inspired by religious faith is far more likely to generate a claim for absolute truth.

THE HISTORY OF LIFE

For many biologists, the most exciting areas in evolution theory focus on the actual mechanism of change, but for the general public, *evolution* still means the development of life on earth, as revealed especially by the fossil record. And that record is still expanding as new discoveries are made, sometimes filling in the details of what we already knew in outline, sometimes dramatically changing our view of a particular episode. Studies of the so-called Cambrian explosion of multicellular animals have raised major questions about the whole course of evolution. But ideas on the relationships between the different living things are also affected by taxonomists who use ever more sophisticated techniques to assess degrees of similarity. In the late twentieth century, the evidence used in these techniques expanded to include molecular biology. We now can estimate degrees of relationship based on the proportion of genes that are common to the genome of two species. In some cases the accumulated evidence has transformed our ideas about the origins of particular species or groups of species. At the same time, too close a focus on taxonomy has led to criticisms that all efforts to chart the actual course of evolution are illusory (see below). To most paleontologists, this restriction seems absurd, and it is significant that the other major development in studies of the history of life—the evidence for catastrophic mass extinctions—bears directly on our understanding of the actual course of evolution.

The continued expansion in our knowledge of the fossil record provides ever more details on the course of evolution. In recent decades the opening up of China to systematic fossil hunting has revealed a wealth of material about the dinosaurs and about the origins of birds and mammals. But paleontologists are not merely concerned with the fossils which throw light on the emergence of the major phyla—one consequence of the modern synthesis was scientists' demand for more complete knowledge of the populations of each period and region, so that they could gain some idea of changes in the diversity of life and in ecological relationships. New ideas also emerged about the character of well-known groups based on more detailed studies of their fossil remains. Much controversy surrounded the claim that the dinosaurs were hot-blooded, popularized by Adrian Desmond (1976) and others. The belief that dinosaurs were much more active creatures than once was supposed has been enshrined in the public imagination by films such as *Jurassic Park*.

One result of the new discoveries has been to push the origin of most major groups further back in geological time. The same point is true for the origin of life itself, since fossil stromatolites (algae mats), indicating the presence of cellular life, have been found at a very early stage in the earth's history. Studies of the structure of modern cells now suggest that the first cells evolved by a process of symbiosis in which originally distinct structures were incorporated together in such a way that they could cooperate with one another (Margulis 1993). The fact that life existed for billions of years at the level of single cells before eventually diversifying into metazoans (multicelled animals and plants) has focused attention on what might have caused this later explosion of more advanced forms. It had been known since Darwin's time that the first major radiation of phyla seemed to occur in the Cambrian period, to which the origins of all living phyla could be traced. We now know that there were some episodes in which metazoa appeared before the Cambrian, but the links between these early, often rather enigmatic creatures and the Cambrian explosion are still debated, as is the cause of the diversification itself.

Stephen Jay Gould's account (1989) of the bizarre creatures preserved in the Burgess shale of British Columbia focused attention on the origin of the major phyla. Although some of these creatures seem to represent early members of the known phyla, Gould insisted that many were so strange that they would have to be classified in phyla unrelated to those still existing. On his model of the history of life, there was a massive diversification in the Cambrian that produced a wide range of basic types, most of which disappeared fairly rapidly. More controversially, Gould also claimed that it seemed to have been a matter of luck which ones survived and which went extinct, because there was no clear evidence that the ancestors of the modern phyla were "fitter" than those that disappeared. On this model, if one could somehow turn the clock back and allow life to evolve again, the end results would be very different and we ourselves certainly would not be here. Although presented as part of Gould's campaign against the emphasis on adaptation in evolution, this argument is not quite as anti-Darwinian as it might seem. Even George Gaylord Simpson insisted that humans were a most unlikely product of evolution, and a good Darwinian would expect a certain amount of unpredictability in the history of life because small changes at crucial points can have major consequences. Critics think that Gould exaggerated the diversity of the Cambrian forms, and a rival account of the Burgess shale fossils by Simon Conway Morris (1998) presents the history of life as a more or less predictable sequence restricted by ecological and other constraints. On this model, even the appearance of human beings was inevitable—something many Darwinians would regard as suspiciously like a reintroduction of the old idea of divinely planned evolution.

When a phylum has living representatives, it is possible to bring in biochemical, and now genetic, evidence to clarify issues not resolved in terms of structural resemblance alone. The expansion of molecular biology in the late twentieth century allowed a wide range of techniques to be brought to bear on these issues, including measurements of the extent of overlap in the genome of the species concerned. Such evidence has confirmed, for instance, the close relationship between humans and chimpanzees, between which only a small percentage of the genes are different. The same evidence has also confirmed the claim that there are no significant genetic differences between the various human races (Cavalli-Sforza 2000). But this kind of evidence is even more useful in dealing with more distant relationships, such as those between the major phyla produced in the Cambrian explosion. The origin of the vertebrates is a case in point (Gee 1996). In the late nineteenth century, William Bateson gave up his study of the origin of the vertebrates in despair at the inability of morphological evidence to decide conclusively among the various theories. Since that time, both biochemical and genetic evidence has confirmed the relationship between the vertebrates and the echinoderms, although the details of how the first creatures with a dorsal chord evolved are still being elucidated (sometimes using more refined versions of the old embryological arguments).

Molecular phylogenies assume that genetic differences build up once two lines of evolution have diverged. This obviously is true for the genes responsible for the adaptive changes which the two lines undergo. But genetics has revealed other underlying changes which have no adaptive significance, because the new gene has exactly the same function as the one it replaces. This is described by the neutral theory of genetic replacement (Kimura 1983; see also Dietrich 1994). Since the replacements are not under the control of natural selection, they take place at a fixed rate determined by how rapidly the equivalent genes can be produced by mutation. If this rate can be determined for a particular group of organisms, then a molecular "clock" can be produced which will date the point of separation of two lines according to the number of neutral replacements by which their modern representatives differ. In some cases, applications of the molecular clock produced results widely divergent from the dates accepted by paleontologists. In human evolution, molecular evidence suggested a much more recent separation between humans and apes than indicated by the estimate derived from fossils (Gribben and Cherfas 1982; Lewin 1998).

Perhaps one of the most striking changes in our ideas about the history of life has been acceptance of the theory that the process has been interrupted from time to time by mass extinctions. The Darwinian paradigm traditionally has been based on the Lyellian notion of uniformity, in which the rate of change in both geology and paleontology is assumed to be slow and gradual. Lyell himself dismissed all evidence for rapid change as illusory, although the majority of geologists remained convinced that there were occasional episodes where transformations were much more abrupt than the principle of uniformity would allow. The main development in twentieth-century geology—the emergence of plate tectonics and the theory of continental drift—tended to revive the Lyellian approach (Hallam 1973; Le Grand 1988; Wood 1985). Although revolutionary in the context of the earth sciences, plate tectonics showed that movements in the earth's crust are slow and gradual, and thus the theory reinforced the Darwinian prejudices of the founders of the modern synthesis for a gradualist view of evolution. There might be episodes of more rapid turnover in evolution caused by unusual combinations of circumstances, as when the great, ancient continent of Pangaea broke up to form the fragments which became the modern continents, dramatically increasing the extent of coastal environments. But there was no room for mass extinctions of the kind envisioned by the old catastrophist theory, in which whole populations were wiped out instantaneously. Paleontologists assumed that the major breaks in the fossil record were due to the accumulated effects of gradual trends, not genuine mass extinctions.

The Russian scholar Immanuel Velikovsky (1950) inspired a popular reaction against this gradualist view. Velikovsky's theory of interplanetary collisions was ridiculed by scientists but made him a hero in the eyes of the counterculture of the 1960s (De Grazia 1966; Goldsmith 1977). It was something of a coincidence that, in the 1980s, L. W. Alvarez and others produced evidence suggesting that one of the most important transitions in the geological record, the Cretaceous-Tertiary boundary was marked by a layer containing a significant amount of iridium, an element that could only have come from an extraterrestrial body such as an asteroid. The claim that the disappearance of the dinosaurs at this point represented a genuine mass extinction due to the climatic disruption caused by an asteroid impact became widely accepted, although some paleontologists protested that the record is not as abrupt as this scenario predicts (Hallam and Wignell 1997; J. Powell 1998). Supporters of the theory argue that the history of life has been interrupted by a series of mass extinctions caused by similar events, although there is evidence of major eruptions of volcanic material that could have had the same environmental effect as an asteroid impact.

The possibility of catastrophic events again brings into question the predictability of the history of life on earth: is extinction due to bad genes or bad luck (Raup 1991)? If catastrophes are involved, then it is bad luck: the casualties of an asteroid impact would depend on where it landed, and the broader environmental effects would depend on whether it hit the land or the ocean. Does this negate the Darwinian paradigm? It certainly introduces a new dimension into evolution, because only the survivors would diversify to repopulate the earth with new species, and in the early stages of the process there would be much less interspecies competition than when diversity had again reached its peak. But if the evidence for the catastrophes is clear, there seems to be no reason why a good Darwinian should be disturbed by the element of unpredictability they add to the story of evolution.

HUMAN ORIGINS

Much attention is still focused on the origin of the human species and the hominid family to which we belong. New fossil discoveries have thrown light on human evolution. But, to a surprising extent, the interpretation of the fossil record and of the significance of our higher faculties is contested by schools of thought reflecting divisions that arose a century or more ago. There are two chief areas of disagreement. One centers on the degree to which human faculties should be seen as a continuous development from characters already present in the higher animals. The original Darwinians tried to minimize the gulf to be bridged by stressing the extent to which animals already exhibit at least the rudiments of all the higher faculties. Their opponents often insisted that the human mind was unique, and that it represented a divine gift that could only have supernatural origins. Few modern scientists endorse the latter view, yet the idea that humans possess unique faculties has been remarkably tenacious, posing problems for the evolutionist seeking to explain how such innovations could have been produced by natural selection. The second area of debate also centers on the issue of continuity, but focuses on whether the later phases of human evolution took place throughout the global population or were confined at first to a small area, from which a new population expanded (for surveys, see Foley 1995; Lewin 1987, 1998).

The supporters of the evolutionary synthesis established a theory of human origins in which the hominid family was first defined by the australopithecines of southern Africa, which had walked upright but still had ape-sized brains. From one species, *Australopithecus africanus*, the first members of the human genus, *Homo erectus*, evolved. Having larger brains, these early humans made use of tools and fire and spread through much of the Old World. Eventually they evolved into *Homo sapiens*, although the details of this transition were unclear, in part because the heavily built Neanderthals of Europe seemed to represent a population that was not only more primitive than anatomically modern humans but also adapted to the cold environment of the ice ages.

The paleontologists of the 1960s estimated that the divergence between the human and ape families had taken place as much as fifteen million years ago, although there was a big gap in the record after this date. The molecular evidence that became available in the 1970s created an uproar when it was used to suggest that the split was no more than five million years ago (Gribben and Cherfas 1982). The more recent date has become more widely accepted, and the position of the ape fossils on which the old estimate was based has been reconsidered. Also in the 1970s, Donald Johanson discovered an earlier form of australopithecine in East Africa which he called *Australopithecus afarensis*, better known as Lucy (Johanson and Edey 1981). This has been widely presented as the ancestor of the whole human family, although later discoveries are throwing more light on this early period and may lead to revisions of the story. Already, in the 1960s, Louis and Mary Leakey had described a fossil which seemed to predate *Homo erectus* as the first true human; this was *Homo habilis*, so called because it seemed to be the maker of the first very primitive stone tools. There was subsequently much rivalry between Johanson and Louis Leakey's son, Richard, over the precise course of human evolution; the Leakeys had always favored the view that the origins of the genus *Homo* lay further back in the record than the australopithecines (Leakey and Lewin 1977, 1992).

It is now clear that the earliest members of the human family stood upright but still had small brains. Because there is no evidence that the australopithecines used tools, the new finds focused attention on the question of what had promoted the expansion of the human brain and, consequently, had led to new faculties such as toolmaking and language. This reopened the question of whether these faculties were genuinely new or merely developed from rudiments already present in the higher animals. The early supporters of the evolutionary synthesis seemed to have been content with the idea that human faculties were genuine novelties which allowed the higher aspects of human behavior to exist independently of biology. But there is a long tradition of evolutionists from Darwin onward who have sought to minimize the novelty by looking for indications that the faculties already exist in a primitive form among the higher animals.

Louis Leakey persuaded Jane Goodall to begin her classic studies of chimpanzee behavior in the hope of throwing light on the origin of human faculties. But these and later observations have undermined many of the traditional assumptions about what makes humans unique. We now know that apes use tools, engage in hunting, and even wage war. From a very different direction, there was an assault on the idea that language is a unique human attribute. Efforts to teach apes to speak had been abandoned because they cannot make the appropriate sounds—but was there a more fundamental mental incapacity that prevented them from engaging in this level of communication? According to the linguist Noam Chomsky, the capacity for language ("deep grammar") is built into the human brain, and there is no equivalent among animals. This was the position urged against the Darwinians of the late nineteenth century, and a century later there was a renewed assault on the question by scientists who felt uneasy at the postulation of such a deep discontinuity. Apes were taught American Sign Language (used by deaf people), and it was soon obvious that they could use symbols to represent objects. There were claims that they could manipulate the symbols to make sentences and thus had the capacity for grammar, that is, the meaningful ordering of signs. The first ape supposedly able to do this was a chimpanzee appropriately named Nim Chimpsky.

These claims have been highly controversial, with critics arguing that the animals have been prompted by their keepers to make the apparently meaningful combinations. Some modern theories accept that the human capacity for language is an evolutionary novelty requiring a significant restructuring of the brain by natural selection. According to Merlin Donald (1991), the first successful members of the genus *Homo* used gesture as a form of communication and enjoyed what he calls a "mimetic culture." This was the species known as *Homo erectus*, which made relatively sophisticated tools, used fire, and spread all over the Old World. Only in the modern *Homo sapiens* was the brain reorganized to allow the use of true language and the emergence of decorative arts. An example of the opposing viewpoint is Robin Dunbar's claim (1996) that brain size increased in proportion to the size of social groups because the primary purpose of higher intelligence is to cope with social interactions ("Machiavellian intelligence"). According to Dunbar, language emerged inevitably as human group sizes increased, because the apes' favorite mechanism of social interaction, body grooming, no longer served adequately as a means for our ancestors to interact with one another successfully.

The other major focus of debate has been the continuity of human evolution from *Homo erectus* to modern *Homo sapiens* and the position of the Neanderthals within this development. In the early twentieth century, there was an active debate between those who saw this as a continuous process across the whole of the territory into which *Homo erectus* expanded—that is, across almost the whole of the Old World—and those who thought that modern humans evolved in a more restricted location and then expanded their territory, displacing the older population without interbreeding with it (Bowler 1986). Supporters of the later view focused in particular on the fate of the Neanderthals in Europe, who were seen as an archaic species exterminated when modern humans invaded their territory. At first the evolutionary synthesis was used to support the rival idea of continuous development. Just as the modern races were seen to be closely related, so even the Neanderthals were labeled a local variant of *Homo sapiens.* The two alternatives have both remained in play, although the replacement theory is now more popular because it seems consistent with evidence from molecular biology (Lewin 1987; Trinkaus and Shipman 1993). Studies of mitochondrial DNA (genetic material contained in small bodies outside the cell nucleus) suggest that all modern humans are closely related and are descended from a comparatively recent ancestor in Africa (the "African Eve"). On this model, the modern population expanded out of Africa and displaced the earlier humans, including the Neanderthals. Since there was no interbreeding, there is no continuity between the ancient inhabitants of most regions and the modern people who occupy the same territory. The rival continuous development theory insists that the morphological similarities between the fossils and the racial variants of modern humans indicate some genetic continuity within each region. The African Eve theory nicely encapsulates the long-standing commitment of the evolutionary synthesists to the unity of modern humanity, denying the reality of the racial types once invoked by physical anthropologists. Its opponents warn that the out-of-Africa theory looks suspiciously like earlier ideas based on race conflict, in which "higher" forms of humanity regularly exterminated their more primitive neighbors.

SOCIOBIOLOGY AND ULTRA-DARWINISM

The debates over the history of life interact with an ongoing feud among evolutionists over how the process operates. Although few biologists now doubt that natural selection is a crucial mechanism of change, there are major disagreements about how it actually works and over the possibility that other factors must also be taken into account. The founders of the modern synthesis took it for granted that natural selection involved a competition between organisms which favored those best fitted to the local environment. Because these scientists saw environmental adaptation as the key, they were convinced that geographical factors affected the outcome of local evolution, and they were open to the possibility that geological changes could significantly influence the course of evolution. They were aware that genes could not be treated as units equivalent to phenotypic structures, because interactions shaped the actual outcome of ontogeny. In the late twentieth century, all these assumptions came under fire, leading to a polarization between what Niles Eldredge (1995) called an ultra-Darwinism, which was characterized by a focus on the gene, and a more pluralistic view.

Ultra-Darwinism arose when it was recognized that the gene, rather than the whole organism, could be treated as the unit of selection. The emphasis was now on the reproductive success as the driving force of selection, leading to a renewed interest in sexual selection (largely ignored by the founders of the synthesis) as opposed to adaptation. By applying the concept of the selfish gene to the origin of social instincts, the new Darwinism opened up a revolutionary approach which became known as sociobiology. The possibility that human behavior might be conditioned by genetically implanted instincts shaped by natural selection generated immense controversy. More generally, ultra-Darwinian philosophers have been led to invoke natural selection as the key to a whole new understanding of nature. In one area after another, they claim, the mechanism of random variation and selection undermines traditional beliefs and shows us a new way to understand our place in the universe.

Traditional Darwinian interests still survive among naturalists, of course. Detailed studies of the Galàpagos finches have shown that even short-term environmental changes are reflected in the character of the population (J. Weiner 1994). Island biogeography more generally has given us insights into the ecological dimensions of evolution, showing how human activity can exert pressures that lead to extinction (Quammen 1996). These studies confirm the significance of adaptation but also show that evolution often depends on geographical factors. Gould's emphasis on the unpredictability of evolution is consistent with such a view, as is the claim that mass extinctions represent important episodes in the history of life. In contrast, the ultra-Darwinians tend to focus so sharply on the mechanistic nature of evolution that they are unwilling to recognize any degree of unpredictability. They accord little significance to biogeography or mass extinctions, instead returning to R. A. Fisher's model of selection as a process which sorts out fitness within large populations, with more or less predictable results. The resulting controversies have been called the Darwin wars (A. Brown 1999; Milkman 1982; and Ruse 1982 are good examples of the literature generated in the opening stages of these debates).

The ultra-Darwinians are convinced that natural selection acting at the most basic level is the only driving force of evolution. This view entails a commitment to extreme gradualism, because adaptive evolution will normally occur as the species tracks gradual changes in its environment caused by geological forces. But even in a stable environment, selection will always produce change, because—as Darwin himself realized—all the species in a region are, in effect, competing with one another for resources. According to Darwin, this effect produced a tendency to specialize for a particular way of life. Nowadays, we think in terms of what L. Van Valen (1973) called the Red Queen hypothesis. In the looking glass world, the Red Queen tells Alice that one has to run as hard as one can merely to stay still; in the evolutionary race, each species has to improve itself constantly or it will be outstripped by its rivals (see Ridley 1993 on this effect and its implications for human sociobiology).

But the most important transformations in ultra-Darwinism have come from a recognition that the theory must deal with other aspects of fitness besides simple adaptation, in particular, with the complexities of animal behavior. Sexual selection, still largely ignored in the 1950s, has become important for perhaps the first time since Darwin proposed the idea (Cronin 1991). With this shift of emphasis comes a different way of conceptualizing selection as a whole: what matters is not survival but reproduction, the successful transfer of genes to the next generation. Because survival seems essential to reproduction, the two processes are complementary rather than antagonistic; but to some ultra-Darwinians it is the reproductive competition which leads to the more interesting questions. In the case of the peacock, sexual selection has produced a tail of such magnificence that it must count as a hindrance in the struggle for existence, yet the advantage of attracting peahens evidently outweighs this adaptive disadvantage. The real puzzle comes in addressing those cases where the males actually fight to obtain access to females, because all too often it seems that the apparent advantages of all-out aggression are not manifested in the actual behavior shown. The British evolutionist John Maynard Smith (a student of J. B. S. Haldane) showed that this problem could be tackled through the application of games theory (Smith 1982). In a population of "doves" who avoid combat, a single "hawk" who fights at every opportunity will be at an advantage, and hawk genes will increase in frequency. But in a population of hawks a single dove may be at an advantage, because the hawks are too busy fighting to reproduce successfully. Thus, there will be an "evolutionary stable strategy" for the species in which a balance of hawks and doves is maintained. In such cases selection does not promote purely aggressive instincts, as some simpleminded critics of Darwinism have assumed, but works to maintain a balance of different characters in the population.

These problems came to a head when the new generation of Darwinists confronted the topic of altruism, or behavior in which the individual sacrifices its own self-interest to promote the survival or reproductive success of another. For example, the worker ant is neuter, spending its whole life helping to raise the offspring of another individual, the queen. How can natural selection have programmed the ant with instincts to behave in this altruistic manner? Even Darwin found this question problematic: for once he abandoned his reliance on individual natural selection and postulated what later became known as group selection. On this model, competition between groups is more important than between individuals: the nest of ants with cooperative instincts is more successful than a rival nest where there is no cooperation, and so the latter is eliminated. In the years following the consolidation of the evolutionary synthesis, V. C. Wynne-Edwards (1962) used the idea of group selection to explain many otherwise puzzling aspects of animal behavior. But to increasingly self-confident Darwinians such as G. C. Williams (1966), this approach was a betrayal of the true spirit of the theory. There is an obvious problem with group selection, spotted immediately by anyone in tune with the thinking of individualistic selectionism: what happens when mutation produces an individual with the instinct to cheat on the reciprocal arrangements within the group? Such an individual would be at an advantage, gaining from the cooperation of others but offering nothing in return. It should therefore be more successful, and the cheating gene would spread into the population, destroying the effects of group selection.

W. D. Hamilton (1964) pioneered what was to become the key insight of sociobiology: the notion of kin selection. As J. B. S. Haldane once had observed, there might be circumstances in which self-sacrifice was the best way of getting one's genes transferred to the next generation, as long as those who benefited were genetic relatives who shared some of the same genes. Kin selection explains apparently altruistic behavior by noting that the altruism is always expressed toward relatives and helps them reproduce. Selection favors the development of altruistic instincts because the genes may be able to reproduce more effectively by encouraging the individual to help relatives breed. In the case of neuter insect castes, scientists knew that the hymenoptera (ants, bees, and wasps) have an unusual reproductive pattern in which males are raised from unfertilized eggs, while among the females, sisters are more closely related than mothers and daughters. This favors the development of sterile female castes, because a female best enhances her genetic representation in the next generation by helping her fertile sisters rather than by reproducing herself. Significantly, a student of ant behavior, Edward O. Wilson, produced the classic survey of sociobiology (1975).

Wilson's book was attacked by social scientists and those on the political left because his final chapter extended the theory to human behavior and thus laid him open to the charge of reviving social Darwinism (Segerstrale 2000; for contemporary literature, see Caplan 1978; Clutton-Brock and Harvey 1978; Ghiselin 1974; Hull 1978; Rose, Kamin, and Lewontin 1984; Ruse 1979b). The theory certainly seems to encapsulate traditional stereotypes about behavior, since it is in the male's reproductive interests to mate as frequently as possible and not stay around to raise the offspring, while the female should look for those males who will be the best fathers for the few offspring she can rear. More specifically, the theory could be used to legitimize our tendency to favor genetic relatives over perfect strangers, so that abstract moral values are dismissed as unworkable because they are incompatible with human nature. Left-wing critics warn that sociobiology thus will enshrine selfishness as natural, something that must be built into the foundations of any workable political philosophy. Human sociobiology does have some evidence in its favor, especially its ability to explain why all cultures have rules discouraging incest. The rules differ among cultures, but the underlying purpose seems to reflect a biological imperative founded on the fact that inbreeding can increase the chances of deformity in the offspring. As Ruse (1979b) argues, we should not rule out on principle the possibility of some biological input into human behavior. We may prefer to believe we behave freely, but it would be remarkable if we had escaped our evolutionary past completely.

A major criticism of sociobiology focused on its more basic claim that human behavior is genetically determined. This diminished the gulf between humans and animals that was based on our supposed development of higher faculties and revived the nature side of the old nature-nurture debate. Here, the new biology challenged the basic assumption on which the social sciences of the twentieth century had been built: the rejection of biology as a determinant of human behavior. In this respect, the sociobiology debate has fed into the much wider discussion of the implications of genetic determinism now being driven by our ability to control the genes directly. Wilson went on to suggest that, by explaining the reasons why we are so attracted to religious beliefs, biology will help us to establish a new, humanistic philosophy by which to live (1978).

To the critics, this deterministic approach is hard to distinguish from the more explicitly right-wing uses of genetics which have begun to flourish in recent years. There are those who argue that individual ability is rigidly pre-determined by the genes. Some would apply the same claim to the alleged differences between the races, openly reintroducing the race biology that the modern evolutionary synthesis was supposed to have destroyed (Herrnstein and Murray 1994; for critiques of the new determinism, see Gould 1981; and Rose, Kamin, and Lewontin 1984). In a less obviously sinister way, the new genetic technology, coupled with the belief that every character has a genetic foundation, offers the prospect of a new eugenics driven not by the state but by individual couples choosing the characters they will pass on to their children (Paul 1998).

Ultra-Darwinism focused attention on the gene as the unit of selection: what drives natural selection is the success of the genes in reproducing themselves, not the success of the individuals who carry the genes. This way of thinking was encapsulated in the title of a book by the British biologist Richard Dawkins, *The Selfish Gene* (1976). Dawkins warns that his use of the term *selfish* is purely metaphorical, since the genes have no feelings and the process of competition between them is totally mechanistic. More recently he has argued that we must take a much broader view of the factors which contribute to successful reproduction, although the bottom line remains the transmission of the genes (1982). In his later career, he has abandoned scientific research to become one of the most vociferous public advocates of the ultra-Darwinian position. He has attacked all efforts to defend any residue of the argument from design (1986), insisting that natural selection is the "blind watchmaker," a purely naturalistic process which replaces Paley's intelligent Creator in the generation of complex structures. The progress which has led to humanity itself is the necessary outcome of the endless competition for reproductive success (1996). In Dawkins's crusade to use Darwinism as a means of dissolving all traditional belief in a purposeful universe, he has been joined by philosophers, such as Daniel Dennett (1995), who apply the mechanism of trial and error to explain everything, including the functioning of the human brain.

OPPONENTS OF ULTRA-DARWINISM

To many religious thinkers, the radical materialism of Dawkins and Dennett encapsulates the most threatening implications of the Darwinian theory. There are also biologists who are by no means religious yet fear that gene selectionism produces an impoverished view of nature. The paleontologist Stephen Jay Gould, also a successful popular-science writer, became the best known of these critics. Some scientists dismissed him as nothing more than a popularizer, and his name has certainly been invoked by a wide range of anti-Darwinians who would go much further than Gould himself. There was an episode in the 1980s when Gould did indeed seem to be toying with explicitly non-Darwinian ideas, including saltationism, but his position became much more moderate before his death in 2002. He had no doubt that natural selection is the main driving force of evolution, and his emphasis on the unpredictability of long-term events in the history of life was an attack on cosmic teleology. But Gould and his colleague Niles Eldredge (1995) felt that the abstract model of evolution developed by the neo-Darwinists left out important elements that explain how evolution works in the real world.

There are a number of points on which the neo-Darwinian theory can be challenged. Gould and Eldredge objected to the assumption that every character has an underlying genetic foundation shaped by natural selection, pointing out that, since there is a complex developmental process leading from the genes to the adult organism, many characters may be merely by-products of natural constraints imposed by that process. This antiadaptationism has broader consequences when applied to the human implications of sociobiology, where it can be used to challenge the claim that every behavioral trait must have a genetic determinant. More generally, the opponents reject neo-Darwinism's tendency to shift the focus of natural selection from the organism to the gene. Older Darwinians such as Ernst Mayr had always insisted that it was the organism which confronted the environment and hence must be treated as the unit of selection. Gould and Eldredge moved in the direction opposite of neo-Darwinism, suggesting that many long-range trends in evolution may be due to selection acting among species, with some species going extinct when confronted with better adapted rivals. From the same perspective comes their interest in the unpredictability of evolution as a result of the hazards of geographical and geological changes, including the possibility of catastrophic mass extinctions. But perhaps the most radical aspect of this modified Darwinism is the hint which occasionally emerges that constraints imposed by the developmental process might generate nonrandom variation and hence actively direct the course of evolution. This is an idea which, in the hands of the more extreme critics, becomes an outright anti-Darwinism.

The public's attention was first drawn to the emerging debates by the publicity centered on the theory of "punctuated equilibria" (Eldredge and Gould 1972; see also Eldredge 1986; Gould and Eldredge 1977). This was a response by the two paleontologists to what they perceived to be the evolutionary synthesists' excessive commitment to gradualism. If evolution is driven solely by individual natural selection enhanced by the Red Queen effect, the fossil record should show continuous evolution as species adapt to gradual changes in their environment. Instead, the record almost always shows species remaining static over geological time spans and then being replaced quite suddenly by related species. This evidence had been dismissed by the synthesis as an artifact of an imperfect record, but Eldredge and Gould now argued that the jumps had to be taken seriously—they were genuine evidence of changes that were abrupt, at least in geological terms (although they might seem quite slow by human standards). To explain these punctuations, Eldredge and Gould drew on an idea that was actually part of the synthesis, Ernst Mayr's claim that new species usually arise from "peripheral isolates" and not the main population. Small populations cut off from the main body of the species in an unusual environment might be expected to evolve quite rapidly, and no evidence of this would be left in the fossil record because of both the rapidity of the change and the restricted area in which it took place. If the new species suddenly got the chance to invade the main territory of the parent species under circumstances where environmental changes gave it an advantage, it rapidly would displace the parent form and thus appear quite abruptly in the fossil record for the area.

At this level it would be possible to claim that the punctuated equilibrium theory was no more than an extension of the synthesis. But to Gould and Eldredge, the element of stasis in species was incompatible with Darwinian gradualism. Over millions of years, there must have been at least slight changes in the environment, yet there was no sign of the species tracking such changes. This suggested that there were internal factors preserving the original form, perhaps constraints imposed by the existing developmental pathway which made it difficult for ontogeny to accommodate genes that would modify the pathway even in adaptively beneficial ways.

It was no accident that, at the same time, Gould teamed up with the geneticist Richard Lewontin to argue that the synthesis was naive in its assumption that every character of the organism had an adaptive purpose (Gould and Lewontin 1979; Lewontin 1974). They argued that many features which we treat as characters and assume to be based on a genetic foundation were in fact merely by-products of the developmental process, necessary for the way the organism is built and not available to the action of selection. They described them as being analogous to the spandrels of San Marco, triangular structures beneath the circular dome of the church of San Marco in Venice, which are highly decorated. One might assume that the spandrels had been designed to provide a prominent space for decoration, but in fact they are simply a structural necessity. Many structures of living things fondly assumed by the Darwinians to be adaptive might be similarly necessary consequences of how the developmental process works, and in this case no amount of selective pressure could change their basic structure (although minor modifications equivalent to the decorations on the spandrels might be possible). Lewontin and Gould suggested that the basic structure of the main phyla (vertebrates, etc.) might be fixed in this way. Thus they form underlying patterns that can be modified, but never fundamentally altered, which explains why no new types have appeared since the Cambrian. Gould was even prepared to use the term *Bauplan* (ground plan)—characteristic of typological thinking and German idealism—to denote this underlying fixity of type. On such a model, the basic plan of each phylum might not be adaptive at all, because only a limited number of fundamental developmental pathways are available. All natural selection can do is tinker with the details.

For a while Gould seemed to be teetering on the edge of outright anti-Darwinism, and there was much discussion about the possibility of a new direction in evolution theory (Gould 1980b, 1982; Milkman 1982). He hinted at support for Richard Goldschmidt's saltationist theory of the hopeful monster, in which new species were produced by sudden transformations of the developmental pathway, perhaps triggered by quite small initial genetic changes. He also wondered if developmental constraints might predispose variation to take place in fixed, and not necessarily adaptive, directions. Some advocates of what became known as epigenetic evolution certainly seized upon these new ideas as sticks with which to beat the Darwinians (see below).

Gould himself soon backed away from these more extreme ideas and returned to an interpretation of punctuated equilibrium theory that could be integrated with a generally Darwinian viewpoint. On the model that he and Eldredge came to favor, evolutionary trends are not due to developmental constraints biasing variation in a particular direction but to species selection (see fig. 34). This retains the idea that the origin of new species may be random with respect to the overall direction of adaptive change, not because the species are formed by nonadaptive trends but because peripheral isolates normally will be confined to small locations with untypical conditions. Only rarely will it turn out that a species formed in this way will have an advantage over the parent species; but when that does happen, competition will allow the newcomer to displace the parent as soon as it is able to invade the main territory. As the very reverse of the ultra-Darwinians' focus on the selfish gene, species selection has generated a major controversy over the question of what is the most appropriate unit of selection: gene, organism,

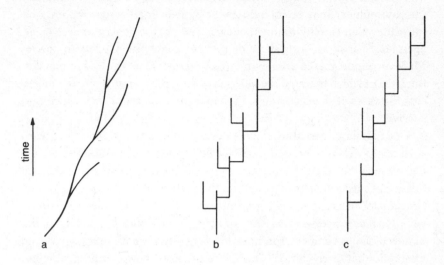

34. Patterns of evolution.
a)   Orthodox Darwinian view of an evolutionary trend in which there is some speciation, but all branches move in the same direction under the pressure of natural selection.
b)   Punctuated equilibrium model with species selection. Each branch remains fixed once it is formed, and speciation is random with respect to the direction of the trend. The trend results from the preferential survival of those branches which diverge to the right, because they are favored by overall changes in the environment.
c)   Punctuated equilibrium model with an "origin bias" influencing the trend. Because of internal factors affecting individual variation, speciation occurs more readily to the right than to the left. The trend is a product not of environmental influence but of built-in, almost orthogenetic, factors.

or species (Brandon and Burian 1984)? Efforts also have been made to revive the theory of group selection as a means of explaining altruistic behavior, thereby checkmating the most important argument of the sociobiologists (Sober and Wilson 1998).

From the perspective of Gould and Eldredge, the overall course of evolution looks much less predetermined than it does within the ultra-Darwinian model of relentless adaptation within the main population. It is a matter of chance whether or not a peripheral environment generates a species better adapted to the main territory, and whether that species gets an opportunity to migrate to the main range. Even then, some less well-adapted populations may remain protected by geographical barriers. Major transformations of the environment may be brought about by geological forces, not only because they alter the environment but also because they create or destroy geographical barriers which allow or prevent competition between species. Massive vulcanism and even extraterrestrial factors might lead to mass extinctions in which many of the old species disappear—but again, the precise circumstances of the catastrophe may make it a matter of luck which ones survive.

One of the more exciting themes developed toward the end of the twentieth century was the recognition that developmental constraints within ontogeny must be taken into account in evolution. Susan Oyama (1985) and others have led a sustained assault on the preformationism of the old unit-gene model, insisting that a better understanding of how the organism develops makes it necessary to devise a new and more flexible understanding of how biological and environmental factors interact both in ontogeny and, by implication, in evolution. This is the proposed new synthesis of epigenetic evolution, or "evo-devo," within which developmental constraints serve as a major limitation on the ability of natural selection to manipulate the organism by simply tinkering with its genotype (Raff 1996; Wilkins 2001). In the eyes of its more extreme advocates, this movement once again looms as a challenge to the underlying Darwinism of the evolutionary synthesis.

ANTI-DARWINIANS

From the beginning, there were some—both inside and, more especially, outside—the scientific community who were uncomfortable with the reemergence of Darwinism. The mainstream critics of ultra-Darwinism often employed arguments similar to those once used to support a non-Darwinian view of evolution. Those with a more active dislike of the selection theory were only too willing to point out what they perceived to be its shortcomings. In some cases, these attacks were presented so vigorously that they seemed to entail a wholesale rejection of evolution as a useful scientific theory, a rejection which seemed to leave a few scientists and academics on the side of the creationists. Other critics seemed to want to turn back the clock to an earlier, less mechanistic form of evolutionism. Yet the moral, philosophical, and theological positions which drive these anti-Darwinian and antievolutionary arguments are genuine enough. They represent fundamentally different worldviews in which nature is perceived as more purposeful or more orderly than seems possible if the natural selection of random genetic variation is held to be the only driving force of evolution.

One of the most emotionally compelling arguments used by the neo-Lamarckians of the late nineteenth century was the claim that Darwinism was a mechanistic theory which reduced living things to puppets driven by heredity. The selection theory made life into a game of Russian roulette, where life or death was predetermined by the genes one inherited. The individual could do nothing to mitigate bad heredity. Lamarckism, in contrast, allowed the individual to choose a new habit when faced with an environmental challenge and shape the whole future course of evolution. This was a view championed relentlessly by literary figures such as Samuel Butler and George Bernard Shaw. In the middle decades of the twentieth century, their mantle was taken up by Arthur Koestler (1967, 1972) as part of a wholesale attack on the mechanistic view of life. Koestler saw the Darwinism of the evolutionary synthesis as a renewal of the mechanist threat, and he portrayed the new selection theory as being as hostile to the notion of individual freedom and creativity as the old. He admitted that earlier selection theory had acknowledged a role for behavioral innovation in directing Darwinian selection—the old Baldwin effect, in which new habits are supposed to determine which genetic variations are most useful. But he insisted that the new generation of Darwinians was blind to this possibility, and it was certainly true that G. G. Simpson had written against the Baldwin effect (1953a). Other mid-twentieth-century biologists, however, took a less rigid view. C. H. Waddington's notion of "genetic assimilation" (1957) allowed a role for individual adaptation without invoking Lamarckism, while Alistair Hardy (1965) openly endorsed the Baldwin effect.

A few biologists still hope to undermine the dogma that a barrier prevents acquired characters from influencing the genes. The discovery of DNA by James Watson and Francis Crick in 1953 (Olby 1974), and the subsequent elucidation of the process by which genes influence the developing organism, reinforced the early geneticists' rejection of Lamarckism. The information flow is one way only: DNA from the parent cells makes RNA, which in turn manufactures the proteins from which the new organism is constructed. There is no feedback process from proteins to DNA, and hence no possibility of a Lamarckian effect. However, the work of Ted Steele (1979; Steele, Lindley, and Blanden 1996) has kept alive the possibility that acquired characters might have a way of evading the barrier. Working with the immune reaction in mice, he produced evidence that acquired immunity might be transmitted to the offspring. New DNA produced by rapid mutations in the cells of the immune system might be able to infect the germ cells and thus be transmitted to the next generation. According to Steele, this opens up the possibility of a more general role for Lamarckian effects, one in which they play a rare but crucial role at key periods in evolution.

Lamarckism assumes that evolutionary changes are adaptive, but the more extreme advocates of epigenetic evolutionism take an antiadaptationist stance in which all substantive developments are internally programmed. This approach draws its intellectual support from a worldview which presupposes that the apparent diversity of living things masks an underlying unity based on fixed archetypes, or *Baupläne* (ground plans). The nineteenth-century idealists supposed that these types were based on eternal patterns in the mind of God. Their modern descendants are more likely to argue that the basic forms are imposed by developmental constraints defined by lawlike processes operating within the fundamental building blocks of life. On this model, the animal kingdom has a fixed number of basic plans, and the species making up each phylum represent distinct modifications of the underlying archetype, each produced by a discrete transformation of the pattern. Evolution is discontinuous and essentially predetermined or at least constrained within well-defined limits.

To those who adopt this position, the whole edifice of modern Darwinism is an illusion based on the unwarranted assumption that the accumulation of small-scale adaptive changes add up to give the major developmental transformations that will generate new types (Ho and Fox 1988, Ho and Saunders 1984; Løvtrup 1977, 1987; see also Gerry Webster's introduction to the reprint of W. Bateson 1894, which nicely identifies the intellectual ancestry of the movement). The Darwinists see this approach as a last-ditch effort to sustain the idealist viewpoint of the nineteenth century. Each finds the other's worldview almost unintelligible. The anti-Darwinians often claim that their perspective will block the efforts of those who would use biology to endorse racist or other right-wing ideologies—forgetting that earlier advocates of theories similar to their own were capable of endorsing the policies they despise.

A related anti-Darwinian viewpoint was developed from the very different perspective of biogeography by Leon Croizat (1958, 1964), whose work has attracted the attention of a small but vociferous group of biologists (Nelson 1978; Nelson and Rosen 1981; see also Hull 1988). Croizat's "panbiogeography" rejects the Darwinians' emphasis on the dispersal of dominant groups from an ancestral point of origin, on the grounds that the observed distribution of species bears no relationship to their dispersal abilities. The theory postulates that all speciation takes place vicariously—

that is, by natural barriers splitting up a once continuous population, so that each fragment then acquires its own distinct character. In his determination to eliminate dispersal, Croizat assumed that all widely distributed groups have occupied more or less the same broad swath of territory as far back as one can trace in the fossil record. To some extent, his insights were compatible with the emergence of plate tectonics and the modern version of continental drift, since this theory endorsed the idea that the breakup of a supercontinent into fragments formed our present landmasses. Any group distributed across the whole of the ancient supercontinent would indeed have had its territory fragmented, leaving separate populations to evolve in isolation. But in order to explain many similarities in later episodes of evolution, Croizat was forced to postulate parallel developments in separate populations, thereby invoking something like the old theory of orthogenesis. Thus his supporters could find some common ground with the critics of Darwinism who invoked developmental constraints on evolution.

DARWINISM NOT SCIENTIFIC?

More fundamental challenges to modern evolutionism claim that it is an illegitimate form of scientific knowledge. Natural selection, critics claim, is a tautology, a play on words, not a genuine proposition with empirical consequences. Some philosophers of science have argued that evolutionary explanations of past events are not scientific because we cannot test them in the way laboratory scientists test their hypotheses. This challenge raises genuine questions about the scientific method. Evolutionists respond that to apply the methodology of physics to any historical science is to miss the point that an empirical study of such questions must adopt different standards. The situation is complicated by the fact that some biologists support the attack on evolutionism, not because they endorse creationism but because they feel that evolutionary relationships cannot be studied scientifically.

The claim that natural selection is a tautology with no explanatory power has been popular among philosophical critics (e.g., Macbeth 1971; Manser 1965). A tautology is a proposition which is true because its components are defined so as to make its truth a logical necessity—such as "all husbands are men"—not because it has empirical content which can be confirmed by observation. The philosophers' argument starts from the definition of natural selection as the survival of the fittest and asks how the fittest are to be recognized. If the only way of identifying the fittest is to see who survives, then natural selection is reduced to the tautology "the survival of those who survive." Darwinians reject this analysis (e.g., Ruse 1982) on the grounds that fitness is defined by how well the organism is adapted to its environment. It is as a consequence of being better adapted that the organism is more likely to survive and reproduce. In the peppered moth, dark coloration provides protection from predators when the surroundings are soot-blackened, and studies have actually shown birds picking out light-colored moths and missing the black ones. Field naturalists who study such interactions in the wild wonder how armchair philosophers who have never done a day's empirical research in their lives can twist logic to make such observations seem an illusion.

Richard Dawkins (1982) concedes that the problem has been exaggerated by the emergence of sociobiology, with its tendency to define fitness in terms of reproductive success. This is necessary if sexual selection is to be included, but it does deflect attention away from the role of adaptation. Even in sexual selection, reproductive success is gained as a result of a behavioral or physical character, such as the peacock's tail. One can still identify in principle the fittest individuals before they actually reproduce—for instance, by predicting that the peacocks with the biggest tails will be more successful in getting mates. The problem, as conceded by Ronald H. Brady (1979), is that in many cases biologists cannot actually identify the character conferring increased reproductive success. The peacock's flamboyant tail is obvious to us as well as to the peahens, but equivalent structures and behavior in the lower animals may not be as easy to identify. In such cases, biologists have tended to assume that, if a gene is maintained at a high level in the population, then it must confer a "fit" character. When they do this, they leave themselves open to the charge of reducing natural selection to a tautology, although they are aware that in such cases they are cutting corners because of the inherent difficulty of the subject matter.

The same problem is evident when evolution theory is used to explain how particular steps in the history of life came about. Karl Popper (1974) argued that the adaptive scenarios postulated to account for the development of new characters are largely untestable and hence unscientific. Popper gained his reputation by promoting a philosophy of science which sought to distinguish between true science and pseudosciences such as astrology. The key, he claimed was the testability, or falsifiability, of their hypotheses: a true science always frames its hypothesis in a way that maximizes the possibility of exposing error through experimental testing. By this standard, he insisted, Darwinism is unscientific because there is no way of rigorously testing the explanations it offers about events in the distant past. A later generation of critics accuses the evolutionists of telling "just so" stories, named after Rudyard Kipling's fanciful tales of how the leopard got his spots, and so on. The target of such an attack is not Darwinism itself but any theory of adaptive evolution. Kipling's own stories were Lamarckian—but in the current situation, Darwinism is the only theory of adaptive evolution on offer. Popper also argued that, since the adaptive scenarios are sequences of individual events, evolutionism offers no general laws and thus cannot predict the future. If there are no predictions, then there can be no effective testing. At best, Popper concluded, Darwinism established a metaphysical framework within which testable theories might eventually be framed.

Ruse (1977, 1982) challenged Popper on this issue by distinguishing between the causal theory of Darwinism and its application to particular events in the history of life. As Gerhard D. Wasserman (1978) also pointed out, the modern Darwinian theory has been tested over and over again against the genetic structure of modern populations. According to Wasserman, microevolution can be tested, but explanations of macroevolutionary events cannot. Dawkins (1982) argued that the different levels of selection postulated by Darwinians generate different and testable conclusions, which is why (in his opinion) group selection has been rejected. Yet most Darwinians would feel that their theory was impotent if it could not address questions about how particular steps in evolution were produced. Ruse conceded that in many cases it is difficult to generate testable explanations of how a species evolved. But there are more general theories about the sequence of events in the past which have been tested and, in some cases, found wanting, so there is no reason why a historical explanation must be dismissed as unscientific. Any explanation based on the existing fossil record can be falsified if new discoveries do not fit in with the scenario it postulates. There are many indirect ways of using modern organisms to throw light on the past—for instance, by testing predictions about the use to which a similar structure in a fossil species might have been put. After some debate, Popper himself softened his opposition to the Darwinian approach (1978).

At the same time, a similar line of attack was being mounted from within biology itself. The new approach to taxonomy known as cladism was pioneered by Willi Hennig (translation 1966), who insisted that the classification of species in terms of their evolutionary origins should focus solely on the points at which branches in the tree of life diverged (the term *clade*, from the Greek for branch, had been introduced by Julian Huxley in 1957). The name *cladism* in fact was coined by Ernst Mayr, a critic of the movement, and reluctantly adopted by its supporters. Although the technique initially was conceived within the framework of evolutionism, its more radical exponents, dubbed the transformed cladists, soon began to argue that the ancestor-descendant relationship so crucial to evolutionary explanations cannot be derived from a scientifically rigorous way of expressing degrees of similarity. From this position they mounted a campaign against Darwinism and against the legitimacy of all efforts to reconstruct the history of life on earth (Hull 1988; Scott-Ram 1990).

Founders of the evolutionary synthesis such as Mayr had always accepted that major branchings in the tree of life were among the most important criteria for defining the groups, or taxa, within the system of classification. In general, each taxon would correspond to a clad—that is, to all the divergent species descended from the form in which the defining features of the taxon first appeared. But for practical purposes they admitted exceptions to this rule when one side branch developed entirely new characters which, they felt, were sufficiently distinctive for all its descendants to be ranked separately. Henning expressed all relationships in the form of a cladogram, in which the only feature displayed was common descent. On this model, by definition all the descendants of the founder species were part of the same clade and hence belonged to the same taxon, however much they subsequently changed. If birds and mammals have both descended from reptiles, then they are part of the same clade and should not be ranked as separate classes. Or, if both are ranked as classes, then the reptiles cannot be a class, because they are merely the residue of a clade after two of its most important divisions have been siphoned off (see fig. 35). To the cladists, the old class Reptilia was a meaningless abstraction.

Hennig's cladograms still represented evolutionary relationships, and he was prepared to use fossil species to help identify the branching points in the tree. A fossil could be treated as the founder member of a clade from which all later species had descended. Later cladists such as Colin Patterson (1980, 1982) argued that, in a cladogram, all species, living and fossil, had to be arranged in a single line at the end of the tree: the branchings represented only degrees of relationship and had no historical significance. A cladogram drawn up in this way did not correspond to an evolutionary tree; indeed, it was consistent with a number of possible trees because it could not represent ancestor-descendant relationships (see fig. 36). The transformed cladists began to argue that evolution was irrelevant to their way of describing relationships. All that could be recognized were groups of sister species descended from a common ancestor which was completely hypothetical and, in principle, could not be identified from the information provided. From this perspective, evolution becomes meaningless because, even

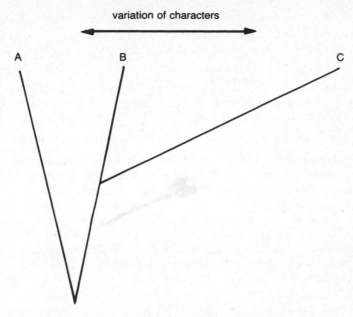

35.   Cladism and evolutionary classification. The diagram
represents two branching points in evolution giving rise to three
species, A, B, and C. The branch leading to C occurs later, but in the
course of subsequent evolution C has changed much more dramatically than either A or B. In an evolutionary classification, A and B
remain closely related and are kept in the same group, while C is
assigned to a new group because it has changed so much. Cladism
links B and C more closely together despite their differences,
because they share a closer common ancestor than either does with
A. To give a specific example, A might represent modern reptiles, B
the dinosaurs, and C the birds. Traditional classifications keep the
modern reptiles and the dinosaurs in the same class but create a
separate class for the birds. Cladistic classification treats the birds
and dinosaurs as a more natural group because they share a closer
common ancestor than either does with the other reptiles. On this
model, the reptiles do not form a natural class, because they can be
defined only by arbitrarily excluding birds (and mammals) from
the natural group formed by descent from a common ancestor.

if one believes it has happened, one cannot use it to generate scientifically
testable hypotheses. Cladists have gone on from this point to attack all efforts to reconstruct the history of life on earth, arguing that the sheer
depth of geological time, and the fragmentary nature of the fossil record,
renders futile all efforts to explain the origin of groups or characters (Gee

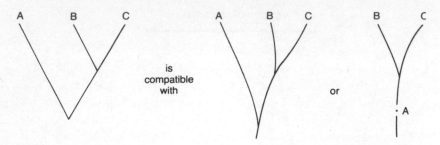

36. Cladograms and evolutionary trees. The cladogram on the left represents a group of three related species in which B and C share some characteristics not possessed by A. The cladogram would be compatible with either of the two evolutionary trees on the right. A might be ancestral to B and C, or it might lie on a side branch of its own. The cladogram thus cannot give information about ancestor-descendant relationships, and this gives rise to the transformed cladists' rejection of evolutionary trees as unfalsifiable. A similar cladogram linking four species would be compatible with twelve possible evolutionary trees.

1999). They are all "just so" stories, with as little foundation in science as Kipling's.

In response, David Hull (1979) pointed out that the cladists' attack was purely methodological: because their technique did not allow evolutionary relationships to be investigated, they assumed that no other techniques were available and dismissed evolutionism as untestable. Information on the fossil record is constantly expanding as new discoveries are made, and the techniques used to investigate it are also being improved. So who can say what will be permanently ruled out as being beyond the scope of investigation? In fact, new ideas are constantly being proposed to reform and sometimes dramatically alter our ideas about particular episodes in evolution, and new evidence is being found to test them. Hypotheses of a historical nature are thus open to testing and refutation, making them as scientific as any other (Nitecki and Nitecki 1992). To rule them out is to deny science access to the question of origins and to risk the consequent danger that nonscientific sources will seize the opportunity to assert their own claims. Relating the whole episode to Popper's philosophy, Hull pointed out that evolutionary trees were actually more falsifiable than cladograms, since the trees specified more details about relationships. Indeed, some of the cladists' literature refutes its own case: the cladists gleefully pick holes in traditional evolutionary scenarios, forgetting that in refuting them they undermine their own claim that such hypotheses are unfalsifiable.

CREATIONISM

The disagreements within the scientific community provide ammunition for those who would dismantle the whole edifice of a scientific approach to understanding the origin and development of life. Modern creationism feeds upon anything it can use to undermine public confidence in scientists' efforts; and to a public conditioned to think of science as offering simple factual information, any sign of theoretical disagreement can be presented as evidence that evolutionism is not science at all. At the same time, creationists are adept at challenging the arguments used to support evolution, often by citing grossly oversimplified versions of those arguments. Scientists all too often find themselves enmeshed in an effort to explain precisely why the attacks are based on misrepresentation, only to find that they are then perceived as being on the defensive or as attempting to hide behind a cloud of technicalities. To some extent, the defense of scientific evolutionism may depend on educating the public to give them a better understanding of what science really is: not a collection of facts but a system for generating hypotheses and testing them against the facts. With such schooling, it becomes evident that the attempt to create an alternative scientific creationism is an illusion. The arguments used are largely negative, and if a modicum of rational analysis is applied to the more extreme form of scientific creationism, its efforts to provide a workable alternative are seen to be inadequate. Even the more sophisticated version based on a revival of the argument from design has no positive research proposals to offer: invoking supernatural creation is merely making a claim that science cannot deal with the question of origins.

The modern creationist movement is largely American in origin (although Islamic countries have experienced a parallel sequence of developments). Europeans look on in amazement at a technologically sophisticated country where, nevertheless, a significant proportion of its citizens still thinks, in accordance with a literal interpretation of the Bible, that the earth is only a few thousand years old. This does not mean that all Europeans are atheists, although organized religion there certainly declined precipitously in the twentieth century. But to European Christians, the challenge is to develop new perspectives on theology which can accommodate all the insights of modern science, not to turn the clock back to a worldview in which the wording of Genesis functions as a straitjacket for knowledge of the material world. Fundamentalists want to hold the line against atheism by treating Darwinism as the agent of militant materialism. It has to be said that there are plenty of atheistical Darwinists about—Dawkins and Dennett are good examples—but there are also evolutionists who still hope to see some evidence of divine purpose in the process of universal history. Some of the most imaginative modern theology seeks to explore ways in which even the suffering and unpredictability of Darwinian evolution can be reconciled with the Christian faith in a God who gave His creation the freedom to evolve as best it could and is willing to suffer along with it.

The first phase of American creationism occurred in the 1920s and has come to be symbolized by the trial of John Thomas Scopes for breaking Tennessee's law against the teaching of evolution (see chap. 8). Contrary to popular belief, this movement—although inspired by fundamentalism—did not seek to establish the literal truth of the Genesis creation story, nor was it totally dominant within the churches of the American South (E. Larson 1998; Numbers 1998). Despite being pilloried in the press, the fundamentalists succeeded in keeping evolution out of the school curriculum for decades. Meanwhile, a more extreme form of the movement, which did hold to the literal truth of Genesis, was working away in the background, waiting for an opportunity (Numbers 1992). This came when the triumph of the evolutionary synthesis in the 1950s gave the scientists a renewed confidence to insist that Darwinism must be recognized at last as an integral part of the modern worldview and, hence, must be taught in the schools (Muller 1959; Simpson 1963).

The most active response came from a resurgent fundamentalism led by what would become known as "young earth" creationists. Their publications ran through the usual litany of anti-Darwinian arguments, many of them dating back to the days of Mivart and the initial response to the *Origin of Species,* and stressed the discontinuity of the fossil record as evidence that new types appeared suddenly rather than by gradual evolution (Gish 1972; Kofahl and Segraves 1975; H. Morris 1974; Parker 1980; Whitcomb and Morris 1961). But in fact, most of these arguments were either unnecessary or deliberately misleading, because the central assumption of the "scientific creationism" being offered as an alternative was that the earth is less than ten thousand years old. The fossil record is not merely discontinuous—according to the arguments, all the layers of sedimentary rock were laid down simultaneously in Noah's flood. For many of these creationists, Noah's ark really had been used to preserve samples of all living species during a flood which had rendered the whole earth temporarily uninhabitable. In this view of earth history, evolution theory would disappear almost as an afterthought when the sciences of geology, paleontology, archaeology, and cosmology were totally rewritten (nuclear physics too, since all forms of radiometric dating are held to be false). Rather than try to ban the teaching of evolution in the schools, the creationists demanded equal time for the teaching of their own alternative version of science (Nelkin 1977, 1983).

Southern states, including Arkansas and Louisiana, tried to pass equal-time legislation, leading to a series of highly publicized trials in which the American Civil Liberties Union sought to get the laws struck down as unconstitutional. The First Amendment of the Bill of Rights, which guarantees freedom of religion, has long been interpreted as forbidding the teaching of religion in the public schools. Although this principle is now again under attack, in the 1980s it was sufficient to allow the ACLU to get creation science dismissed as a smokescreen for presenting a Genesis-based account of earth history. Many eminent scientists and philosophers were called to give evidence; all sought to drive home the point that there is a crucial difference between a worldview in which the evidence is shoehorned into a preexisting pattern defined by the sacred text and the scientific methodology of putting forward hypotheses to be tested by the facts. The purely negative emphasis of most creationist arguments was stressed: creation science consisted mostly of arguments against evolution, leaving it to be assumed that, if that theory were discredited, the only alternative was divine creation as described in Genesis. There was no real effort to use creationist theory as a guide to future research.

The equal-time laws were all struck down; but on platforms around the country, creationist speakers such as Duane T. Gish were embarrassing scientists who found it difficult to present a complex case for evolution to a public that had been led to assume that a scientific truth should be immediately demonstrable. Unlike most scientists, the creationists were skilled debaters, and they knew that once they had put the scientists on the defensive by challenging their theories, any effort to respond must founder on the difficulty of explaining to ordinary people how their perceptions of geology and paleontology had been manipulated. In the long term, the best response was to encourage an improvement in the public's understanding of science, and at this level the evolutionists began to fight back (Eldredge 1982; Futuyma 1982; Godfrey 1983; Kitcher 1982; Montagu 1982; Ruse 1982; Strahler 1988). But long-term education is of little help to a debater faced with an audience of laypersons who have not yet had access to better information. The most effective rebuttals to creationism came from those who sought to expose the true foundations of creation science and subject it to critical scrutiny. Here, as in any debate, the best form of defense is attack. It may be hard to prove from the fossil record that evolution has occurred, but it is much easier to show that the complex structure of the earth's crust is incompatible with the claim that all the strata were laid down in a single event. Any attempt to explain the current geographical distribution of species on the assumption that they all migrated out of Noah's ark also generates numerous absurdities. How, for instance, did the koala bear, which moves laboriously and feeds only on eucalyptus leaves, get from Mount Ararat to Australia?

It was the problems generated by the link with Genesis, rather than the efforts of the evolutionists, which forced the creationists to change tack in the 1990s. The new assault comes from the "intelligent design" theory proposed by the biochemist Michael Behe (1996) and the lawyer Philip E. Johnson (1991). This abandons the young-earth position and concentrates purely on a revival of the argument from design. As with the analogy of William Paley's watch, the claim is that living systems are so complex that they cannot have been created by a natural process, least of all one based on trial and error. Behe points to biochemical processes in the body which, he insists, show "irreducible complexity" in the sense that (like the watch) the whole mechanism has to be in place for the system to function at all. Take away any part, and the whole ceases to function—so how can it have been built up from a more primitive state? Scientists have responded by pointing to cases where we can in fact see intermediate stages which do function.

There have been signs of disagreements within the creationist camp, with one advocate of intelligent design conceding that the argument is compatible with a theistic evolutionism, that is, with a version of evolution in which God's designing hand directs the process (Denton 1998). The argument that many scientists would bring against this whole program is that, once again, the creationist position is purely negative. It can never be used as the basis for a research program because all it can do is set up the claim that something is impossible, leaving the evolutionists to work out how nature could have circumvented the alleged barrier. The supernatural can never be the basis for a scientific investigation because in principle we cannot use natural evidence to throw light on how a miracle, or a supernatural guiding hand, works (Pennock 1999).

Whatever the limitations of the intelligent design movement as perceived by the majority of scientists who actually investigate origins, it has been remarkably successful in sustaining creationist views among educated Americans. Public protests against the teaching of evolution multiply, and in 1999, the Kansas State Board of Education voted to remove all references to evolution from the school curriculum (although the decision was subsequently reversed). The fact that the Kansas board also removed all references to the age of the earth and the big bang theory of cosmology makes it clear that, whatever the public rhetoric, the creationist movement is still being driven at the grassroots level by the young-earth position. Surveys suggest that about 35 percent of Americans accept the literal truth of the Genesis creation story.

To some scientists, it is equally disturbing to find that their enemies among what is called the academic left actually support the creationists' demand to have their theory taught in the schools. The academic left flourishes in humanities departments, where it teaches a relativist view of knowledge in which there can be no citation of objective evidence: all texts, including those written by scientists, are produced as attempts to gain cultural power over others (for critiques, see Gross and Levitt 1994; Sokal and Bricmont 1998). On such a philosophy, objective science is an illusion and the creationists' texts are just as valid as the evolutionists', since each side is merely bidding for influence over the public (Eve and Harrold 1991; Smout 1998). Scientists no longer seek to deny that external factors do have some influence over the development of theories, if only because such factors can determine which areas are seen as priorities for study. But scientists are naturally concerned at the suggestion that any hypothesis, however bizarre, can be proposed with impunity because all efforts at empirical testing are to be dismissed as worthless. The fact that many creationists think they have an alternative source of legitimacy for their theories, a divine imprimatur, seems to have escaped the attention of the scholars who are so anxious to challenge what they perceive as the cultural hegemony of modern science.

Thus, evolutionism has become sucked into the science wars of the academic community while still being viewed with suspicion by a substantial number of Americans. Clearly, the Darwinian revolution is not yet completed. The assumption of Christian creationists that an attack on evolution must automatically generate support for a biblical view of human origins cannot be left unchallenged in any multicultural society. Islam too has its fundamentalists who are suspicious of evolution because it is inconsistent with the word of the Koran. Evolutionary biologists are currently under attack in Turkey from Muslim creationists. The proponents of exotic alternatives, such as Erich von Däniken (1970, 1977), use almost identical arguments against evolution. Von Däniken invokes the discontinuity of the fossil record to support a theory which postulates episodes of genetic engineering by extraterrestrial visitors at several points in the earth's history. His books were extremely popular, and although he no longer has a large following today, there are many who share his suspicion of the orthodox scientific story of earth history without wanting to see it replaced by biblical literalism. Unfortunately for the evolutionists, the supporters of the rival positions seldom bother to attack one other, even though their alternatives are radically incompatible. Evolution is thus the target for attacks from a multitude of hostile forces, all of which use the same weapons and seem to be willing to forget their differences in their anxiety to resist a perceived threat from a dogmatic scientific orthodoxy.

Many critics charge that evolutionism is the agent of a militant materialism determined to overthrow all traditional values. They take no account of the fact that many different theories of evolution have been proposed over the last two hundred years and considerable differences of opinion arise even within modern science. There are indeed militant atheists such as Dennett and Dawkins who use Darwinism to undermine belief in the traditional foundations of religious and moral values. But there are also many scientists, including some evolutionary biologists, who resist this interpretation of the theory's implications. When this view of science is linked to a liberal Christianity, the whole notion of a conflict between them can be overturned (Peacocke 1980). Admittedly, there has been a long-standing preference among religious evolutionists for non-Darwinian theories which preserve a more obvious role for order and purpose in the world. But with the continuing support for at least some form of the selection theory in modern biology, liberal Christian thinkers are struggling to develop a theology of creation which can cope with Darwinism. They see God not as the Creator who designed everything from the start, but as a Power struggling to articulate its purpose within the world and participating in the suffering that is a necessary consequence of that struggle. They point to the open-endedness of Darwinian evolution to show that the result is not preordained and, hence, to argue that the freedom of the human will is not an illusion (K. Miller 1999). These efforts demonstrate the falsity of the claim that serious engagement with religion must necessarily require a rejection of Darwinism. A middle ground does exist, for all the efforts of extremists on both sides to conceal it.

The public hostility to science which has been exploited by the creationists also points to a pressing need for a more realistic public understanding of how science works, in order to dispel the simple image which supposes it to be a collection of established facts. It obviously would help if people simply had more scientific knowledge. Anyone familiar with the structure of the earth's crust, or the distribution of animals and plants around the globe, would be less likely to take seriously a theory based on Noah's flood and the ark as a center of dispersal. But equally important is a clearer understanding of how science operates. The various opponents of evolutionism have thrived on the claim that, if evolution were a scientific fact, no opposition would be possible. Since they can show there are difficulties, then it cannot be a fact and their own alternative must be better. The defenders of evolution have to some extent played into the hands of their opponents by actually claiming that it is a fact. But clearly it is not, at least not in the sense that a layperson understands the term: it is a theory better supported by the facts than any of the alternatives. And the theory itself is evolving by the normal scientific process of throwing out new ideas and testing them against an ever-expanding realm of evidence. The controversies among scientists over the details of the evolutionary process are not, as the critics claim, a sign of weakness. On the contrary, they are signs of healthy debate and ongoing research. The creationists would be able to claim that they had an alternative form of science only if they could demonstrate how their own theory was generating similar research programs. Instead, they continue to focus on negative arguments against evolution.

If better understanding of science is the key, then perhaps evolutionists can draw some comfort from the fact that the public's awareness of science's limitations has been enlarged by controversies in the biomedical and environmental sciences. We are used to seeing scientists in these areas explain that they do not have immediate answers to complex questions because more research is needed. We are also getting used to seeing scientists disagree over the details of, say, the effect of greenhouse gases on global warming. Few educated people can still be trapped by the old image of science as a collection of facts. The controversies within evolution theory thus come to seem far more normal, part of the complex process by which science advances on a wide range of fronts. This book displays the immensely complex sequence of new ideas and factual discoveries which have shaped the emergence of evolution theory. That process is ongoing, and we should welcome the continuing debates as a sign that this area of science is as active as any other. To turn the clock back to a theory of the earth and a natural theology constructed by theologians in the seventeenth century would be a betrayal not only of science but of Western culture itself.

Bibliography

Despite its size, this is by no means an exhaustive bibliography. Its aim is to provide an outline of the most easily accessible primary sources and a good introduction to the secondary literature. Wherever possible, I have included English translations of works published originally in a foreign language and modern reprints of books published before 1900. Facsimile reprints of primary sources are cited under the original date of publication, but modern editions of older works are listed under the later date. In general, the original publisher is not given for books printed before 1920. Some books are issued by different publishers in Britain and America, and, because I have worked on both sides of the Atlantic, it is a matter of chance which edition is listed here. The bibliography has been extensively updated for this new edition, and in the process some older books, and foreign-language books recently translated, have been deleted.

Biographical information on many of the scientists mentioned in the text may be found in C. C. Gillispie's *Dictionary of Scientific Biography* (1970–80).

Adams, Mark B. 1968. "The Foundations of Population Genetics: Contributions of the Chetverikov School, 1924–34." *J. Hist. Biology* 1: 23–39.

———. 1970. "Toward a Synthesis: Population Concepts in Russian Evolutionary Thought." *J. Hist. Biology* 3: 107–209.

———. 1979. "From 'Gene Fund' to 'Gene Pool': On the Evolution of Evolutionary Language." *Stud. Hist. Biology* 3: 241–285.

———. 2000. "Last Judgement: The Visionary Biology of J. B. S. Haldane." *J. Hist. Biology* 33: 457–491.

———, ed. 1990. *The Wellborn Science: Eugenics in Germany, France, Brazil, and Russia.* New York: Oxford University Press.

———. 1994. *The Evolution of Theodosius Dobzhansky.* Princeton: Princeton University Press.

Adanson, Michel. 1763. *Familles de plantes.* Reprint, Paris: Lehre, Cramer.

Adelmann, Howard B. 1966. *Marcello Malpighi and the Evolution of Embryology.* 5 vols. Ithaca, N.Y.: Cornell University Press.

Agassiz, Louis. 1833–43. *Recherches sur les poissons fossiles.* 5 vols. and plates. Neuchatel.

————. 1842. "On the Succession and Development of Organized Beings at the Surface of the Terrestrial Globe." *Edinburgh New Phil. J.* 33: 388–399.

————. 1962. *Essay on Classification.* Edited by Edward Lurie. Cambridge, Mass.: Harvard University Press.

————. 1967. *Studies on Glaciers.* Translated by Albert V. Carozzi. New York: Hafner.

Alaya, Flavia. 1977. "Victorian Science and the 'Genius' of Women." *J. Hist. Ideas* 38: 261–280.

Alexander, Samuel. 1920. *Space, Time, and Deity.* 2 vols. London: Macmillan.

Allan, Mea. 1967. *The Hookers of Kew, 1785–1911.* London: Joseph.

Allen, David C. 1978. *The Naturalist in Britain: A Social History.* Harmondsworth, Middlesex: Penguin Books.

Allen, Garland E. 1968. "Thomas Hunt Morgan and the Problem of Natural Selection." *J. Hist. Biology* 1: 113–139.

————. 1969a. "Thomas Hunt Morgan and the Emergence of a New American Biology." *Quart. Rev. Biology* 44: 168–188.

————. 1969b. "Hugo De Vries and the Reception of the Mutation Theory." *J. Hist. Biology* 2: 55–87.

————. 1974. "Opposition to the Mendelian-Chromosome Theory: The Physiological and Developmental Genetics of Richard Goldschmidt." *J. Hist. Biology* 7: 49–92.

————. 1975a. *Life Science in the Twentieth Century.* New York: Wiley.

————. 1975b. "Genetics, Eugenics, and Class Struggle." *Genetics* 79: 29–45.

————. 1976. "Genetics, Eugenics, and Society: Internalists and Externalists in Contemporary History of Science." *Social Studies of Science* 6: 105–122.

————. 1978. *Thomas Hunt Morgan: The Man and His Science.* Princeton: Princeton University Press.

————. 1986. "The Eugenics Records Office at Cold Spring Harbor, 1910–1940." *Osiris*, n.s., 2: 225–264.

Alter, Stephen G. 1999. *Darwinism and the Linguistic Image: Language, Race, and Natural Theology in the Nineteenth Century.* Baltimore: Johns Hopkins University Press.

Amrine, F., F. J. Zucker, and H. Wheeler, eds. 1987. *Goethe and the Sciences: A Reappraisal.* Dordrecht: Kluwer.

Anderson, Lorin. 1982. *Charles Bonnet and the Order of the Known.* Dordrecht: D. Reidel.

Appel, Toby A. 1987. *The Cuvier-Geoffroy Debate: French Biology in the Decades before Darwin.* Oxford: Oxford University Press.

————. 1988. "Jeffries Wyman, Philosophical Anatomy, and the Scientific Reception of Darwinism." *J. Hist. Biology* 21: 69–94.

Appleman, Philip, ed. 1970. *Darwin: A Norton Critical Edition.* New York: Norton. 3d ed., New York: Norton, 2001.

Ardrey, Robert. 1966. *The Territorial Imperative: A Personal Inquiry into the Animal Origins of Property and Nations.* New York: Athenaeum.

Argyll, George Douglas Campbell, Duke of. 1867. *The Reign of Law.* London. 5th ed., London, 1868.

———. 1868. *Primeval Man: An Examination of Some Recent Speculations.* London.

———. 1898. *Organic Evolution Cross-Examined.* London.

Arrhenius, Svante. 1908. *Worlds in the Making: The Evolution of the Universe.* Translated by H. Borns. New York: Harper.

Ashworth, William B., Jr. 1990. "Natural History and the Emblematic World View." In *Reappraisals of the Scientific Revolution,* edited by D. Lindberg and R. Westman, pp. 303–332. Chicago: University of Chicago Press.

Atran, Scott. 1990. *Cognitive Foundations of Natural History: Towards an Anthropology of Science.* Cambridge: Cambridge University Press.

Augstein, Hannah, ed. 1996. *Race: The Origins of an Idea.* Bristol: Thoemmes Press.

Babbage, Charles. 1838. *The Ninth Bridgewater Treatise: A Fragment.* 2d ed. London. Reprint, London: Cass, 1968.

Baer, Karl Ernst von. 1828. *Über Entwickelungsgeschichte der Thiere: Beobachtung und Reflexion. Erster Theil.* Konigsburg. Reprint, Brussels: Culture et Civilization, 1967.

Bagehot, Walter. 1872. *Physics and Politics: Or Thoughts on the Application of the Principles of "Natural Selection" and "Inheritance" to Political Society.* London. Reprint, Farnborough: Gregg International, 1971.

Bajema, Carl Jay. 1977. *Eugenics: Then and Now.* Stroudsburg, Penn.: Dowden, Hutchinson, and Ross.

Baker, J. A. 1952. *Abraham Trembley of Geneva: Scientist and Philosopher.* London: Arnold.

Baker, Keith M. 1975. *Condorcet: From Natural Philosophy to Social Mathematics.* Chicago: University of Chicago Press.

Baker, Leo D. 1998. *From Savage to Negro: Anthropology and the Construction of Race, 1896–1954.* Berkeley and Los Angeles: University of California Press.

Baldwin, James Mark. 1902. *Development and Evolution: Including Psychophysical Evolution, Evolution by Orthoplasy, and the Theory of Genetic Modes.* New York.

Bannister, Robert C. 1979. *Social Darwinism: Science and Myth in Anglo-American Social Thought.* Philadelphia: Temple University Press.

Banton, Michael. 1987. *Racial Theories.* Cambridge: Cambridge University Press.

Barkan, Elazar. 1992. *The Retreat of Scientific Racism: Changing Concepts of Race in Britain and the United States between the World Wars.* Cambridge: Cambridge University Press.

Barker, David. 1989. "The Biology of Stupidity: Genetics, Eugenics, and Mental Deficiency in the Inter-war Years." *Brit. J. Hist. Sci.* 89: 347–378.

Barlow, Nora. 1946. *Charles Darwin and the Voyage of the* Beagle. New York: Philosophical Library.

Barnes, Barry, David Bloor, and John Henry. 1996. *Scientific Knowledge: A Sociological Analysis.* London: Athlone.

Barnes, Barry, and Stephen Shapin, eds. 1979. *Natural Order: Historical Studies of Scientific Culture.* Beverly Hills: Sage Publications.

Barr, Alan P., ed. 1997. *Thomas Henry Huxley's Place in Science and Letters.* Athens: University of Georgia Press.

Barrett, Paul H. 1974. "The Sedgwick-Darwin Geological Tour of North Wales." *Proc. Am. Phil. Soc.* 118: 146–164.

Barthélemy-Madaule, Madeleine. 1982. *Lamarck the Mythical Precursor: A Study of the Relations between Science and Ideology.* Cambridge, Mass.: MIT Press.

Bartholomew, Michael. 1973. "Lyell and Evolution: An Account of Lyell's Response to the Prospect of an Evolutionary Ancestry for Man." *Brit. J. Hist. Sci.* 6: 261–303.

———. 1975. "Huxley's Defence of Darwinism." *Annals of Science* 32: 525–535.

Barton, Ruth. 1998. " 'Huxley, Lubbock, and Half a Dozen Others': Professionals and Gentlemen in the Formation of the X Club, 1851–1864." *Isis* 89: 410–444.

Barzun, Jacques. 1958. *Darwin, Marx, Wagner: Critique of a Heritage.* 2d ed. Garden City, N.Y.: Doubleday.

———. 1965. *Race: A Study in Superstition.* New York: Harcourt Brace.

Bates, Henry Walter. 1862. "Contributions to an Insect Fauna of the Amazon Valley: Lepidoptera: Heliconidae." *Trans. Linn. Soc. Lond.* 23: 495–515.

———. 1863. *The Naturalist on the River Amazons.* 2 vols. London.

Bateson, Beatrice. 1928. *William Bateson, F.R.S.: Naturalist.* Cambridge: Cambridge University Press.

Bateson, William. 1894. *Materials for the Study of Variation: Treated with Especial Regard to Discontinuity in the Origin of Species.* London. Reprint, with a foreword by Peter J. Bowler and introduction by Gerry Webster, Baltimore: Johns Hopkins University Press, 1992.

———. 1902. *Mendel's Principles of Heredity.* Cambridge.

———. 1914. "President's Address." *Report of the British Association for the Advancement of Science,* 3–28.

Beatty, John. 1982. "What's in a Word? Coming to Terms in the Darwinian Revolution." *J. Hist. Biology* 15: 215–239.

Beddall, Barbara C. 1968. "Wallace, Darwin, and the Theory of Natural Selection." *J. Hist. Biology* 1: 261–323.

———. 1969. *Wallace and Bates in the Tropics: An Introduction to the Theory of Natural Selection.* London: Macmillan.

———. 1972. "Wallace, Darwin, and Edward Blyth: Further Notes on the Development of Evolution Theory." *J. Hist. Biology* 5: 153–158.

———. 1973. "Notes for Mr. Darwin: Letters to Charles Darwin from Edward Blyth at Calcutta: A Study in the Process of Discovery." *J. Hist. Biology* 6: 69–95.

————. 1988. "Darwin and Divergence: The Wallace Connection." *J. Hist. Biology* 21: 1–68.

Beer, Gillian. 1983. *Darwin's Plots: Evolutionary Narrative in Darwin, George Eliot, and Nineteenth-Century Fiction*. London: Routledge and Kegan Paul.

Behe, Michael. 1996. *Nature's Black Box: The Biochemical Challenge to Evolution*. New York: Simon and Schuster.

Bellomy, Donald C. 1984. "Social Darwinism Revisited." *Perspectives in American History*, n.s., 1: 1–129.

Bender, Bert. 1996. *The Descent of Love: Darwin and the Theory of Sexual Selection in American Fiction, 1871–1926*. Philadelphia: University of Pennsylvania Press.

Bennett, J. H., ed. 1983. *Natural Selection, Heredity, and Eugenics*. Oxford: Oxford University Press.

Berg, Leo S. 1926. *Nomogenesis: Or Evolution Determined by Law*. Translated by I. N. Rostovtsov, introduced by D'Arcy Wentworth Thomson. London. Reprint, with a preface by Theodosius Dobzhansky, Cambridge, Mass.: MIT Press, 1969.

Bergson, Henri. 1911. *Creative Evolution*. Translated by Arthur Mitchell. New York.

Bernasconi, Robert, ed. 2001. *Concepts of Race in the Eighteenth Century*. 8 vols. Bristol: Thoemmes Press.

Bilberg, Isaac. 1752. "Oeconomia naturae." In *Amoenitates Academicae, seu Dissertationes variae physicae, medicae et botanicae*, edited by Carolus Linnaeus, 2:1–58. Leiden and Amsterdam, 1749–90. Translation: "The Economy of Nature." In *Miscellaneous Tracts Relating to Natural History*, edited by Benjamin Stillingfleet, pp. 39–126. London, 1762.

Blacker, C. 1952. *Eugenics: Galton and After*. Cambridge, Mass.: Harvard University Press.

Blaisdell, Muriel. 1982. "Natural Theology and Nature's Disguises." *J. Hist. Biology* 15: 163–189.

Blinderman, Charles. 1986. *The Piltdown Inquest*. Buffalo, N.Y.: Prometheus Books.

Blitz, David. 1992. *Emergent Evolution: Qualitative Novelty and the Levels of Reality*. Dordrecht: Kluwer.

Blumenbach, Johann Friedrich. 1865. *On the Natural Varieties of Mankind*. Translated by Thomas Bendyshe. London. Reprint, New York: Bergman, 1969.

Blunt, Wilfred. 1971. *The Compleat Naturalist: A Life of Linnaeus*. With the assistance of William T. Stearn. New York: Viking Press.

Bocking, Stephen. 1988. "Alpheus Spring Packard and Cave Fauna in the Evolution Debate." *J. Hist. Biology* 21: 425–456.

Bölsche, Wilhelm. 1906. *Haeckel: His Life and Work*. Translated by Joseph McCabe. London: T. Fisher Unwin.

Bolt, Christine. 1971. *Victorian Attitudes to Race*. London: Routledge and Kegan Paul.

Bonnet, Charles. 1779. *Oeuvres d'histoire naturelle et de philosophie.* 19 vols. Neuchatel.

Borelli, Giovano Alfonso. 1989. *On the Movement of Animals.* Translated by Paul Maquet. Berlin: Springer-Verlag.

Botting, Douglas. 1973. *Humboldt and the Cosmos.* London: Sphere.

Boule, Marcellin. 1923. *Fossil Man: Elements of Human Palaeontology.* Edinburgh: Oliver and Boyd.

Bourdier, Frank. 1969. "Geoffroy Saint-Hilaire versus Cuvier: The Campaign for Paleontological Evolution." In *Toward a History of Geology,* edited by Cecil J. Schneer, pp. 36–61. Cambridge, Mass.: MIT Press.

Bowlby, John. 1990. *Charles Darwin: A Biography.* London: Hutchinson.

Bowler, Peter J. 1971. "Preformation and Pre-existence in the Seventeenth Century: A Brief Analysis." *J. Hist. Biology* 4: 221–244.

———. 1973. "Bonnet and Buffon: Theories of Generation and the Problem of Species." *J. Hist. Biology* 6: 259–281.

———. 1974a. "Evolutionism in the Enlightenment." *History of Science* 12: 159–183.

———. 1974b. "Darwin's Concepts of Variation." *J. Hist. Medicine* 29: 196–212.

———. 1975. "The Changing Meaning of 'Evolution.' " *J. Hist. Ideas* 36: 95–114.

———. 1976a. *Fossils and Progress: Paleontology and the Idea of Progressive Evolution in the Nineteenth Century.* New York: Science History Publications.

———. 1976b. "Malthus, Darwin, and the Concept of Struggle." *J. Hist. Ideas* 37: 631–650.

———. 1976c. "Alfred Russel Wallace's Concepts of Variation." *J. Hist. Medicine* 31: 17–29.

———. 1977. "Darwinism and the Argument from Design: Suggestions for a Re-evaluation." *J. Hist. Biology* 10: 29–43.

———. 1983. *The Eclipse of Darwinism: Anti-Darwinian Evolution Theories in the Decades around 1900.* Baltimore: Johns Hopkins University Press.

———. 1984. "E. W. MacBride's Lamarckian Eugenics and Its Implications for the Social Construction of Scientific Knowledge." *Annals of Science* 41: 245–260.

———. 1985. "Scientific Attitudes to Darwinism in Britain and America." In *The Darwinian Heritage: A Centennial Retrospect,* edited by David Kohn, pp. 641–682. Princeton: Princeton University Press.

———. 1986. *Theories of Human Evolution: A Century of Debate, 1844–1944.* Baltimore: Johns Hopkins University Press; Oxford: Basil Blackwell.

———. 1988. *The Non-Darwinian Revolution: Reinterpreting a Historical Myth.* Baltimore: Johns Hopkins University Press.

———. 1989a. *The Invention of Progress: The Victorians and the Past.* Oxford: Basil Blackwell.

———. 1989b. *The Mendelian Revolution: The Emergence of Hereditarian Concepts in Modern Science and Society.* London: Athlone; Baltimore: Johns Hopkins University Press.

————. 1989c. "Holding Your Head Up High: Degeneration and Orthogenesis in Theories of Human Evolution." In *History, Humanity, and Evolution: Essays for John C. Greene*, edited by James R. Moore, pp. 329–353. Cambridge: Cambridge University Press.

————. 1990. *Charles Darwin: The Man and His Influence*. Oxford: Basil Blackwell. Reprint, Cambridge: Cambridge University Press, 1996.

————. 1992. *The Fontana History of the Environmental Sciences*. London: Fontana. American edition published as *The Norton History of the Environmental Sciences*. New York: Norton. Subsequently retitled *The Earth Encompassed*.

————. 1993. *Biology and Social Thought, 1850–1914*. Berkeley: Office for the History of Science and Technology, University of California.

————. 1996. *Life's Splendid Drama: Evolutionary Biology and the Reconstruction of Life's Ancestry, 1860–1940*. Chicago: University of Chicago Press.

————. 1998. "Evolution and the Eucharist: Bishop E. W. Barnes on Science and Religion in the 1920s and 1930s." *Brit. J. Hist. Sci.* 31: 453–467.

————. 2001. *Reconciling Science and Religion: The Debates in Early Twentieth-Century Britain*. Chicago: University of Chicago Press.

Box, Joan Fisher. 1978. *R. A. Fisher: The Life of a Scientist*. New York: Wiley.

Brace, C. Loring. 1964. "The Fate of the 'Classic' Neanderthals: A Study in Hominid Catastrophism." *Current Anthropology* 5: 3–43.

Brackman, Arnold C. 1980. *A Delicate Arrangement: The Strange Case of Charles Darwin and Alfred Russel Wallace*. New York: Times Books.

Brady, Ronald H. 1979. "Natural Selection and the Criteria by Which a Theory Is Judged." *Systematic Zoology* 28: 600–621.

Brandon, Robert N., and Richard M. Burian, eds. 1984. *Genes, Organisms, Populations: Controversies over the Units of Selection*. Cambridge, Mass.: MIT Press.

Brewster, Sir David. 1855. *Memoirs of the Life, Writings, and Discoveries of Sir Isaac Newton*. 2 vols. London. Reprint, New York: Johnson Reprint Corporation, 1965.

Broberg, Gunnar, ed. 1980. *Linnaeus: Progress and Prospects in Linnaean Research*. Stockholm: Almquist and Wiksell International.

Brock, W. H., and R. M. MacLeod. 1976. "The Scientists' Declaration: Reflections on Science and Belief in the Wake of *Essays and Reviews*." *Brit. J. Hist. Sci.* 9: 39–66.

Brongniart, Adolphe. 1828. *Prodome d'une histoire des végétaux fossiles*. Paris and Strasburg.

————. 1829. "General Considerations on the Nature of the Vegetation Which Covered the Earth at the Different Epochs of the Formation of Its Crust." *Edinburgh New Phil. J.* 6: 349–371.

Bronn, Heinrich Georg. 1859. "On the Laws of Evolution of the Organic World during the Formation of the Crust of the Earth." *Ann. & Mag. of Nat. Hist.*, 3d ser., 4: 81–89, 175–184.

Brooke, John Hedley. 1977. "Richard Owen, William Whewell, and the *Vestiges*." *Brit. J. Hist. Sci.* 10: 132–145.

———. 1985. "The Relations between Darwin's Science and His Religion." In *Darwinism and Divinity: Essays on Evolution and Religious Belief*, edited by John R. Durant, pp. 40–75. Oxford: Basil Blackwell.

———. 1991. *Science and Religion: Some Historical Perspectives.* Cambridge: Cambridge University Press.

Brooks, John Langdon. 1983. *Just before the Origin: Alfred Russel Wallace's Theory of Evolution.* New York: Columbia University Press.

Broom, Robert. 1932. *The Mammal-Like Reptiles of South Africa.* London: H. and F. Witherby.

———. 1950. *Finding the Missing Link.* London: Watts.

Brown, Andrew. 1999. *The Darwin Wars: How Stupid Genes Became Selfish Gods.* New York: Simon and Schuster.

Brown, F. B. 1986. "The Evolution of Darwin's Theism." *J. Hist. Biology* 19: 1–45.

Browne, Janet. 1980. "Darwin's Botanical Arithmetic and the 'Principle of Divergence,' 1854–1858." *J. Hist. Biology* 13: 53–89.

———. 1983. *The Secular Ark: Studies in the History of Biogeography.* New Haven: Yale University Press.

———. 1989. "Botany for Gentlemen: Erasmus Darwin and *The Loves of the Plants*." *Isis* 80: 593–621.

———. 1995. *Charles Darwin: Voyaging.* London: Jonathan Cape.

———. 2002. *Charles Darwin: The Power of Place.* London: Jonathan Cape.

Brundell, Barry. 2001. "Catholic Church Politics and Evolution Theory, 1894–1902." *Brit. J. Hist. Sci.* 34: 81–96.

Brush, Stephen G. 1978. *The Temperature of History: Phases of Science and Culture in the Nineteenth Century.* New York: Burt Franklin.

Buckland, William. 1820. *Vindiciae Geologicae: Or the Connexion of Geology and Religion Explained.* Oxford.

———. 1823. *Reliquiae Diluvianae: Or Observations of the Organic Remains contained in Caves, Fissures and Diluvial Gravel, and other Geological Phenomena, attesting the Action of a Universal Deluge.* London. Reprint, New York: Arno Press, 1977.

———. 1836. *Geology and Mineralogy Considered with Reference to Natural Theology.* 2 vols. London.

Buffetaut, Eric. 1986. *A Short History of Vertebrate Paleontology.* Bekenham: Croom Helm.

Buffon, Georges Louis Leclerc, Comte de. 1749–67. *Histoire naturelle, générale et particulière.* 15 vols. Paris.

———. 1774–89. *Histoire naturelle: Supplement.* 7 vols. Paris.

———. 1791. *Natural History, General and Particular.* Translated by William Smellie. 3d ed. 9 vols. London. Reprint, Bristol: Thoemmes Press, 2000.

———. 1962. *Les époques de la nature.* Edited by Jacques Roger. Paris: Museum d'histoire naturelle.

———. 1976. "The 'Critical Discourse' to Buffon's *Histoire Naturelle:* The First Complete English Translation." By John Lyon. *J. Hist. Biology* 9: 133–181.

———. 1981. *From Natural History to the History of Nature: Readings from Buffon and His Critics.* Edited by Philip R. Sloan and J. Lyon. London: University of Notre Dame Press.

Bulmer, Michael. 1999. "The Development of Francis Galton's Ideas on the Mechanism of Heredity." *J. Hist. Biology* 32: 263–292.

Burchfield, Joe D. 1974. "Darwin and the Dilemma of Geological Time." *Isis* 65: 301–321.

———. 1975. *Lord Kelvin and the Age of the Earth.* New York: Science History Publications.

Burian, Richard, Jean Gayon, and Doris Zallen. 1988. "The Singular Fate of Genetics in the History of French Biology." *J. Hist. Biology* 21: 357–402.

Burkhardt, Richard W., Jr. 1970. "Lamarck, Evolution, and the Politics of Science." *J. Hist. Biology* 3: 275–296.

———. 1972. "The Inspiration of Lamarck's Belief in Evolution." *J. Hist. Biology* 5: 413–438.

———. 1977. *The Spirit of System: Lamarck and Evolutionary Biology.* Cambridge, Mass.: Harvard University Press.

Burnet, Thomas. 1691. *The Sacred Theory of the Earth.* London. Reprint, with an introduction by Basil Willey, London: Centaur Press, 1965.

Burr, Clinton Stoddard. 1922. *America's Race Heritage.* New York. Reprint, New York: Arno, 1977.

Burrow, J. W. 1966. *Evolution and Society: A Study in Victorian Social Theory.* Cambridge: Cambridge University Press.

Burstyn, H. C. 1975. "If Darwin Wasn't the *Beagle*'s Naturalist, Why Was He on Board?" *Brit. J. Hist. Sci.* 8: 62–69.

Bury, J. B. 1932. *The Idea of Progress: An Inquiry into Its Growth and Origins.* Reprint, New York: Dover, 1955.

Buss, Allan R. 1976. "Galton and the Birth of Differential Psychology and Eugenics: Social, Political, and Economic Forces." *J. Hist. Behavioral Sci.* 12: 47–58.

Butler, Samuel. 1879. *Evolution, Old and New: Or the theories of Buffon, Dr. Erasmus Darwin, and Lamarck, as Compared with that of Mr. Charles Darwin.* London.

———. 1890. *Essays on Life, Art, and Science.* London. Reprint, Port Washington, N.Y.: Kennikat Press, 1970.

———. 1916. *Life and Habit.* 2d ed. London.

———. 1920a. *Luck, or Cunning, as the Main Means of Organic Modification?* 2d ed. London.

———. 1920b. *Unconscious Memory.* 3d ed. London.

Bynum, William F. 1975. "The Great Chain of Being." *History of Science* 13: 1–28.

———. 1984. "Charles Lyell's *Antiquity of Man* and Its Critics." *J. Hist. Biology* 17: 153–187.

Cabanis, Pierre. 1981. *On the Relations between the Physical and the Moral Aspects of Man.* Edited by George Mora. Baltimore: Johns Hopkins University Press.

Cain, Joe. 2000. "Towards a 'Greater Degree of Integration': The Society for the Study of Speciation, 1939–1941." *Brit. J. Hist. Sci.* 33: 85–108.

Call, Lewis. 1998. "Anti-Darwin, Anti-Spencer: Friedrich Nietzsche's Critique of Darwin on 'Darwinism.' " *History of Science* 36: 1–22.

Camerini, Jane R. 1994. "Evolution, Biogeography, and Maps: The Early History of Wallace's Line." In *Darwin's Laboratory: Evolutionary Thought and Natural History in the Pacific,* edited by Roy M. MacLeod and Philip F. Rehbock, pp. 70–109. Honolulu: University of Hawaii Press.

Campbell, M. 1980. "Did De Vries Discover the Law of Segregation Independently?" *Annals of Science* 37: 639–655.

Camper, Petrus. 1779. "Account of the Organs of Speech of the Orang Outang." *Phil. Trans. Roy. Soc.* 69: 155–156.

Cannon, Susan F. 1978. *Science in Culture: The Early Victorian Period.* New York: Science History Publications.

Cannon, Walter F. 1960a. "The Uniformitarian-Catastrophist Debate." *Isis* 51: 38–55.

———. 1960b. "The Problem of Miracles in the 1830s." *Victorian Studies* 4: 5–32.

———. 1961a. "The Bases of Darwin's Achievement: A Reevaluation." *Victorian Studies* 5: 109–132.

———. 1961b. "The Impact of Uniformitarianism: Two Letters from John Herschel to Charles Lyell, 1836–37." *Proc. Am. Phil. Soc.* 105: 301–314.

Caplan, Arthur L., ed. 1978. *The Sociobiology Debate.* New York: Harper and Row.

Carlson, Elof Axel. 1966. *The Gene: A Critical History.* Philadelphia: Saunders.

Caron, Joseph A. 1988. " 'Biology' and the Life Sciences: A Historiographical Contribution." *History of Science* 26: 223–268.

Carozzi, Albert V. 1969. "De Maillet's *Telliamed* (1748): An Ultra-Neptunian Theory of the Earth." In *Toward a History of Geology,* edited by Cecil J. Schneer, pp. 80–91. Cambridge, Mass.: MIT Press.

Carpenter, William Benjamin. 1851. *Principles of Physiology: General and Comparative.* 3d ed. London.

———. 1888. *Nature and Man: Essays Scientific and Philosophical. With an Introductory Memoir by J. Erstlin Carpenter.* London. Reprint, Farnborough: Gregg International, 1970.

Carroll, Joseph. 1995. *Evolution and Literary Theory.* Columbia: University of Missouri Press.

Cartmill, Matt. 1993. *A View to a Death in the Morning: Hunting and Nature through History.* Cambridge, Mass.: Harvard University Press.

Cassirer, Ernst. 1945. *Rousseau, Kant, Goethe: Two Essays.* Translated by James Gutman et al. Princeton: Princeton University Press.

———. 1951. *The Philosophy of the Enlightenment.* Translated by Fritz Koelln and James Pettegrove. Princeton: Princeton University Press.

Castle, W. E. 1911. *Heredity in Relation to Evolution and Animal Breeding.* New York: Appleton.

Cavalli-Sforza, Luigi L. 2000. *Genes, Peoples, and Languages.* London: Allen Lane.

Chadwick, Owen. 1966. *The Victorian Church.* London: A. C. Black.

———. 1975. *The Secularization of the European Mind in the Nineteenth Century.* Cambridge: Cambridge University Press.

Chamberlin, J. Edward, and Sander L. Gilman, eds. 1985. *Degeneration: The Dark Side of History.* New York: Columbia University Press.

Chambers, Robert. 1844. *Vestiges of the Natural History of Creation.* London. Reprint, with an introduction by Sir Gavin de Beer, Leicester: Leicester University Press.

———. 1846. *Explanations: A Sequel to the Vestiges of the Natural History of Creation.* 2d ed. London.

———. 1994. *Vestiges of the Natural History of Creation and Other Evolutionary Writings.* Edited by James Secord. Chicago: University of Chicago Press.

Chetverikov, S. S. 1961. "On Certain Aspects of the Evolutionary Process from the Standpoint of Modern Genetics." Translated by M. Barker, edited by I. M. Lerner. *Proc. Am. Phil. Soc.* 105: 167–195.

———. 1997. "On Certain Aspects of the Evolutionary Process from the Standpoint of Modern Genetics." Edited by Charles David Mellon. Placitas, N.M.: Genetics Heritage Press.

Churchill, Frederick B. 1968. "August Weismann and a Break from Tradition." *J. Hist. Biology* 1: 91–112.

———. 1974. "William Johannsen and the Genotype Concept." *J. Hist. Biology* 7: 5–30.

———. 1976. "Rudolph Virchow and the Pathologist's Criterion for the Inheritance of Acquired Characteristics." *J. Hist. Medicine* 31: 117–148.

———. 1986. "Weismann, Hydromedusae, and the Biogenetic Imperative: A Reconsideration." In *A History of Embryology,* edited by T. J. Horder, J. A. Witkowski, and C. C. Wylie, pp. 7–34. Cambridge: Cambridge University Press.

———. 1999. "August Weismann: A Developmental Evolutionist." In *Ausgewahlte Briefe und Dokumente/Selected Letters and Documents,* by August Weismann, edited F. Churchill and H. Risler, 2:749–798. Freiburg i. Br.: Universitatsbibliothek.

Clark, Linda L. 1984. *Social Darwinism in France.* University, Ala.: University of Alabama Press.

Clark, Ronald W. 1969. *JBS: The Life and Work of J. B. S. Haldane.* New York: Coward-McCann.

Clark, Wilfrid Le Gros. 1949. *History of the Primates: An Introduction to the Study of Fossil Man.* London: British Museum.

———. 1955. *The Fossil Evidence for Human Evolution.* Chicago: University of Chicago Press.

Clark, William, Jan Golinski, and Simon Schaffer. 1999. *The Sciences in Enlightened Europe*. Chicago: University of Chicago Press.

Clutton-Brock, T. H., and Paul H. Harvey. 1978. *Readings in Sociobiology*. San Francisco: W. H. Freeman.

Cock, A. G. 1973. "William Bateson, Mendelism, and Biometry." *J. Hist. Biology* 6: 1–36.

Cohen, H. Floris. 1994. *The Scientific Revolution: A Historiographical Inquiry*. Chicago: University of Chicago Press.

Colbert, E. H. 1971. *Men and Dinosaurs: The Search in Field and in Laboratory*. Reprint, Harmondsworth: Penguin Books.

Coleman, William. 1962. "Lyell and the Reality of Species." *Isis* 53: 325–338.

———. 1964. *Georges Cuvier, Zoologist: A Study in the History of Evolution Theory*. Cambridge, Mass.: Harvard University Press.

———. 1965. "Cell, Nucleus, and Inheritance: A Historical Study." *Proc. Am. Phil. Soc.* 109: 124–158.

———. 1966. "Science and Symbol in the Turner Frontier Hypothesis." *Am. Hist. Rev.* 72: 22–49.

———. 1970. "Bateson and Chromosomes: Conservative Thought in Science." *Centaurus* 15: 228–315.

———. 1971. *Biology in the Nineteenth Century: Problems of Form, Function, and Transmutation*. New York: Wiley.

———. 1973. "Limits of the Recapitulation Theory: Carl Freidrich Kielmeyer's Critique of the Presumed Parallelism of Earth History, Ontogeny, and the Present Order of Organisms." *Isis* 64: 341–350.

———. 1976. "Morphology between Type Concept and Descent Theory." *J. Hist. Medicine* 31: 149–175.

Colp, Ralph, Jr. 1974. "The Contacts between Karl Marx and Charles Darwin." *J. Hist. Ideas* 35: 329–338.

———. 1977. *To Be an Invalid: The Illness of Charles Darwin*. Chicago: University of Chicago Press.

———. 1982. "The Myth of the Darwin-Marx Letter." *History of Political Economy* 14: 461–482.

———. 1986. "Confessing a Murder: Darwin's First Revelations about Transmutation." *Isis* 77: 9–32.

Comte, Auguste. 1975. *August Comte and Positivism: The Essential Writings*. Edited by Gertrud Lenzer. New York: Harper and Row.

Condillac, Etienne Bonnot de. 1756. *An Essay on the Origins of Human Knowledge*. Reprint, with an introduction by James H. Stam, New York: AMS Press, 1974.

Condorcet, Marie-Jean-Antoine-Nicholas. 1955. *Sketch for a Historical Picture of the Progress of the Human Mind*. Translated by I. Barraclough. London: Weidenfeld and Nicholson.

Conry, Yvette. 1974. *L'introduction du Darwinisme en France au XIXe siècle*. Paris: Vrin.

Conway, Jill. 1973. "Stereotypes of Femininity in a Theory of Sexual

Evolution." In *Suffer and Be Still: Women in the Victorian Age,* edited by Martha Vicinus, pp. 140–157. Bloomington: Indiana University Press.

Cook, George M. 1999. "Neo-Lamarckian Experimentalism in America: Origins and Consequences." *Quart. Rev. Biology* 74: 417–437.

Cooter, Roger. 1985. *The Cultural Meaning of Popular Science: Phrenology and the Organization of Consent in Nineteenth-Century Britain.* Cambridge: Cambridge University Press.

Cope, Edward Drinker. 1887. *The Origin of the Fittest: Essays in Evolution.* Reprinted with Cope, *The Primary Factors of Organic Evolution.* New York: AMS Press, 1974.

———. 1896. *The Primary Factors of Organic Evolution.* Chicago.

Copleston, Frederick. 1963. *A History of Philosophy.* Vol. 7: *Fichte to Nietzsche.* London: Burns and Oates.

———. 1966. *A History of Philosophy.* Vol. 8: *Bentham to Russell.* London: Burns and Oates.

Corbey, Raymond, and Wil Roebroeks. 2001. *Studying Human Origins: Disciplinary History and Epistemology.* Amsterdam: University of Amsterdam Press.

Corbey, Raymond, and Bert Theunissen, eds. 1995. *Ape, Man, Ape-Man: Changing Views since 1600.* Leiden: Department of Prehistory, University of Leiden.

Cornell, John F. 1983. "From Creation to Evolution: Sir William Dawson and the Idea of Design in the Nineteenth Century." *J. Hist. Biology* 16: 137–170.

———. 1984. "Analogy and Technology in Darwin's Vision of Nature." *J. Hist. Biology* 17: 303–344.

Corsi, Pietro. 1988a. *The Age of Lamarck: Evolutionary Theories in France, 790–1830.* Berkeley and Los Angeles: University of California Press.

———. 1988b. *Science and Religion: Baden Powell and the Anglican Debate, 1800–1860.* Cambridge: Cambridge University Press.

Cosans, Christopher. 1994. "Anatomy, Metaphysics, and Value: The Ape Brain Debate Reconsidered." *Biology and Philosophy* 9: 129–166.

Cowan, Ruth Schwartz. 1972a. "Francis Galton's Contributions to Genetics." *J. Hist. Biology* 5: 389–412.

———. 1972b. "Francis Galton's Statistical Ideas: The Influence of Eugenics." *Isis* 63: 509–528.

———. 1977. "Nature and Nurture: The Interplay of Biology and Politics in the Work of Francis Galton." *Stud. Hist. Biology* 1: 133–208.

Cravens, Hamilton. 1978. *The Triumph of Evolution: American Scientists and the Heredity-Environment Controversy, 1900–1941.* Philadelphia: University of Pennsylvania Press.

Crocker, Lester G. 1959. "Diderot and Eighteenth-Century French Transformism." In *Forerunners of Darwin, 1745–1859,* edited by Bentley Glass, Owsei Temkin, and William Strauss Jr., pp. 114–143. Baltimore: Johns Hopkins University Press.

———. 1963. *Nature and Culture: Ethical Thought in the French Enlightenment.* Baltimore: Johns Hopkins University Press.

Croizat, Leon. 1958. *Panbiogeography*. 3 vols. Caracas: Leon Croizat.

——. 1964. *Space, Time, Form: The Biological Synthesis*. Caracas: Leon Croizat.

Cronin, Helena. 1991. *The Ant and the Peacock: Altruism and Sexual Selection from Darwin to Today*. Cambridge: Cambridge University Press.

Crook, Paul. 1994. *Darwinism, War, and History: The Debate over the Biology of War from the* Origin of Species *to the First World War*. Cambridge: Cambridge University Press.

Crow, James F. 1990. "Sewall Wright's Place in Twentieth-Century Biology." *J. Hist. Biology* 29: 57–89.

Cunningham, Andrew, and Nicholas Jardine, eds. 1990. *Romanticism and the Sciences*. Cambridge: Cambridge University Press.

Cuvier, Georges. 1805. *Leçons d'anatomie comparée*. 5 vols. Paris. Reprint, Brussels: Culture et Civilisation, 1969.

——. 1812. *Recherches sur les ossemens fossiles de quadrupèdes* . . . 4 vols. Paris. Reprint, Brussels: Culture et Civilisation, 1969. 3d ed., 5 vols., Paris, 1825.

——. 1817a. *Le règne animal distribué d'après son organization* . . . 4 vols. Paris. Reprint, Brussels: Culture et Civilisation, 1969.

——. 1817b. *An Essay on the Theory of the Earth*. Translated by Robert Kerr, with notes by Robert Jameson. 3d ed. Edinburgh. Reprint, New York: Arno Press, 1977.

——. 1825. *Discours sur les révolutions de la surface du globe* . . . 2d ed. Paris. Reprint, Brussels: Culture et Civilization, 1969.

——. 1863. *The Animal Kingdom Arranged after Its Organization* . . . New ed. London. Reprint, Millwood, N.Y.: Kraus Reprints.

Cuvier, Georges, and Alexandre Brongniart. 1825. *Description géologiques des environs de Paris*. 2d ed. Paris. Reprint, Brussels: Culture et Civilisation, 1969.

Dalrymple, G. Brent. 1991. *The Age of the Earth*. Stanford: Stanford University Press.

Daniel, Glyn. 1975. *A Hundred and Fifty Years of Archaeology*. London: Duckworth.

Daniels, George. 1968. *Darwinism Comes to America*. Waltham, Mass.: Blaisdell Publishing.

Däniken, Erich von. 1970. *Chariots of the Gods? Unsolved Mysteries of the Past*. Translated by Michael Heron. New York: C. P. Putnam's Sons.

——. 1977. *According to the Evidence: My Proof of Man's Extraterrestrial Origins*. Translated by Michael Heron. London: Souvenir Press.

Darden, Lindley. 1976. "Reasoning in Scientific Change: Charles Darwin, Hugo De Vries, and the Rediscovery of Segregation." *Stud. Hist. & Phil. Sci.* 7: 127–169.

——. 1977. "William Bateson and the Promise of Mendelism." *J. Hist. Biology* 10: 87–106.

——. 1991. *Theory Change in Science: Strategies for Mendelian Genetics*. New York: Oxford University Press.

Darwin, Charles Robert. 1842. *The Structure and Distribution of Coral Reefs.* London. Reprint, Brussels: Culture et Civilisation, 1969. 1897 ed. reprinted New York: AMS Press, 1972.

———. 1845. *Journal of Researches into the Natural History and Geology of the Various Countries Visited by H.M.S. Beagle.* London. Reprint, London: Everyman.

———. 1851–53. *A Monograph on the Sub-Class Cirripedia.* 2 vols. London. Reprint, New York: Johnson Reprint Corporation, 1964.

———. 1859. *On the Origin of Species by Means of Natural Selection: Or the Preservation of Favoured Races in the Struggle for Life.* London. Reprint, with an introduction by Ernst Mayr, Cambridge, Mass.: Harvard University Press, 1964.

———. 1868. *The Variation of Animals and Plants under Domestication.* 2 vols. London. Reprint, Brussels: Culture et Civilisation, 1969. 1883 ed., edited by H. Ritvo, reprinted Baltimore: Johns Hopkins University Press, 1998.

———. 1871. *The Descent of Man and Selection in Relation to Sex.* 2 vols. London. Reprint, Brussels: Culture et Civilisation, 1969. 1895 ed. reprinted New York: AMS Press, 1972.

———. 1872. *The Expression of the Emotions in Man and the Animals.* London. Reprint, Brussels: Culture et Civilisation, 1969. Reprint, with an introduction by Paul Ekman, London: Harper Collins, 1999.

———. 1958. *The Autobiography of Charles Darwin: With the Original Omissions Restored.* Edited by Nora Barlow. New York: Harcourt Brace.

———. 1959. *The Origin of Species . . . A Variorum Text.* Edited by Morse Peckham. Philadelphia: University of Pennsylvania Press.

———. 1975. *Charles Darwin's Natural Selection: Being the Second Part of His Big Species Book Written from 1856 to 1858.* Edited by Robert C. Stauffer. London: Cambridge University Press.

———. 1977. *The Collected Papers of Charles Darwin.* Edited by Paul H. Barrett. 2 vols. Chicago: University of Chicago Press.

———. 1981. *A Concordance to Darwin's Origin of Species, First Edition.* Edited by Paul H. Barrett et al. Ithaca, N.Y.: Cornell University Press.

———. 1984–. *The Correspondence of Charles Darwin.* Edited by Frederick Burkhardt and Sydney Smith. 12 vols. published to 2001. Cambridge: Cambridge University Press.

———. 1985. *A Calendar of the Correspondence of Charles Darwin.* Edited by Frederick Burkhardt et al. New York: Garland Publishing.

———. 1987. *Charles Darwin's Notebooks (1836–1844).* Edited by Paul H. Barrett et al. Cambridge: Cambridge University Press.

———. 1988. *Charles Darwin's* Beagle *Diary.* Edited by R. D. Keynes. Cambridge: Cambridge University Press.

———. 1990a. *Charles Darwin's Marginalia.* Edited by Mario di Gregorio and Nick Gill. Pt. 1. New York: Garland.

———. 1990b. *A Concordance to Charles Darwin's Notebooks, 1836–1844.* Edited by Donald J. Weinshank et al. Ithaca, N.Y.: Cornell University Press.

Darwin, Charles, and Alfred Russel Wallace. 1958. *Evolution by Natural Selection.* With a foreword by Sir Gavin de Beer. Cambridge: Cambridge University Press.

Darwin, Erasmus. 1791. *The Botanic Garden: A Poem in Two Parts. Part One: Containing the Economy of Nature, Part Two: The Loves of the Plants.* 2 vols. London. Reprint, Elmsford, N.Y.: Pergamon (British Book Center).

———. 1794–96. *Zoonomia: Or the Laws of Organic Life.* 2 vols. London. Reprint, New York: AMS Press, 1974.

———. 1803. *The Temple of Nature.* London. Reprint, Elmsford, N.Y.: Pergamon (British Book Center).

———. 1968. *The Essential Erasmus Darwin.* Edited by Desmond King-Hele. London: McGibbon and Kee.

Darwin, Francis, ed. 1887. *The Life and Letters of Charles Darwin.* 3 vols. London. Reprint, New York: Johnson Reprint Corporation, 1969.

———. 1903. *More Letters of Charles Darwin.* 2 vols. London. Reprint, New York: Johnson Reprint Corporation, 1972.

Daudin, Henri. 1926. *Études d'histoire des sciences naturelles.* I. *De Linné à Jussieu: Méthodes de la classification et l'idée de série en botanique et en zoologie.* II. *Cuvier et Lamarck: Les classes zoologiques et l'idée de série animale.* Paris: Alcan.

Davies, Gordon L. 1969. *The Earth in Decay: A History of British Geomorphology, 1578–1878.* New York: Science History Publications.

Dawkins, R. 1976. *The Selfish Gene.* Oxford: Oxford University Press.

———. 1982. *The Extended Phenotype: The Gene as the Unit of Selection.* Oxford: W. H. Freeman.

———. 1986. *The Blind Watchmaker.* Harlow: Longman Scientific and Technical.

———. 1996. *Climbing Mount Improbable.* New York: Norton.

Dawson, Sir John William. 1890. *Modern Ideas of Evolution.* London. Reprint, edited by William R. Shea and John F. Cornell, New York: Prodist.

Dawson, Virginia P. 1988. *Nature's Enigma: The Problem of the Polyp in the Letters of Bonnet, Trembley, and Reaumur.* Vol. 174. Philadelphia: Memoirs of the American Philosophical Society.

Dean, Dennis R. 1981. "The Age of the Earth Controversy: Beginnings to Hutton." *Annals of Science* 38: 435–456.

———. 1992. *James Hutton and the History of Geology.* Ithaca, N.Y.: Cornell University Press.

———. 1999. *Gideon Mantell and the Discovery of Dinosaurs.* Cambridge: Cambridge University Press.

de Beer, Gavin. 1930. *Embryology and Evolution.* Oxford: Oxford University Press.

———. 1940. *Embryos and Ancestors.* Oxford: Oxford University Press.

———. 1963. *Charles Darwin: Evolution by Natural Selection.* London: Nelson.

———. 1964. "Mendel, Darwin, and Fisher." *Notes and Records Roy. Soc. London* 19: 192–226.

De Camp, L. S. 1968. *The Great Monkey Trial.* Garden City, N.Y.: Doubleday.

De Grazia, A. 1966. *The Velikovsky Affair.* New Hyde Park, N.Y.: University Books.

Delages, Yves, and Marie Goldsmith. 1912. *The Theories of Evolution.* Translated by Andre Tridon. New York.

Delisle, Richard G. 2001. "Adaptationism versus Cladism in Human Evolution Studies." In *Studying Human Origins: Disciplinary History and Epistemology,* edited by Raymond Corbey and Wil Roebroeks, pp. 107–121. Amsterdam: University of Amsterdam Press.

de Maillet, Benoît. 1968. *Telliamed, or Conversations between an Indian Philosopher and a French Missionary on the Diminution of the Sea . . .* Translated by Albert V. Carozzi. Urbana: University of Illinois Press.

De Marrais, Robert. 1974. "The Double-Edged Effect of Sir Francis Galton: A Search for the Motives of the Biometrician-Mendelian Debate." *J. Hist. Biology* 7: 141–174.

de Mortillet, Gabriel. 1883. *Le préhistorique: L'antiquité de l'homme.* Paris.

Dempster, W. J. 1996. *Natural Selection and Patrick Matthew: Evolutionary Concepts in the Nineteenth Century.* Edinburgh: Pentland Press.

Dennert, E. 1904. *At the Deathbed of Darwinism.* Translated by E. V. O'Hara and John H. Peschges. Burlington, Iowa.

Dennett, Daniel. 1995. *Darwin's Dangerous Idea: Evolution and the Meaning of Life.* New York: Simon and Schuster.

Denton, Michael J. 1998. *Nature's Destiny: How the Laws of Biology Reveal Purpose in the Universe.* New York: Free Press.

Depew, David J., and Bruce H. Weber. 1995. *Darwinism Evolving: Systems Dynamics and the Genealogy of Natural Selection.* Cambridge, Mass.: MIT Press.

Desmond, Adrian. 1976. *The Hot-Blooded Dinosaurs: A Revolution in Paleontology.* New York: Dial Press.

———. 1982. *Archetypes and Ancestors in Victorian London, 1850–1875.* London: Blond and Briggs. Reprint, Chicago: University of Chicago Press, 1984.

———. 1984. "Robert E. Grant: The Social Predicament of a Pre-Darwinian Evolutionist." *J. Hist. Biology* 17: 180–223.

———. 1987. "Artisan Resistance and Evolution in Britain, 1819–1848." *Osiris,* n.s., 3: 72–110.

———. 1989. *The Politics of Evolution: Morphology, Medicine, and Reform in Radical London.* Chicago: University of Chicago Press.

———. 1994. *Huxley: The Devil's Disciple.* London: Michael Joseph.

———. 1997. *Huxley: Evolution's High Priest.* London: Michael Joseph.

———. 2001. "Redefining the X-Axis: 'Professionals,' 'Amateurs,' and the Making of Mid-Victorian Biology." *J. Hist. Biology* 34: 3–50.

Desmond, Adrian, and James R. Moore. 1991. *Darwin.* London: Michael Joseph.

De Vries, Hugo. 1904. *Species and Varieties: Their Origin by Mutation.* Edited by D. T. MacDougal. Chicago. New ed., 1910.

————. 1910a. *Intracellular Pangenesis*. Translated by C. Stuart Gager. Chicago.

————. 1910b. *The Mutation Theory: Experiments and Observations on the Origin of Species in the Vegetable Kingdom*. Translated by J. B. Farmer and A. D. Darbyshire. 2 vols. London.

Dewey, John. 1910. *The Impact of Darwin on Philosophy: And Other Essays in Contemporary Thought*. Reprint, Bloomington: Indiana University Press, 1965.

Dexter, Ralph W. 1979. "The Impact of Evolutionary Theories on the Salem Group of Agassiz Zoologists (Morse, Hyatt, Packard, Putnam)." *Essex Institute Historical Collections* 115: 144–171.

d'Holbach, Paul Henri Thiry, Baron. 1821. *Système de la nature: Ou des lois du monde physique et du monde moral*. 2 vols. Reprint, with an introduction by Yvon Belavel, Hildesheim: Georg Olms, 1966.

————. 1868. *The System of Nature: Or Laws of the Moral and Physical World*. Translated by H. D. Robinson. Reprint, New York: Burt Franklin, 1970.

Diderot, Denis. 1964. *Oeuvres philosophiques*. Edited by Paul Vernière. Paris: Classiques Garnier.

————. 1966. *D'Alembert's Dream and Rameau's Nephew*. Translated by L. W. Tancock. Harmondsworth, Middlesex: Penguin Books.

Dietrich, Michael R. 1994. "The Origins of the Neutral Theory of Molecular Evolution." *J. Hist Biology* 27: 21–60.

————. 1995. "Richard Goldschmidt's 'Heresies' and the Evolutionary Synthesis." *J. Hist. Biology* 28: 431–461.

————. 1998. "Paradox and Persuasion: Negotiating the Place of Molecular Evolution within Evolutionary Biology." *J. Hist. Biology* 31: 85–111.

Di Gregorio, Mario A. 1982. "In Search of the Natural System: Problems of Zoological Classification in Victorian Britain." *Hist. & Phil. of Life. Sciences* 4: 225–254.

————. 1984. *T. H. Huxley's Place in Natural Science*. New Haven: Yale University Press.

————. 1992. "Entre Mephistopheles et Luther: Ernst Haeckel et la reforme de l'univers." In *Darwinisme et société*, edited by P. Tort, pp. 237–284. Paris: PUF.

————. 1995. "A Wolf in Sheep's Clothing: Carl Gegenbaur, Ernst Haeckel, the Vertebral Theory of the Skull, and the Survival of Richard Owen." *J. Hist Biology* 28: 247–280.

Dobzhansky, Theodosius. 1937. *Genetics and the Origin of Species*. New York: Columbia University Press.

————. 1956. *The Biological Basis of Human Freedom*. New York: Columbia University Press.

————. 1962. *Mankind Evolving*. New Haven: Yale University Press.

————. 1981. *Dobzhansky's Genetics of Natural Populations. I–XLIII*. Edited by R. C. Lewontin et al. New York: Columbia University Press.

Dohrn, Anton. 1993. "The Origin of the Vertebrates on the Principle of the Succession of Functions." Translated by Michael T. Ghiselin. *Hist. & Phil. of Life Sciences* 16: 1–98.

Donald, Merlin. 1991. *Origins of the Modern Mind: Three Stages in the Evolution of Culture and Cognition.* Cambridge, Mass.: Harvard University Press.

Dorlodot, Henri de. 1925. *Darwinism and Catholic Thought.* Translated by E. Messenger. New York: Benziger Brothers.

Draper, J. W. 1875. *History of the Conflict between Religion and Science.* London.

Drummond, Henry. 1894. *The Ascent of Man.* New York.

Duffin, Lorna. 1978. "Prisoners of Progress: Women and Evolution." In *The Nineteenth-Century Woman: Her Cultural and Physical World,* edited by Sara Delamont and Lorna Duffin, pp. 57–91. London: Croom Helm.

Dunbar, Robin. 1996. *Grooming, Gossip, and the Evolution of Language.* London: Faber and Faber.

Duncan, David, ed. 1911. *The Life and Letters of Herbert Spencer.* Reissue, London.

Dunn, Leslie Clarence. 1965. *A Short History of Genetics.* New York: McGraw-Hill.

Dupree, A. Hunter. 1959. *Asa Gray.* Cambridge, Mass.: Harvard University Press. Reprint, Baltimore: Johns Hopkins University Press, 1988.

Durant, John R. 1979. "Scientific Naturalism and Social Reform in the Thought of Alfred Russel Wallace." *Brit. J. Hist. Sci.* 12: 31–58.

———, ed. 1985. *Darwinism and Divinity: Essays on Evolution and Religious Belief.* Oxford: Basil Blackwell.

Easlea, Brian. 1981. *Science and Sexual Oppression: Patriarchy's Confrontation with Women and Nature.* London: Weidenfeld and Nicolson.

East, Edward M. 1910. "A Mendelian Interpretation of Variation That Is Apparently Continuous." *Am. Naturalist* 44: 65–82.

Eddy, J. H., Jr. 1984. "Buffon, Organic Alterations, and Man." *Stud. Hist. Biology* 7: 1–46.

Egerton, Frank N. 1968. "Studies of Animal Population from Lamarck to Darwin." *J. Hist. Biology* 1: 225–259.

———. 1970a. "Humboldt, Darwin, and Population." *J. Hist. Biology* 3: 325–360.

———. 1970b. "Refutation and Conjectures: Darwin's Response to Sedgwick's Attack on Chambers." *Stud. Hist. & Phil. Sci.* 1: 176–183.

———. 1973. "Changing Concepts of the Balance of Nature." *Quart. Rev. Biology* 48: 322–350.

———. 1979. "Hewett C. Watson: Great Britain's First Phytogeographer." *Huntia* 3: 87–102.

Eigen, Edward A. 1997. "Overcoming First Impressions: Georges Cuvier's Types." *J. Hist. Biology* 30: 179–209.

Eimer, Gustav Heinrich Theodor. 1890. *Organic Evolution as the Result of the*

Inheritance of Acquired Characters According to the Laws of Organic Growth. Translated by James T. Cunningham. London.

———. 1898. *On Orthogenesis and the Impotence of Natural Selection in Species Formation.* Translated by J. M. McCormack. Chicago.

Eiseley, Loren. 1958. *Darwin's Century: Evolution and the Men Who Discovered It.* New York: Doubleday.

———. 1959. "Charles Darwin, Edward Blyth, and the Theory of Natural Selection." *Proc. Am. Phil. Soc.* 103: 94–158.

Elder, Gregory P. 1996. *Chronic Vigor: Darwin, Anglicans, Catholics, and the Development of a Doctrine of Providential Evolution.* Lanham, Md.: University Press of America.

Eldredge, N. 1982. *The Monkey Business; a Scientist Looks at Creationism.* New York: Washington Square Press.

———. 1985. *Unfinished Synthesis: Biological Hierarchies and Modern Evolutionary Thought.* New York: Oxford University Press.

———. 1986. *Time Frames: The Rethinking of Darwinian Evolution and the Theory of Punctuated Equilibria.* London: Heinemann.

———. 1995. *Reinterpreting Darwin: The Great Evolutionary Debate.* London: Weidenfeld and Nicolson.

Eldredge, N., and S. J. Gould. 1972. "Punctuated Equilibria: An Alternative to Phyletic Gradualism." In *Models in Paleobiology,* edited by T. J. M. Schopf, pp. 82–115. San Francisco: Freeman, Cooper, and Company.

Eliade, Mircea. 1951. *The Myth of the Eternal Return.* London: Routledge and Kegan Paul.

Ellegård, Alvar. 1958. *Darwin and the General Reader: The Reception of Darwin's Theory of Evolution in the British Periodical Press, 1859–1872.* Goteburg: Acta Universitatis Gothenburgensis. Reprint, Chicago: University of Chicago Press, 1990.

Ellis, I. 1980. *Seven against Christ: A Study of "Essays and Reviews."* Leiden: Brill.

Emmet, Dorothy. 1932. *Whitehead's Philosophy of Organism.* New ed. London: Macmillan, 1966.

England, Richard. 1997. "Natural Selection before the *Origin:* Public Reaction of Some Naturalists to the Darwin-Wallace Papers." *J. Hist. Biology* 30: 267–290.

Evans, Brian, and Bernard Waites. 1981. *IQ and Mental Testing: An Unnatural Science and Its Social History.* London: Macmillan.

Evans, L. T. 1984. "Darwin's Use of the Analogy between Artificial and Natural Selection." *J. Hist. Biology* 17: 113–140.

Eve, Raymond, and Francis Harrold. 1991. *The Creationist Movement in Modern America.* New York: Twayne.

Everdell, William R. 1997. *The First Moderns: Origins of Twentieth-Century Thought.* Chicago: University of Chicago Press.

Eyles, Joan M. 1969. "William Smith: Some Aspects of His Life and Work." In *Toward a History of Geology,* edited by Cecil J. Schneer, pp. 142–158. Cambridge, Mass.: MIT Press.

Fairchild, Henry Pratt. 1926. *The Melting Pot Mistake.* Boston. Reprint, New York: Arno, 1977.

Fancher, Raymond E. 1983. "Francis Galton's African Ethnography and Its Role in the Development of His Psychology." *Brit. J. Hist. Sci.* 16: 67–79.

Farber, Paul Lawrence. 1972. "Buffon and the Problem of Species." *J. Hist. Biology* 5: 259–284.

———. 1975. "Buffon and Daubenton: Divergent Traditions in the *Histoire Naturelle.*" *Isis* 66: 63–74.

Farley, John. 1974. "The Initial Reaction of French Biologists to Darwin's *Origin of Species.*" *J. Hist. Biology* 7: 275–300.

———. 1977. *The Spontaneous Generation Controversy: From Descartes to Oparin.* Baltimore: Johns Hopkins University Press.

———. 1982. *Gametes and Spores: Ideas about Sexual Reproduction, 1750–1914.* Baltimore: Johns Hopkins University Press.

Farrall, Lyndsay A. 1979. "The History of Eugenics: A Bibliographical Review." *Annals of Science* 36: 111–123.

Fay, Margaret A. 1978. "Did Marx Offer to Dedicate *Capital* to Darwin?" *J. Hist. Ideas* 39: 133–146.

Fellows, Otis T., and Stephen Milliken. 1972. *Buffon.* Boston: Twayne.

Feuer, Lewis S. 1975. "Is the Darwin-Marx Correspondence Authentic?" *Annals of Science* 32: 1–12.

Fichman, Martin. 1977. "Wallace's Zoogeography and the Problem of Land Bridges." *J. Hist. Biology* 10: 45–63.

———. 1981. *Alfred Russel Wallace.* Boston: Twayne.

———. 1984. "Ideological Factors in the Dissemination of Darwinism in England, 1860–1900." In *Transformation and Tradition in the Sciences: Essays in Honor of I. Bernard Cohen,* edited by E. Mendelsohn, pp. 471–485. Cambridge: Cambridge University Press.

———. 1997. "Biology and Politics: Defining the Boundaries." In *Victorian Science in Context,* edited by Bernard Lightman, pp. 94–118. Chicago: University of Chicago Press.

Fisch, Menachem, and Simon Schaffer, eds. 1990. *William Whewell: A Composite Portrait.* Oxford: Clarendon Press.

Fisher, R. A. 1918. "The Correlation between Relatives on the Supposition of Mendelian Inheritance." *Trans. Roy. Soc. Edinb.* 52: 399–433.

———. 1930. *The Genetical Theory of Natural Selection.* Oxford: Clarendon Press. Reprint, New York: Dover, 1958.

———. 1936. "Has Mendel's Work Been Rediscovered?" *Annals of Science* 1: 115–137.

Fiske, John. 1874. *Outlines of Cosmic Philosophy.* Boston. Reprint, with an introduction by David W. Noble, New York: Johnson Reprint Corporation, 1969.

Foley, Robert. 1995. *Humans before Humanity: An Evolutionary Perspective.* Oxford: Basil Blackwell.

Ford, E. B. 1931. *Mendelism and Evolution.* London: Methuen.

Forrest, D. 1974. *Francis Galton: The Life and Work of a Victorian Genius*. New York: Tapplinger.

Foucault, Michel. 1970. *The Order of Things: The Archaeology of the Human Sciences*. New York: Pantheon Books.

Fox, R., ed. 1976. "Lyell Centenary Issue: Papers Delivered at the Charles Lyell Centenary Symposium, London, 1975." *Brit. J. Hist. Sci.* 9: pt. 2.

Frangsmyr, Tore, ed. 1984. *Linnaeus: The Man and His Work*. Berkeley and Los Angeles: University of California Press.

Frazer, James George. 1924. *The Golden Bough: A Study of Magic and Religion*. Abridged ed. London: Macmillan.

Frederickson, George. 1971. *The Black Image in the White Mind: The Debate on Afro-American Character and Destiny, 1817–1914*. New York: Harper and Row.

Freeman, Derek. 1974. "The Evolutionary Theories of Charles Darwin and Herbert Spencer." *Current Anthropology* 15: 211–237.

———. 1983. *Margaret Mead and Samoa*. Cambridge, Mass.: Harvard University Press.

Froggatt, P., and N. C. Nevin. 1971a. "The 'Law of Ancestral Heredity' and the Mendelian-Ancestrian Controversy in England, 1889–1900." *J. Med. Genetics* 8: 1–36.

———. 1971b. "Galton's 'Law of Ancestral Heredity': Its Impact on the Early Development of Human Genetics." *History of Science* 10: 1–27.

Futuyma, D. 1982. *Science on Trial: The Case for Evolution*. New York: Pantheon.

Fyfe, Aileen. 1997. "The Reception of William Paley's Natural Theology in the University of Cambridge." *Brit. J. Hist. Sci.* 30: 321–335.

Gaissinovitch, A. E. 1980. "The Origins of Soviet Genetics and the Struggle with Lamarckism, 1922–1929." *J. Hist. Biology* 13: 1–52.

Gale, Barry C. 1972. "Darwin and the Concept of the Struggle for Existence: A Study in the Extra-Scientific Origins of Scientific Ideas." *Isis* 63: 321–344.

Gallagher, Idella J. 1970. *Morality in Evolution: The Moral Philosophy of Henri Bergson*. The Hague: Martinus Nijhoff.

Galton, Francis. 1889. *Natural Inheritance*. London.

———. 1892. *Hereditary Genius*. Rev. ed. London. Reprint, with an introduction by H. J. Eysenck, London: Julian Friedman, 1978.

Gascoyne, Robert M. 1991. "Julian Huxley and Biological Progress." *J. Hist. Biology* 24: 433–455.

Gasking, Elizabeth. 1959. "Why Was Mendel's Work Ignored?" *J. Hist. Ideas* 20: 60–84.

———. 1967. *Investigations into Generation, 1651–1828*. London: Hutchinson.

Gasman, Daniel. 1971. *The Scientific Origins of National Socialism: Social Darwinism in Ernst Haeckel and the Monist League*. New York: American Elsevier.

———. 1998. *Haeckel's Monism and the Birth of Fascist Ideology*. New York: Peter Lang.

Gay, Peter. 1966, 1969. *The Enlightenment: An Interpretation.* Vol. 1: *The Rise of Modern Paganism.* Vol. 2: *The Science of Freedom.* New York: Alfred A. Knopf.

Gayon, Jean. 1998. *Darwinism's Struggle for Survival: Heredity and the Hypothesis of Natural Selection.* Cambridge: Cambridge University Press.

Geddes, Patrick, and J. Arthur Thomson. 1889. *The Evolution of Sex.* London.

Gee, Henry. 1996. *Before the Backbone: Views on the Origin of the Vertebrates.* London: Chapman and Hall.

———. 1999. *In Search of Deep Time: Beyond the Fossil Record to a New History of Life.* New York: Free Press.

Geison, Gerald. 1969. "Darwin and Heredity: The Evolution of His Hypothesis of Pangenesis." *J. Hist. Medicine* 24: 375–411.

Geoffroy Saint-Hilaire, Etienne. 1818–22. *Philosophie anatomique: Les organes réspiratoires sous le rapport de la détermination et de l'identité de leur pièces osseuses.* 2 vols. Paris. Reprint, Brussels: Culture et Civilisation, 1968.

———. 1833. "Le degré d'influence du monde ambiant pour modifier les formes animales." *Mem. Acad. Roy. des Sciences* 12: 63–92.

George, Wilma. 1964. *Biologist-Philosopher: A Study of the Life and Writings of Alfred Russel Wallace.* New York: Abelard-Schuman.

Ghiselin, Michael T. 1969. *The Triumph of the Darwinian Method.* Berkeley and Los Angeles: University of California Press.

———. 1974. *The Economy of Nature and the Evolution of Sex.* Berkeley and Los Angeles: University of California Press.

Ghiselin, Michael T., and L. Jaffe. 1973. "Phylogenetic Classification in Darwin's *Monograph on the Sub-Class Cirripedia." Systematic Zoology* 22: 132–140.

Gilbert, Scott F. 1978. "The Embryological Origins of the Gene Theory." *J. Hist. Biology* 11: 307–351.

Gilham, Nicholas Wright. 2001. *A Life of Sir Francis Galton: From African Exploration to the Birth of Eugenics.* Oxford: Oxford University Press.

Gillespie, Neal C. 1977. "The Duke of Argyll, Evolutionary Anthropology, and the Art of Scientific Controversy." *Isis* 68: 40–54.

———. 1979. *Charles Darwin and the Problem of Creation.* Chicago: University of Chicago Press.

———. 1987. "Natural History, Natural Theology, and Social Order: John Ray and the 'Newtonian Ideology.' " *J. Hist. Biology* 20: 1–50.

———. 1990. "Divine Design and the Industrial Revolution: William Paley's Abortive Reform of Natural Theology." *Isis* 81: 214–229.

Gillispie, Charles Coulston. 1951. *Genesis and Geology: A Study in the Relations of Scientific Thought, Natural Theology, and Social Opinions in Great Britain, 1790–1859.* Reprint, New York: Harper, 1959.

———. 1956. "The Formation of Lamarck's Evolutionary Theory." *Arch. Internat. Hist. Sci.* 9: 323–338.

———. 1959. "Lamarck and Darwin in the History of Science." In *Forerunners of Darwin, 1745–1859,* edited by Bentley Glass, Owsei Temkin, and William Strauss Jr., pp. 265–291. Baltimore: Johns Hopkins University Press.

————. 1960. *The Edge of Objectivity: An Essay in the History of Scientific Ideas.* Princeton: Princeton University Press.

————, ed. 1970–80. *Dictionary of Scientific Biography.* 16 vols. New York: Charles Scribner's Sons.

Ginger, Ray. 1958. *Six Days or Forever.* Boston: Beacon Press.

Ginsberg, Morris. 1953. *The Idea of Progress: A Revaluation.* Reprint, Westport, Conn.: Greenwood Press, 1972.

Gish, Duane T. 1972. *Evolution: The Fossils Say No!* San Diego: Creation Life Publishers.

Glass, Bentley. 1959a. "The Germination of the Biological Species Concept." In *Forerunners of Darwin, 1745–1859,* edited by Bentley Glass, Owsei Temkin, and William Strauss Jr., pp. 30–48. Baltimore: Johns Hopkins University Press.

————. 1959b. "Maupertuis, Pioneer of Genetics." In *Forerunners of Darwin, 1745–1859,* edited by Bentley Glass, Owsei Temkin, and William Strauss Jr., pp. 51–83. Baltimore: Johns Hopkins University Press.

————. 1959c. "Heredity and Variation in the Eighteenth-Century Concept of the Species." In *Forerunners of Darwin, 1745–1859,* edited by Bentley Glass, Owsei Temkin, and William Strauss Jr., pp. 144–172. Baltimore: Johns Hopkins University Press.

Glass, Bentley, Owsei Temkin, and William Strauss Jr., eds. 1959. *Forerunners of Darwin, 1745–1859.* Baltimore: Johns Hopkins University Press.

Gliboff, Sander. 1999. "Gregor Mendel and the Laws of Evolution." *History of Science* 37: 217–235.

Glick, Thomas F., ed. 1974. *The Comparative Reception of Darwinism.* Austin: University of Texas Press. New ed., Chicago: University of Chicago Press, 1988.

Godfrey, L. R., ed. 1983. *Scientists Confront Creationism.* New York: W. W. Norton.

Goldschmidt, Richard. 1940. *The Material Basis of Evolution.* New Haven: Yale University Press. Reprint, with an introduction by S. J. Gould, New Haven: Yale University Press, 1982.

Goldsmith, Donald, ed. 1977. *Scientists Confront Velikovsky.* Ithaca, N.Y.: Cornell University Press.

Gould, Stephen Jay. 1974a. "On Biological and Social Determinism." *History of Science* 12: 212–220.

————. 1974b. "The Origin and Function of 'Bizarre' Structures: Antler Size and Skull Size in the 'Irish Elk,' *Megaloceros giganteus.*" *Evolution* 28: 191–220.

————. 1977a. "The Eternal Metaphors of Paleontology." In *Patterns of Evolution,* edited by Anthony Hallam, pp. 1–26. Amsterdam: Elsevier.

————. 1977b. *Ontogeny and Phylogeny.* Cambridge, Mass.: Harvard University Press.

————. 1980a. "G. G. Simpson, Paleontology, and the Modern Synthesis." In *The Evolutionary Synthesis: Perspectives on the Unification of Biology,* edited by Ernst Mayr and William B. Provine, pp. 153–172. Cambridge, Mass.: Harvard University Press.

———. 1980b. "Is a New and General Theory of Evolution Emerging?" *Paleobiology* 6: 119–130.

———. 1981. *The Mismeasure of Man.* New York: Norton.

———. 1982. "The Meaning of Punctuated Equilibrium and Its Role in Validating a Hierarchical Approach to Macroevolution." In *Perspectives on Evolution,* edited by Roger Milkman, pp. 83–104. Sunderland, Mass.: Sinauer.

———. 1983. "The Hardening of the Modern Synthesis." In *Dimensions of Darwinism: Themes and Counterthemes in Twentieth-Century Evolutionary Theory,* edited by Marjorie Grene, pp. 71–93. Cambridge: Cambridge University Press.

———. 1987. *Time's Arrow, Time's Cycles: Myth and Metaphor in the Discovery of Geological Time.* Cambridge, Mass.: Harvard University Press.

———. 1989. *Wonderful Life: The Burgess Shale and the Nature of History.* London: Hutchinson Radius.

———. 1993. *Eight Little Piggies: Reflections on Natural History.* New York: Norton.

———. 2002. *The Structure of Evolutionary Theory.* Cambridge, Mass.: Harvard University Press.

Gould, Stephen J., and Niles Eldredge. 1977. "Punctuated Equilibria: The Tempo and Mode of Evolution Reconsidered." *Paleobiology* 3: 115–151.

Gould, Stephen J., and R. C. Lewontin. 1979. "The Spandrels of San Marco and the Panglossian Paradigm: A Critique of the Adaptationist Programme." *Proc. Roy. Soc. London,* ser. B, 205: 581–598.

Grabiner, J. V., and P. D. Miller. 1974. "Effects of the Scopes Trial: Was It a Victory for Evolutionists?" *Science* 185: 832–837.

Grant, Madison. 1918. *The Passing of the Great Race.* New York. Reprint, New York: Arno, 1977.

Gray, Asa. 1876. *Darwiniana: Essays and Reviews Pertaining to Darwinism.* New York. Reprint, edited by A. Hunter Dupree, Cambridge, Mass.: Harvard University Press, 1963.

Grayson, Donald K. 1983. *The Establishment of Human Antiquity.* New York: Academic Press.

Greenaway, A. P. 1973. "The Incorporation of Action into Associationism: The Psychology of Alexander Bain." *J. Hist. Behavioral Sci.* 9: 42–52.

Greene, John C. 1959a. *The Death of Adam: Evolution and Its Impact on Western Thought.* Ames: Iowa State University Press.

———. 1959b. "Biology and Social Theory in the Nineteenth Century: Auguste Comte and Herbert Spencer." In *Critical Problems in the History of Science,* edited by M. Clagett, pp. 419–466. Madison: University of Wisconsin Press. Reprinted in *Science, Ideology, and World View: Essays in the History of Evolutionary Ideas,* by Greene, pp. 60–94. Berkeley and Los Angeles: University of California Press, 1981.

————. 1961. *Darwin and the Modern World View*. Baton Rouge: Louisiana State University Press.

————. 1971. "The Kuhnian Paradigm and the Darwinian Revolution in Natural History." In *Perspectives in the History of Science and Technology*, edited by Duane H. D. Roller, pp. 3–25. Norman: University of Oklahoma Press. Reprinted in *Science, Ideology, and World View: Essays in the History of Evolutionary Ideas*, by Greene, pp. 30–59. Berkeley and Los Angeles: University of California Press, 1981.

————. 1975. "Reflections on the Progress of Darwin Studies." *J. Hist. Biology* 8: 243–273.

————. 1977. "Darwin as a Social Evolutionist." *J. Hist. Biology* 10: 1–27. Reprinted in *Science, Ideology, and World View: Essays in the History of Evolutionary Ideas*, by Greene, pp. 95–127. Berkeley and Los Angeles: University of California Press, 1981.

————. 1981. *Science, Ideology, and World View: Essays in the History of Evolutionary Ideas*. Berkeley and Los Angeles: University of California Press.

————. 1990. "The Interaction of Science and World View in Sir Julian Huxley's Evolutionary Biology." *J. Hist. Biology* 23: 39–55. Reprinted in *Science, Ideology, and World View: Essays in the History of Evolutionary Ideas*, by Greene, pp. 71–90. Berkeley and Los Angeles: University of California Press, 1981.

————. 1999. *Debating Darwin: Adventures of a Scholar*. Claremont, Calif.: Regina Books.

Greene, Mott T. 1982. *Geology in the Nineteenth Century: Changing Views of a Changing World*. Ithaca, N.Y.: Cornell University Press.

Greenwood, Davydd J. 1984. *The Taming of Evolution: The Persistence of Nonevolutionary Views in the Study of Humans*. Ithaca, N.Y.: Cornell University Press.

Gregory, Frederick. 1977. *Scientific Materialism in Nineteenth-Century Germany*. Dordrecht: D. Reidel.

Grene, Marjorie, ed. 1983. *Dimensions of Darwinism: Themes and Counterthemes in Twentieth-Century Evolutionary Theory*. Cambridge: Cambridge University Press.

Gribben, John, and Jeremy Cherfas. 1982. *The Monkey Puzzle: A Family Tree*. London: Bodley Head.

Grogin, R. C. 1988. *The Bergsonian Controversy in France, 1900–1914*. Calgary: University of Calgary Press.

Gross, Paul R., and Norman Levitt. 1994. *Higher Superstition: The Academic Left and Its Quarrels with Science*. Baltimore: Johns Hopkins University Press.

Gruber, Howard E. 1968. "Who Was the *Beagle*'s Naturalist?" *Brit. J. Hist. Sci.* 4: 266–282.

————. 1974. *Darwin on Man: A Psychological Study of Scientific Creativity: Together with Darwin's Early and Unpublished Notebooks*. New York: E. P. Dutton.

Gruber, Jacob W. 1960. *A Conscience in Conflict: The Life of St. George Jackson Mivart*. New York: Columbia University Press.

Gulick, Addison. 1932. *Evolutionist and Missionary: John Thomas Gulick*. Chicago: University of Chicago Press.

Gulick, John T. 1888. "Divergent Evolution through Cumulative Segregation." *J. Linn. Soc. (Zool.)* 20: 189–274.

Gutting, Gary. 1989. *Michel Foucault's Archaeology of Scientific Reason*. Cambridge: Cambridge University Press.

———, ed. 1994. *The Cambridge Companion to Foucault*. Cambridge: Cambridge University Press.

Haber, Francis C. 1959. *The Age of the World: Moses to Darwin*. Baltimore: Johns Hopkins University Press.

Haeckel, Ernst. 1866. *Generelle Morphologie der Organismen*. 2 vols. Reprint, Berlin: Walter de Gruyter, 1988.

———. 1876. *The History of Creation: Or the Development of the Earth and Its Inhabitants by the Action of Natural Causes. A Popular Exposition of the Doctrine of Evolution in General and of that of Darwin, Goethe and Lamarck in Particular*. 2 vols. New York.

———. 1879. *The Evolution of Man: A Popular Exposition of the Principal Points of Human Ontogeny and Phylogeny*. New York.

———. 1898. *The Last Link: Our Present Knowledge of the Descent of Man*. London.

Hagberg, Knut. 1953. *Carl Linnaeus*. Translated by Alan Blair. New York: Dutton.

Hagen, Joel B. 1999a. "Retelling Experiments: H. B. D. Kettlewell's Studies of Industrial Melanism in Peppered Moths." *Biology and Philosophy* 14: 39–54.

———. 1999b. "Naturalists, Molecular Biologists, and the Challenges of Molecular Evolution." *J. Hist. Biology* 32: 321–341.

Haldane, J. B. S. 1932. *The Causes of Evolution*. London. Reprint, Ithaca, N.Y.: Cornell University Press, 1966.

———. 1938. *Heredity and Politics*. London: Allen and Unwin.

Halévy, Elie. 1955. *The Growth of Philosophic Radicalism*. Translated by Mary Morris. Boston: Beacon Press.

Hall, G. Stanley. 1904. *Adolescence: Its Psychology and Its Relation to Physiology, Anthropology, Sociology, Sex, Crime, Religion, and Education*. 2 vols. New York.

Hall, Thomas S. 1969. *Ideas of Life and Matter*. 2 vols. Chicago: University of Chicago Press.

Hallam, Anthony. 1973. *A Revolution in the Earth Sciences: From Continental Drift to Place Tectonics*. Oxford: Clarendon Press.

———. 1983. *Great Geological Controversies*. Oxford: Oxford University Press.

Hallam, Anthony, and P. B. Wignell. 1997. *Mass Extinctions and Their Aftermath*. Oxford: Oxford University Press.

Haller, John S. 1975. *Outcasts from Evolution: Scientific Attitudes of Racial Inferiority, 1859–1900.* Urbana: University of Illinois Press.

Haller, Mark H. 1963. *Eugenics: Hereditarian Attitudes in American Thought.* New Brunswick, N.J.: Rutgers University Press.

Halliday, R. J. 1971. "Social Darwinism: A Definition." *Victorian Studies* 14: 389–405.

Hamilton, W. D. 1964. "The Genetical Evolution of Social Behavior, I and II." *J. Theoretical Biology* 7: 1–32.

Hammond, Michael. 1980. "Anthropology as a Weapon of Social Combat in Late Nineteenth-Century France." *J. Hist. Behavioral Sci.* 16: 118–132.

———. 1982. "The Expulsion of the Neanderthals from Human Ancestry: Marcellin Boule and the Social Context of Scientific Research." *Social Studies of Science* 12: 1–36.

Hampson, Norman. 1968. *The Enlightenment.* Harmondsworth, Middlesex: Penguin Books.

Hanks, Leslie. 1966. *Buffon avant l'Histoire naturelle.* Paris: Presses Universitaires de France.

Haraway, Donna. 1990. *Primate Visions: Gender, Race, and Nature in the World of Modern Science.* London: Routledge.

Hardy, A. C. 1965. *The Living Stream: A Restatement of Evolution Theory.* London: Collins.

Harmsen, H. 1955. "The German Sterilization Act of 1933." *Eugenics Review* 46: 227–232.

Harris, Marvin. 1968. *The Rise of Anthropological Theory: A History of Theories of Culture.* New York: Thomas Y. Crowell.

Harrison, James. 1972. "Erasmus Darwin's Views on Evolution." *J. Hist. Ideas* 32: 247–264.

Harrison, Peter. 1998. *The Bible, Protestantism, and the Rise of Natural Science.* Cambridge: Cambridge University Press.

Hartley, David. 1749. *Observations upon Man: His Frame, His Duty, and His Expectations.* London. Reprint, Gainesville, Fla.: Scholars' Reprint Corporation, 1966.

———. 1775. *Hartley's Theory of the Human Mind.* Edited by Joseph Priestley. London. Reprint, New York: AMS, 1972.

Hartmann, Johann. 1756. "Plantae Hybridae." In *Amoenitates Academicae, seu Dissertationes variae physicae, medicae et botanicae,* edited by Carolus Linnaeus, 3:28–62. Leiden and Amsterdam, 1749–90.

Harwood, Jonathan. 1985. "Genetics and the Evolutionary Synthesis in Interwar Germany." *Annals of Science* 42: 279–301.

———. 1993. *Styles of Scientific Thought: The German Genetics Community, 1900–1937.* Chicago: University of Chicago Press.

———. 1994. "Metaphysical Foundations of the Evolutionary Synthesis: A Historiographical Note." *J. Hist. Biology* 27: 1–20.

Hatch, Elvin. 1973. *Theories of Man and Culture.* New York: Columbia University Press.

Hawkins, Mike. 1997. *Social Darwinism in European and American Thought, 1860–1945: Nature as Model, Nature as Threat.* Cambridge: Cambridge University Press.

Hazard, Paul. 1953. *The European Mind, 1680–1715.* Translated by J. Lewis May. London: Hollis and Carter.

———. 1963. *European Thought in the Eighteenth Century: From Montesquieu to Lessing.* Translated by J. Lewis May. Cleveland: Meridian Books.

Hegel, G. W. F. 1953. *Reason in History: A General Introduction to the Philosophy of History.* Translated by Robert S. Hartman. New York: Bobbs Merrill.

Heimann, P. M., and J. E. McGuire. 1971. "Newtonian Forces and Lockean Powers: Concepts of Matter in Eighteenth Century Thought." *Hist. Stud. Phys. Sci.* 3: 233–306.

Helfand, M. S. 1977. "T. H. Huxley's 'Evolution and Ethics': The Politics of Evolution and the Evolution of Politics." *Victorian Studies* 20: 159–177.

Helvétius, Claude Adrien. 1810. *A Treatise on Man: His Intellectual Faculties and His Education.* Translated by W. Hooper. Reprint, New York: Franklin, 1969.

Hempel, Carl. 1966. *Philosophy of Natural Science.* Englewood Cliffs, N.J.: Prentice-Hall.

Henfrey, Arthur, and T. H. Huxley, eds. 1853. *Scientific Memoirs: Selected from the Transactions of Foreign Academies of Science and from Foreign Journals: Natural History.* London. Reprint, New York: Johnson Reprint Corporation, 1966.

Henig, Robin Maranz. 2000. *A Monk and Two Peas: The Story of Gregor Mendel and the Discovery of Genetics.* New York: Houghton Mifflin; London: Weidenfeld and Nicolson.

Henkin, Leo J. 1963. *Darwinism in the English Novel, 1860–1910.* New York: Russell and Russell.

Hennig, Willi. 1966. *Phylogenetic Systematics.* Translated by D. Dwight Davis and Rainer Zangerl. Urbana: University of Illinois Press.

Henslow, George. 1888. *The Origin of Floral Structures through Insect and Other Agencies.* London.

———. 1895. *The Origin of Plant Structures by Self-Adaptation to the Environment.* London.

Herbert, Sandra. 1971. "Darwin, Malthus, and Selection." *J. Hist. Biology* 4: 209–218.

———. 1974–77. "The Place of Man in the Development of Darwin's Theory of Transmutation. Part I, to July 1837." *J. Hist. Biology* 7: 217–258. "Part II." Ibid., 10: 155–227.

Herder, Johann Gottfried von. 1968. *Reflections on the Philosophy of the History of Mankind.* Translated by T. O. Churchill, edited by Frank E. Manuel. Chicago: University of Chicago Press.

Herrnstein, Richard J., and Charles Murray. 1994. *The Bell Curve.* New York: Free Press.

Herschel, Sir J. F. W. 1830. *A Preliminary Discourse on the Study of Natural Philosophy.* London. Reprint, with an introduction by Michael Partridge, New York: Johnson Reprint Corporation, 1966.

———. 1861. *Physical Geography.* London.

Heyer, Paul. 1982. *Nature, Human Nature, and Society: Marx, Darwin, Biology, and the Human Sciences.* Westport, Conn.: Greenwood Press.

Hill, Emita. 1968. "Materialism and Monsters in *Le rêve de d'Alembert.*" *Diderot Studies* 10: 69–94.

Himmelfarb, Gertrude. 1959. *Darwin and the Darwinian Revolution.* Reprint, New York: Norton.

Ho, Mae-Wan, and S. W. Fox, eds. 1988. *Evolutionary Processes and Metaphors.* New York: John Wiley.

Ho, Mae-Wan, and Peter T. Saunders, eds. 1984. *Beyond Neo-Darwinism: An Introduction to the New Evolutionary Paradigm.* London: Academic Press.

Hobbes, Thomas. 1957. *Leviathan: Or the Matter, Forme, and Power of a Commonwealth, Ecclesiastical and Civil.* Edited by Michael Oakeshott. Oxford: Blackwell.

Hodge, M. J. S. 1971. "Lamarck's Science of Living Bodies." *Brit. J. Hist. Sci.* 5: 323–352.

———. 1972. "The Universal Gestation of Nature: Chambers' *Vestiges* and *Explanations.*" *J. Hist. Biology* 5: 127–152.

———. 1977. "The Structure and Strategy of Darwin's 'Long Argument.' " *Brit. J. Hist. Sci.* 10: 237–246.

———. 1982. "Darwin and the Laws of the Animate Part of the Terrestrial System (1835–1837): On the Lyellian Origins of His Zoonomical Explanatory Programme." *Stud. Hist. Biology* 6: 1–106.

———. 1985. "Darwin as a Lifelong Generation Theorist." In *The Darwinian Heritage: A Centennial Retrospect,* edited by David Kohn, pp. 207–243. Princeton: Princeton University Press.

———. 1989. "Generation and the Origin of Species (1837–1937): A Historiographical Suggestion." *Brit. J. Hist. Sci.* 22: 267–282.

———. 1992. "Biology and Philosophy (Including Ideology): A Study of Fisher and Wright." In *The Founders of Evolutionary Genetics,* edited by S. Sarker, pp. 231–293. Dordrecht: Kluwer.

Hodge, M. J. S., and D. Kohn. 1985. "The Immediate Origins of Natural Selection." In *The Darwinian Heritage: A Centennial Retrospect,* edited by David Kohn, pp. 185–206. Princeton: Princeton University Press.

Hofstadter, Richard. 1959. *Social Darwinism in American Thought.* Rev. ed. New York: George Braziller.

Holt, Niles R. 1971. "Ernst Haeckel's Monist Religion." *J. Hist. Ideas* 32: 265–280.

Hooke, Robert. 1705. *The Posthumous Works of Robert Hooke.* London. Reprint, New York: Johnson Reprint Corporation, 1969.

———. 1977. *Lectures and Discourses of Earthquakes and Subterraneous Eruptions.* Reprint, New York: Arno Press, 1977.

Hooker, Joseph Dalton. 1860. "On the Origination and Distribution of Vegetable Species: Introductory Essay to the Flora of Tasmania." *Am. J. Sci.*, 2d ser., 29: 1–25, 303–326.

Hooykaas, R. 1957. "The Parallel between the History of the Earth and the History of the Animal World." *Arch. Internat. Hist. Sci.* 10: 3–18.

———. 1959. *Natural Law and Divine Miracle: The Principle of Uniformity in Geology, Biology, and History.* Leiden: Brill.

———. 1966. "Geological Uniformitarianism and Evolution." *Arch. Internat. Hist. Sci.* 19: 3–19.

———. 1970. *Catastrophism in Geology: Its Scientific Character in Relation to Actualism and Uniformitarianism.* Amsterdam: North Holland Publishing.

Horder, T. J., J. A. Witkowski, and C. C. Wylie, eds. 1986. *A History of Embryology.* Cambridge: Cambridge University Press.

Hoskin, M. A. 1964. *William Herschel and the Construction of the Heavens.* New York: W. W. Norton.

Hull, David L. 1973a. "Charles Darwin and Nineteenth Century Philosophies of Science." In *Foundations of Scientific Thought: The Nineteenth Century,* edited by N. Giere and R. S. Westfall. Bloomington: Indiana University Press.

———. 1973b. *Darwin and His Critics: The Reception of Darwin's Theory of Evolution by the Scientific Community.* Cambridge, Mass.: Harvard University Press.

———. 1978. "Sociobiology: A Scientific Bandwagon or a Travelling Medicine Show?" In *Sociobiology and Human Nature,* edited by M. S. Gregory et al., pp. 136–163. San Francisco: Jossey-Bass.

———. 1979. "The Limits of Cladism." *Systematic Zoology* 28: 416–440.

———. 1985. "Darwinism as a Historical Entity: A Historiographical Proposal." In *The Darwinian Heritage: A Centennial Retrospect,* edited by David Kohn, pp. 773–812. Princeton: Princeton University Press.

———. 1988. *Science as Process: An Evolutionary Account of the Growth and Conceptual Development of Science.* Chicago: University of Chicago Press.

Hull, David L., Peter D. Tessner, and Arthur M. Diamond. 1978. "Plank's Principle: Do Younger Scientists Accept New Scientific Ideas with Greater Alacrity Than Older Scientists?" *Science* 202: 717–723.

Humboldt, Alexander von. 1814–29. *Personal Narrative of Travels to the Equinoctal Regions of the New Continent during the Years 1799–1804.* Translated by Helen Maria Williams. 7 vols. London. Reprint, New York: AMS Press, 1966.

Hutton, James. 1795. *Theory of the Earth, with Proofs and Illustrations.* 2 vols. Edinburgh. Reprint, Weinheim/Bergstr.: H. R. Engelmann (J. Cramer) and Codiocote, Herts., Wheldon and Wesley, 1960.

Huxley, Aldous. 1955. *Brave New World.* Reprint, Harmondsworth: Penguin Books.

Huxley, Julian S. 1932. *Problems of Relative Growth.* London: Methuen.

———. 1942. *Evolution: The Modern Synthesis.* London: Allen and Unwin.

————. 1953. *Evolution in Action; Based on the Pattern Foundation Lectures Delivered at Indiana University in 1951*. London: Chatto and Windus.

————. 1970. *Memories*. London: Allen and Unwin.

————, ed. 1961. *The Humanist Frame*. London: Allen and Unwin.

Huxley, Leonard. 1900. *The Life and Letters of Thomas Henry Huxley*. 2 vols. London. Reprint, Farnborough: Gregg International, 1969.

————. 1918. *The Life and Letters of Sir Joseph Dalton Hooker*. 2 vols. London.

Huxley, T. H. 1854. "Vestiges of the Natural History of Creation." *Brit. & Foreign Med. Chirurg. Rev.* 13: 332–343.

————. 1863. *Man's Place in Nature*. London. Reprinted in *Collected Essays*, by T. H. Huxley. Vol. 7. London. Reprint, Bristol: Thoemmes Press, 2001.

————. 1888. *American Addresses: With a Lecture on the Study of Biology*. New York.

————. 1893–94. *Collected Essays*. 9 vols. London. Reprint, Bristol: Thoemmes Press, 2001.

————. 1894. *Evolution and Ethics: Collected Essays*. Vol. 9. Reprint, New York: AMS Press, 1970.

————. 1997. *The Major Prose of Thomas Henry Huxley*. Edited by Alan P. Barr. Athens: University of Georgia Press.

Hyatt, Alpheus. 1866. "On the Parallelism between the Different Stages of Life in the Individual and Those in the Entire Group of the Molluscous Order Tetrabranchiata." *Mem. Boston Soc. Nat. Hist.* 1: 193–209.

————. 1880. "Genesis of the Tertiary Species of Planorbis at Steinheim." *Boston Soc. Nat. Hist., Anniversary Memoir*. Abstracted *Proc. A.A.A.S.* 1880, 527–550; and *Am. Naturalist* 16 (1882): 441–453.

————. 1884. "Evolution of the Cephalopods." *Science* 3: 122–127 and 145–149.

————. 1889. *Genesis of the Arietidae*. Washington.

Iltis, Hugo. 1932. *Life of Mendel*. Translated by Eden and Cedar Paul. Reprint, New York: Hafner, 1966.

Ingold, Tim. 1987. *Evolution and Social Life*. Cambridge: Cambridge University Press.

Irvine, William. 1955. *Apes, Angels, and Victorians: The Story of Darwin, Huxley, and Evolution*. London. Reprint, Cleveland: Meridian Books, 1959.

Israel, Jonathan I. 2001. *Radical Enlightenment: Philosophy and the Making of Modernity, 1650–1750*. Oxford: Oxford University Press.

Jacob, Margaret C. 1981. *The Radical Enlightenment: Pantheists, Freemasons, and Republicans*. London: Allen and Unwin.

Jaki, Stanley. 1978a. *The Road of Science and the Way to God*. Chicago: University of Chicago Press.

————. 1978b. *Planets and Planetarians: A History of Theories of the Origin of Planetary Systems*. Edinburgh: Scottish University Press.

James, Patricia. 1979. *Population Malthus: His Life and Times*. London: Routledge.

Jameson, Robert. 1808. *The Wernerian Theory of the Neptunian Origin of Rocks*. Introduced by Jesse M. Sweet. New York: Hafner; London: Collier Macmillan, 1976.

Jarvis, Adrian. 1997. *Samuel Smiles and the Reconstruction of Victorian Values.* London: Alan Sutton.

Jenkin, Fleeming. 1867. "The Origin of Species." *North British Review* 46: 277–318.

Jensen, J. Vernon. 1988. "Return to the Wilberforce-Huxley Debate." *Brit. J. Hist. Sci.* 21: 161–179.

————. 1991. *Thomas Henry Huxley: Communicating for Science.* Newark: University of Delaware Press.

Johannsen, Wilhelm. 1955. "Concerning Heredity in Populations and in Pure Lines." Translated by Harold Call and Elga Putsch. In *Selected Readings in Biology for Natural Sciences,* pp. 172–215. Chicago: University of Chicago Press.

Johanson, D. C., and M. A. Edey. 1981. *Lucy: The Beginnings of Humankind.* New York: Simon and Schuster.

Johnson, Phillip E. 1991. *Darwin on Trial.* New York: Intervarsity Press and Regnery Gateway.

Jones, Greta. 1980. *Social Darwinism and English Thought: The Interaction between Biological and Social Theory.* London: Harvester Press.

————. 1986. *Social Hygiene in Twentieth-Century Britain.* London: Croom Helm.

————. 2002. "Alfred Russel Wallace, Robert Owen, and the Theory of Natural Selection." *Brit. J. Hist. Sci.* 35: 73–96.

Jones, Richard Foster. 1965. *Ancients and Moderns: A Study of the Rise of the Scientific Movement in Seventeenth-Century England.* 2d ed. Berkeley and Los Angeles: University of California Press.

Joravsky, D. 1970. *The Lysenko Affair.* Cambridge, Mass.: Harvard University Press.

Jordanova, L. 1984. *Lamarck.* Oxford: Oxford University Press.

Jordanova, L., and R. S. Porter, eds. 1979. *Images of the Earth: Essays in the History of the Environmental Sciences.* Chalfont St. Giles, Bucks.: British Society for the History of Science.

Kammerer, Paul. 1923. "Breeding Experiments on the Inheritance of Acquired Characters." *Nature* 111: 637–640.

————. 1924. *The Inheritance of Acquired Characteristics.* New York: Boni and Liveright.

Kant, Immanuel. 1969. *Universal Natural History and Theory of the Heavens.* Introduction by Milton K. Munitz. Ann Arbor: University of Michigan Press.

Keith, Sir Arthur. 1915. *The Antiquity of Man.* London.

————. 1949. *A New Theory of Human Evolution.* New York: Philosophical Library.

————. 1955. *Darwin Revalued.* London: Watts.

Kellner, L. 1963. *Alexander von Humboldt.* London: Oxford University Press.

Kellogg, Vernon L. 1907. *Darwinism Today: A Discussion of Present Day Scientific Criticism of the Darwinian Selection Theories.* New York: Henry Holt.

Kelly, Alfred. 1981. *The Descent of Darwinism: The Popularization of Darwinism in Germany, 1860–1914.* Chapel Hill: University of North Carolina Press.

Kelly, Suzanne. 1969. "Theories of the Earth in Renaissance Cosmologies." In *Toward a History of Geology,* edited by Cecil J. Schneer, pp. 214–225. Cambridge, Mass.: MIT Press.

Kennedy, James C. 1978. *Herbert Spencer.* Boston: Twayne Publishers.

Kettlewell, H. B. D. 1955. "Selection Experiments on Industrial Melanism in the Lepidoptera." *Heredity* 9: 323–342.

———. 1973. *The Evolution of Melanism.* Oxford: Clarendon Press.

Kevles, Daniel. 1985. *In the Name of Eugenics: Genetics and the Uses of Human Heredity.* New York: Knopf.

Keynes, Richard Darwin. 1979. *The Beagle Record: Selections from the Original Pictorial Records and Written Accounts of the Voyage of H.M.S.* Beagle. Cambridge: Cambridge University Press.

Kimura, Motoo. 1983. *The Neutral Theory of Molecular Evolution.* Cambridge: Cambridge University Press.

Kinch, Michael Paul. 1980. "Geographical Distribution and the Origin of Life: The Development of Early Nineteenth-Century British Explanations." *J. Hist. Biology* 13: 91–119.

King-Hele, Desmond. 1963. *Erasmus Darwin.* New York: Scribner.

Kirwan, Richard. 1799. *Geological Essays.* London. Reprint, New York: Arno Press, 1977.

Kitcher, Philip. 1982. *Abusing Science: The Case against Creationism.* Cambridge, Mass.: MIT Press.

Knight, Isobel F. 1968. *The Geometric Spirit: The Abbé de Condillac and the French Enlightenment.* New Haven: Yale University Press.

Knox, Robert. 1862. *The Races of Man: A Philosophical Enquiry into the Influence of Race on the Destiny of Nations.* 2d ed. London.

Koerner, Lisbet. 1999. *Linnaeus: Nature and Nation.* Cambridge, Mass.: Harvard University Press.

Koestler, Arthur. 1967. *The Ghost in the Machine.* New York: Macmillan.

———. 1971. *The Case of the Midwife Toad.* London: Hutchinson.

———. 1972. *The Roots of Coincidence.* London: Hutchinson.

Kofahl, R. E., and K. L. Segraves. 1975. *The Creation Explanation: A Scientific Alternative to Evolution.* Wheaton, Ill.: Harold Shaw.

Kohler, Robert E. 1994. *Lords of the Fly:* Drosophila *Genetics and the Experimental Life.* Chicago: University of Chicago Press.

Kohn, David. 1980. "Theories to Work By: Rejected Theories, Reproduction, and Darwin's Path to Natural Selection." *Stud. Hist. Biol.* 4: 67–170.

———. 1985. "Darwin's Principle of Divergence as Internal Dialogue." In *The Darwinian Heritage: A Centennial Retrospect,* edited by David Kohn, pp. 245–257. Princeton: Princeton University Press.

———. 1989. "Darwin's Ambiguity: The Secularization of Biological Meaning." *Brit. J. Hist. Sci.* 22: 215–240.

————, ed. 1985. *The Darwinian Heritage: A Centennial Retrospect*. Princeton: Princeton University Press.

Kottler, Malcolm Jay. 1974. "Alfred Russel Wallace, the Origin of Man, and Spiritualism." *Isis* 65: 145–192.

————. 1978. "Charles Darwin's Biological Species Concept and Theory of Geographic Speciation: The Transmutation Notebooks." *Annals of Science* 35: 275–297.

————. 1979. "Hugo De Vries and the Rediscovery of Mendel's Laws." *Annals of Science* 36: 517–538.

————. 1980. "Darwin, Wallace, and the Origin of Sexual Dimorphism." *Proc. Am. Phil. Soc.* 124: 203–226.

————. 1985. "Charles Darwin and Alfred Russel Wallace: Two Decades of Debate over Natural Selection." In *The Darwinian Heritage: A Centennial Retrospect*, edited by David Kohn, pp. 367–432. Princeton: Princeton University Press.

Krause, Ernst. 1879. *Erasmus Darwin*. Translated by W. S. Dallas, with a preliminary notice by Charles Darwin. London. Reprint, Farnborough: Gregg International, 1971.

Kroeber, A. L. 1917. "The Superorganic." *Am. Anthropologist*, n.s., 19: 163–213.

Kropotkin, Peter. 1902. *Mutual Aid: A Factor in Evolution*. London. Reprint, with an introduction by Ashley Montagu, Boston: Extending Horizon Books.

Kuehl, Stefan. 1994. *The Nazi Connection: Eugenics, American Racism, and German National Socialism*. New York: Oxford University Press.

Kuhn, Thomas S. 1962. *The Structure of Scientific Revolutions*. Chicago: University of Chicago Press. Reprint, 1969.

Kuklick, Henrika. 1991. *The Savage Within: A Social History of British Anthropology*. Cambridge: Cambridge University Press.

Kuper, Adam. 1972. *Anthropologists and Anthropology: The British School, 1922–1972*. London: Allan Lane.

————. 1985. "The Development of Lewis Henry Morgan's Evolutionism." *J. Hist. Behavioral Sci.* 21: 3–21.

————. 1988. *The Invention of Primitive Society: Transformations of an Illusion*. London: Routledge.

Lack, David. 1947. *Darwin's Finches*. Cambridge: Cambridge University Press.

Lakatos, Imre, and Alan Musgrave. 1970. *Criticism and the Growth of Knowledge*. Cambridge: Cambridge University Press.

Lamarck, Jean-Baptiste Pierre Antoine de Monet, Chevalier de. 1815–22. *Histoire naturelle des animaux sans vertèbres*. 6 vols. Paris. Reprint, Brussels: Culture et Civilisation, 1969.

————. 1914. *Zoological Philosophy*. Translated by Hugh Elliot. London. Reprint, New York: Hafner, 1963.

————. 1964. *Hydrogeology*. Translated by Albert V. Carozzi. Urbana: University of Illinois Press.

La Mettrie, Julien Offray de. 1774. *Oeuvres philosophiques*. 2 vols. Berlin.

———. 1991. Man a Machine *and* Man a Plant. Translated by Richard A. Watson and Maya Rybalka. Indianapolis: Hackett.

Landau, Misia. 1990. *Narratives of Human Evolution.* New Haven: Yale University Press.

Lanham, Url. 1973. *The Bone Hunters.* New York: Columbia University Press.

Lankester, E. Ray. 1880. *Degeneration: A Chapter in Darwinism.* London. Reprinted in *The Interpretation of Animal Form,* edited by W. Coleman. New York: Johnson Reprint Corporation, 1967.

Laplace, Pierre-Simon, Marquis de. 1830. *The System of the World.* Translated by H. H. Harte. 2 vols. Dublin and London.

Laporte, Leo F. 1990. "The World into Which Darwin Led Simpson." *J. Hist. Biology* 23: 499–516.

———. 2000. *George Gaylord Simpson: Paleontologist and Evolutionist.* New York: Columbia University Press.

Largent, Margaret A. 1999. "Bionomics: Vernon Lyman Kellogg and the Defence of Darwinism." *J. Hist. Biology* 32: 465–488.

Larson, Edward J. 1998. *Summer for the Gods: The Scopes Trial and America's Continuing Debate over Science and Religion.* New York: Basic Books; Cambridge: Harvard University Press.

———. 2001. *Evolution's Workshop: God and Science in the Galapagos Islands.* New York: Basic Books.

Larson, James L. 1971. *Reason and Experience: The Representation of Natural Order in the Work of Carl von Linné.* Berkeley and Los Angeles: University of California Press.

Latour, Bruno, and Steve Woolgar. 1979. *Laboratory Life: The Social Construction of Scientific Facts.* Beverly Hills: Sage Publications.

Laudan, Larry. 1977. *Progress and Its Problems: Toward a Theory of Scientific Growth.* Berkeley and Los Angeles: University of California Press.

Laudan, Rachel. 1982. "The Role of Methodology in Lyell's Science." *Stud. Hist. & Phil. Sci.* 13: 215–249.

———. 1987. *From Mineralogy to Geology: The Foundations of a Science, 1650–1830.* Chicago: University of Chicago Press.

Laurent, Goulven, ed. 1997. *Jean-Baptiste Lamarck: 1744–1828.* Paris: Comité des Travaux Historiques et Scientifiques.

Leakey, Richard, and Robert Lewin. 1977. *Origins: What New Discoveries Reveal about the Emergence of Our Species and Its Possible Future.* London: Macdonald and Jane's.

———. 1992. *Origins Reconsidered: In Search of What Makes Us Human.* London: Little, Brown.

Le Conte, Joseph. 1899. *Evolution: Its Nature, Its Evidences, and Its Relation to Religious Thought.* 2d ed. New York. Reprint, New York: Kraus, 1970.

Lecourt, Dominique. 1977. *Proletarian Science? The Case of Lysenko.* Introduced by Louis Althusser. London: NLB Books.

Lederman, Muriel. 1989. "Genes on Chromosomes: The Conversion of Thomas Hunt Morgan." *J. Hist. Biology* 22: 163–176.

Le Grand, H. E. 1988. *Drifting Continents and Shifting Theories: The Modern Revolution in Geology and Scientific Change*. Cambridge: Cambridge University Press.

Le Mahieu, D. L. 1976. *The Mind of William Paley: A Philosopher of His Age*. Lincoln: University of Nebraska Press.

Lenoir, Timothy. 1978. "Generational Factors in the Origin of *Romantische Naturphilosophie*." *J. Hist. Biology* 11: 57–100.

———. 1982. *The Strategy of Life: Teleology and Mechanics in Nineteenth-Century German Biology*. Dordrecht: D. Reidel.

Lesch, John E. 1975. "The Role of Isolation in Evolution: George J. Romanes and John T. Gulick." *Isis* 66: 483–503.

Lester, Joe. 1995. *E. Ray Lankester and the Making of Modern British Biology*. Edited by Peter J. Bowler. Stanford in the Vale: British Society for the History of Science.

Levine, George. 1988. *Darwin and the Novelists: Patterns of Science in Victorian Fiction*. Cambridge, Mass.: Harvard University Press.

Levine, J. E. 1977. *Dr. Woodward's Shield: History, Science, and Satire in Augustan England*. Cambridge, Mass.: Harvard University Press.

Lewin, Roger. 1987. *Bones of Contention: Controversies in the Search for Human Origins*. New York: Simon and Schuster.

———. 1998. *Principles of Human Evolution: A Core Textbook*. Oxford: Blackwell Scientific.

Lewis, Cherry. 2000. *The Dating Game: One Man's Search for the Age of the Earth*. Cambridge: Cambridge University Press.

Lewontin, Richard C. 1974. *The Genetic Basis of Evolutionary Change*. New York: Columbia University Press.

———. 1983. "The Organism as the Subject and as the Object of Evolution." *Scientia* 118: 65–80.

Lewontin, Richard C., and Richard Levins. 1976. "The Problem of Lysenkoism." In *The Radicalization of Science: Ideology of/in the Natural Sciences*, edited by Hilary and Steven Rose, pp. 32–64. London: Macmillan.

Lightman, Bernard. 1987. *The Origins of Agnosticism: Victorian Unbelief and the Limits of Knowledge*. Baltimore: Johns Hopkins University Press.

———. 1997. " 'Fighting Even with Death': Balfour, Scientific Naturalism, and Thomas Henry Huxley's Final Battle." In *Thomas Henry Huxley's Place in Science and Letters*, edited by Alan P. Barr, pp. 323–350. Athens: University of Georgia Press.

———, ed. 1997. *Victorian Science in Context*. Chicago: University of Chicago Press.

Limoges, Camille. 1970. *Le sélection naturelle: Étude sur le première construction d'un concept*. Paris: Presses Universitaires de France.

———. 1976. "Natural Selection, Phagocytosis and Preadaptation: Lucien Cuénot, 1886–1901." *J. Hist. Medicine* 31: 176–214.

Lindberg, David C., and Ronald L. Numbers, eds. 1986. *God and Nature:*

Historical Essays on the Encounter between Science and Religion. Berkeley and Los Angeles: University of California Press.

Lindberg, David C., and Robert Westman, eds. 1990. *Reappraisals of the Scientific Revolution.* Cambridge: Cambridge University Press.

Linnaeus, Carolus (Karl von Linné). 1735. *Systema Naturae.* Reprint, edited by M. S. J. Engel-Ledeboer and H. Engel, Nieuwkoop: B. de Graff 1964.

———. 1753. *Species Plantarum.* 2 vols. Reprint, London: Ray Society, 1955–59.

———. 1754. *Genera Plantarum.* 5th ed., reprint, Weinheim: J. Cramer, 1960.

———. 1758–59. *Systema Naturae.* 10th ed., reprint, vol. 1, London: Ray Society, 1956; vol. 2, Weinheim: J. Cramer, 1964.

———, ed. 1749–90. *Amoenitates Academicae, seu Dissertationes variae physicae, medicae et botanicae.* 10 vols. Leiden and Amsterdam.

Livingstone, David N. 1987. *Darwin's Forgotten Defenders: The Encounter between Evangelical Theology and Evolutionary Thought.* Edinburgh: Scottish Universities Press; Grand Rapids, Mich.: Eerdmans.

Livingstone, David N., D. G. Hart, and Mark A. Knoll, eds. 1999. *Evangelicalism and Science in Historical Perspective.* New York: Oxford University Press.

Locke, John. 1960. *Two Treatises of Government.* Edited by Peter Laslett. Cambridge: Cambridge University Press.

———. 1975. *An Essay Concerning Human Understanding.* Edited by P. Nidditch. Oxford: Clarendon Press.

Loewenberg, Bert James. 1959. *Darwin, Wallace, and the Theory of Natural Selection: Including the Linnean Society Papers.* Cambridge, Mass.: Arlington Books.

———. 1965. "Darwin and Darwin Studies." *History of Science* 4: 15–54.

———. 1969. *Darwin Comes to America: 1859–1900.* Philadelphia: Fortress Press.

Lorenz, Konrad. 1966. *On Aggression.* Translated by Marjorie Kerr Wilson. New York: Harcourt, Brace, and World.

Lorimer, Douglas A. 1988. "Theoretical Racism in Late Victorian Anthropology, 1870–1900." *Victorian Studies* 31: 405–430.

———. 1997. "Science and the Secularization of Victorian Images of Race." In *Victorian Science in Context,* edited by Bernard Lightman, pp. 212–235. Chicago: University of Chicago Press.

Lovejoy, Arthur O. 1936. *The Great Chain of Being: A Study in the History of an Idea.* Reprint, New York: Harper, 1960.

———. 1959a. "Buffon and the Problem of Species." In *Forerunners of Darwin, 1745–1859,* edited by Bentley Glass, Owsei Temkin, and William Strauss Jr., pp. 84–113. Baltimore: Johns Hopkins University Press.

———. 1959b. "Herder: Progressionism without Transformism." In *Forerunners of Darwin, 1745–1859,* edited by Bentley Glass, Owsei Temkin, and William Strauss Jr., pp. 207–221. Baltimore: Johns Hopkins University Press.

———. 1959c. "The Argument for Organic Evolution before the *Origin of*

Species, 1830–1858." In *Forerunners of Darwin, 1745–1859,* edited by Bentley Glass, Owsei Temkin, and William Strauss Jr., pp. 356–414. Baltimore: Johns Hopkins University Press.

———. 1959d. "Schopenhauer as an Evolutionist." In *Forerunners of Darwin, 1745–1859,* edited by Bentley Glass, Owsei Temkin, and William Strauss Jr., pp. 415–437. Baltimore: Johns Hopkins University Press.

Løvtrup, Søren. 1977. *The Phylogeny of the Vertebrata.* London: John Wiley.

———. 1987. *Darwinism: The Refutation of a Myth.* London: Croom Helm.

Lubbock, John. 1865. *Prehistoric Times.* London.

———. 1870. *The Origin of Civilization and the Primitive Condition of Man.* London.

Lucas, J. R. 1979. "Wilberforce and Huxley: A Legendary Encounter." *Historical Journal* 22: 313–330.

Ludmerer, Kenneth M. 1972. *Genetics and American Society: A Historical Appraisal.* Baltimore: Johns Hopkins University Press.

Lurie, Edward. 1960. *Louis Agassiz: A Life in Science.* Chicago: University of Chicago Press.

Lyell, Charles. 1830–33. *Principles of Geology: Being an Attempt to Explain the Former Changes of the Earth's Surface by Reference to Causes Now in Operation.* 3 vols. London. Reprint, with an introduction by M. J. S. Rudwick, Chicago: University of Chicago Press, 1990–91.

———. 1851. Presidential Address. *Quart. J. Geol. Soc. Lond.* 7: 25–76.

———. 1863. *Geological Evidences of the Antiquity of Man: With Remarks on Theories of the Origin of Species by Variation.* London. 1873 ed. reprinted New York: AMS Press, 1973.

———. 1970. *Sir Charles Lyell's Journals on the Species Question.* Edited by Leonard J. Wilson. New Haven: Yale University Press.

Lyell, Mrs. K. M., ed. 1881. *The Life, Letters, and Journals of Sir Charles Lyell.* 2 vols. London. Reprint, Farnborough: Gregg International, 1970.

Lynch, John M., ed. 2000. *Vestiges and the Debate before Darwin.* 7 vols. Bristol: Thoemmes Press.

———. 2001. *Darwin's Theory of Natural Selection: British Responses, 1859–1871.* 4 vols. Bristol: Thoemmes Press.

Lyons, Cherrie L. 1999. *Thomas Henry Huxley: The Evolution of a Scientist.* New York: Prometheus Books.

Macbeth, Norman. 1971. *Darwin Retried: An Appeal to Reason.* Boston: Gambit.

MacBride, E. W. 1914. *Textbook of Embryology.* Vol. 1: *Invertebrata.* London: Macmillan.

———. 1924. *An Introduction to the Study of Heredity.* London: Williams and Norgate.

McDougall, William. 1927. "An Experiment for the Testing of the Hypothesis of Lamarck." *Brit. J. Psychology* 17: 267–304.

Mackenzie, Donald. 1976. "Eugenics in Britain." *Social Studies of Science* 6: 499–532.

———. 1982. *Statistics in Britain, 1865–1930: The Social Construction of Scientific Knowledge.* Edinburgh: Edinburgh University Press.

McKinney, H. Lewis. 1972. *Wallace and Natural Selection.* New Haven: Yale University Press.

———, ed. 1971. *Lamarck to Darwin: Contributions to Evolutionary Biology.* Lawrence, Kans.: Coronado Press.

MacLeod, Roy M. 1965. "Evolutionism and Richard Owen." *Isis* 56: 259–280.

———. 1970. "The X-Club: A Scientific Network in Late Victorian England." *Notes and Records Roy. Soc. Lond.* 24: 305–322.

MacLeod, Roy M., and Philip F. Rehbock, eds. 1994. *Darwin's Laboratory: Evolutionary Thought and Natural History in the Pacific.* Honolulu: University of Hawaii Press.

McNeil, M. 1987. *Under the Banner of Science: Erasmus Darwin and His Age.* Manchester: Manchester University Press.

McOuat, Gordon, and Mary P. Winsor. 1995. "J. B. S. Haldane's Darwinism in Its Religious Context." *Brit. J. Hist. Sci.* 28: 227–231.

Magnello, Eileen. 1996. "Karl Pearson's Gresham Lectures: W. F. R. Weldon, Speciation, and the Origins of Pearsonian Statistics." *Brit. J. Hist. Sci.* 29: 43–63.

———. 1998. "Karl Pearson's Mathematization of Inheritance: From Ancestral Heredity to Mendelian Genetics (1895–1090)." *Annals of Science* 55: 35–94.

———. 1999. "The Non-Correlation of Biometrics and Eugenics: Rival Forms of Laboratory Work in Karl Pearson's Career at University College, London." *History of Science* 37: 79–106, 123–150.

Maienschein, Jane. 1978. "Cell Lineage, Ancestral Reminiscence, and the Biogenetic Law." *J. Hist. Biology* 11: 129–158.

———. 1984. "What Determines Sex: A Study of Converging Research Approaches." *Isis* 75: 457–480.

———. 1991. *Transforming Traditions in American Biology, 1880–1915.* Baltimore: Johns Hopkins University Press.

Malthus, Thomas Robert. 1959. *Population: The First Essay.* Ann Arbor: University of Michigan Press.

———. 1990. *An Essay on the Principle of Population.* 2d ed. 2 vols. Reprint, edited by Patricia James, Cambridge: Cambridge University Press.

Mandelbaum, Maurice. 1971. *History, Man, and Reason: A Study in Nineteenth Century Thought.* Baltimore: Johns Hopkins University Press.

Manier, Edward. 1978. *The Young Darwin and His Cultural Circle: A Study of the Influences Which Shaped the Language and Logic of the Theory of Natural Selection.* Dordrecht: D. Reidel.

———. 1980. "History, Philosophy, and Sociology of Biology: A Family Romance." *Stud. Hist. & Phil. Sci.* 11: 1–24.

Manser, A. R. 1965. "The Concept of Evolution." *Philosophy* 40: 18–34.

Manuel, F. E. 1956. *The New World of Henri de Saint-Simon.* Cambridge, Mass.: Harvard University Press.

Marchant, James. 1916. *Alfred Russel Wallace: Letters and Reminiscences.* London. Reprint, New York: Arno Press.

Margulis, Lynn. 1993. *Symbiosis in Cell Evolution.* 2d ed. New York: W. H. Freeman.

Marsh, Othniel C. 1880. *Odontornithes: A Monograph on the Extinct Toothed Birds of North America.* Vol. 8. Washington: Report of the Geological Exploration of the Fortieth Parallel.

Maupertuis, Pierre Louis Moreau de. 1768. *Oeuvres.* 4 vols. Reprint, Hildesheim: Georg Olms, 1968.

———. 1968. *The Earthly Venus.* Translated by Simon Brangier Boas. New York: Johnson Reprint Corporation.

Mayr, Ernst. 1942. *Systematics and the Origin of Species.* New York: Columbia University Press. Reprint, with a new introduction by Mayr, New York: Columbia University Press, 1999.

———. 1954. "Wallace's Line in the Light of Recent Zoogeographic Studies." *Quart. Rev. Biology* 29: 1–14. Reprinted in *Evolution and the Diversity of Life,* by Mayr, pp. 626–645. Cambridge, Mass.: Harvard University Press.

———. 1955. "Karl Jordan's Contribution to Current Concepts in Systematics and Evolution." *Trans. Roy. Entomological Soc. London* 107: 45–66. Reprinted in *Evolution and the Diversity of Life,* by Mayr, pp. 135–143, 297–306, and 485–492. Cambridge, Mass.: Harvard University Press.

———. 1959a. "Agassiz, Darwin, and Evolution." *Harvard Library Bulletin* 12: 165–194. Reprinted in *Evolution and the Diversity of Life,* by Mayr, pp. 251–276. Cambridge, Mass.: Harvard University Press.

———. 1959b. "Isolation as an Evolutionary Factor." *Proc. Am. Phil. Soc.* 103: 221–230. Reprinted in *Evolution and the Diversity of Life,* by Mayr, pp. 120–134. Cambridge, Mass.: Harvard University Press.

———. 1959c. "Where Are We?" In *Evolution and the Diversity of Life,* by Mayr, pp. 307–328. Cambridge, Mass.: Harvard University Press.

———. 1964. Introduction to *On the Origin of Species,* by Charles Darwin. Facsimile reprint, Cambridge, Mass.: Harvard University Press.

———. 1972. "Lamarck Revisited." *J. Hist. Biology* 5: 55–94. Reprinted in *Evolution and the Diversity of Life,* by Mayr, pp. 222–250. Cambridge, Mass.: Harvard University Press.

———. 1976. *Evolution and the Diversity of Life.* Cambridge, Mass.: Harvard University Press.

———. 1977. "Darwin and Natural Selection: How Darwin May Have Discovered His Highly Unconventional Theory." *Am. Scientist* 65: 321–377.

———. 1982. *The Growth of Biological Thought: Diversity, Evolution, and Inheritance.* Cambridge, Mass.: Harvard University Press.

———. 1985. "Weismann and Evolution." *J. Hist. Biology* 18: 259–322.

———. 1991. *One Long Argument: Charles Darwin and the Genesis of Evolutionary Thought.* Cambridge, Mass.: Harvard University Press.

———. 2001. *What Evolution Is.* New York: Basic Books.

Mayr, Ernst, and William B. Provine, eds. 1980. *The Evolutionary Synthesis: Perspectives on the Unification of Biology.* Cambridge, Mass.: Harvard University Press.

Mazumdar, Pauline. 1992. *Eugenics, Human Genetics, and Human Failings: The Eugenics Society, Its Sources, and Its Critics in Britain.* London: Routledge.

Medawar, P. B. 1961. Review of *The Phenomenon of Man,* by Teilhard de Chardin. *Mind* 70: 99–106.

Medvedev, Zhores. 1969. *The Rise and Fall of T. D. Lysenko.* Translated by I. Michael Lerner. New York: Columbia University Press.

Meijer, Onno. 1985. "Hugo De Vries No Mendelian?" *Annals of Science* 42: 189–232.

Mendel, Gregor Johann. 1965. *Experiments on Plant Hybridization.* Foreword by Paul C. Mangelsdorf. Cambridge, Mass.: Harvard University Press.

Merz, John Theodore. 1896–1903. *A History of European Thought in the Nineteenth Century.* 2 vols. Edinburgh. Reprint (in 4 vols.), New York: Dover.

Messenger, Ernest. 1931. *Evolution and Theology: The Problem of Man's Origin.* London: Burns, Oates, and Washbourne.

Messerly, John G. 1996. *Piaget's Conception of Evolution: Beyond Darwin and Lamarck.* Lanham, Md.: Rowan and Littlefield.

Metzger, Helene. 1930. *Newton, Stahl, Boerhaave et la doctrine chimique.* Paris: Alcan.

Meyer, A. W. 1956. *Human Generation: The Conclusions of Burdach, Dollinger, and von Baer.* Stanford: Stanford University Press.

Milkman, Roger, ed. 1982. *Perspectives on Evolution.* Sunderland, Mass.: Sinauer.

Mill, John Stuart. 1950. *Mill on Bentham and Coleridge.* London: Chatto and Windus.

———. 1981–91. *Collected Works.* 23 vols. Toronto: University of Toronto Press.

Millar, Ronald. 1972. *The Piltdown Men: A Case of Archaeological Fraud.* London: Victor Gollancz.

Miller, Hugh. 1841. *The Old Red Sandstone: Or New Walks in an Old Field.* Edinburgh. Boston, 1857 ed. reprinted New York: Arno Press, 1977.

———. 1850. *Footprints of the Creator: Or the Asterolepis of Stromness.* 3d ed. Edinburgh. Edinburgh, 1861 ed. reprinted Farnborough: Gregg International, 1971.

Miller, Kenneth R. 1999. *Finding Darwin's God: A Scientist's Search for Common Ground between God and Evolution.* New York: HarperCollins.

Miller, Stanley. 1953. "A Production of Amino Acids under Possible Primitive Earth Conditions." *Science* 117: 528.

Millhauser, Milton. 1954. "The Scriptural Geologists." *Osiris* 11: 65–86.

———. 1959. *Just Before Darwin: Robert Chambers and Vestiges.* Middletown, Conn.: Wesleyan University Press.

Mitman, Greg. 1990. "Evolution as Gospel: William Pattern, the Language of Democracy, and the Great War." *Isis* 81: 446–463.

Mivart, St. George Jackson. 1871. *The Genesis of Species*. London.

Montagu, Ashley, ed. 1982. *Evolution and Creation*. New York: Oxford University Press.

Montesquieu, Charles Louis Secondat, Baron de. 1989. *The Spirit of the Laws*. Edited by Anne M. Cohler. Cambridge: Cambridge University Press.

Montgomery, William M. 1974. "Germany." In *The Comparative Reception of Darwinism*, edited by Thomas F. Glick, pp. 81–116. Austin: University of Texas Press. New ed., Chicago: University of Chicago Press, 1988.

Moore, James R. 1979. *The Post-Darwinian Controversies: A Study of the Protestant Struggle to Come to Terms with Darwin in Great Britain and America, 1870–1900*. New York: Cambridge University Press.

———. 1982. "Charles Darwin Lies in Westminster Abbey." *Biological J. Linn. Soc.* 17: 97–113.

———. 1985a. "Herbert Spencer's Henchmen: The Evolution of Protestant Liberals in Late-Nineteenth-Century America." In *Darwinism and Divinity: Essays on Evolution and Religious Belief*, edited by John R. Durant, pp. 76–100. Oxford: Basil Blackwell.

———. 1985b. "Evangelicals and Evolution: Henry Drummond, Herbert Spencer, and the Naturalization of the Spiritual World." *Scottish Journal of Theology* 38: 383–417.

———. 1989. "Of Love and Death: Why Darwin 'Gave up Christianity.' " In *History, Humanity, and Evolution: Essays for John C. Greene*, edited by James R. Moore, pp. 195–230. Cambridge: Cambridge University Press.

———. 1991. "Deconstructing Darwinism: The Politics of Evolution in the 1860s." *J. Hist. Biology* 24: 353–408.

———. 1994. *The Darwin Legend*. Grand Rapids, Mich.: Baker Books.

———. 1997. "Wallace's Malthusian Moment: The Common Context Revisited." In *Victorian Science in Context*, edited by Bernard Lightman, pp. 290–311. Chicago: University of Chicago Press.

———, ed. 1989. *History, Humanity, and Evolution: Essays for John C. Greene*. Cambridge: Cambridge University Press.

Moorehead, Alan. 1969. *Darwin and the* Beagle. London: Hamish Hamilton.

Moravia, Sergio. 1978. "From *Homme Machine* to *Homme Sensible:* Changing Eighteenth-Century Models of Man's Image." *J. Hist. Ideas* 39: 45–60.

Morgan, Conwy Lloyd. 1923. *Emergent Evolution: The Gifford Lectures Delivered at the University of St. Andrews in the Year 1922*. London.

Morgan, Lewis Henry. 1877. *Ancient Society: Or Researches into the Lines of Human Progress from Savagery through Barbarism to Civilization*. Reprint, Cambridge, Mass.: Harvard University Press, 1964.

Morgan, Thomas Hunt. 1903. *Evolution and Adaptation*. New York. Reprint, 1908.

———. 1916. *A Critique of the Theory of Evolution*. Princeton: Princeton University Press.

Morgan, Thomas Hunt, A. H. Sturtevant, H. J. Muller, and C. B. Bridges. 1915. *The Mechanism of Mendelian Heredity*. New York. Reprint, with an introduction by Garland E. Allen, New York: Johnson Reprint Corporation, 1972.

Morris, Desmond. 1967. *The Naked Ape: A Zoologist's Study of the Human Animal*. London: Jonathan Cape.

Morris, H. M. 1974. *Scientific Creationism*. San Diego: Creation Life.

Morris, Paul J. 1997. "Louis Agassiz's Arguments against Darwinism in His Additions to the French Translation of the *Essay on Classification*." *J. Hist. Biology* 30: 121–134.

Morris, Simon Conway. 1998. *The Crucible of Creation: The Burgess Shale and the Rise of Animals*. Oxford: Oxford University Press.

Morss, J. R. 1990. *The Biologizing of Childhood: Developmental Psychology and the Darwinian Myth*. Hove: Laurence Erlbaum.

Morton, Peter. 1984. *The Vital Science: Biology and the Literary Imagination*. London: Allen and Unwin.

Morus, Iwan R. 1998. *Frankenstein's Children: Electricity, Exhibition, and Experiment in Early Nineteenth-Century London*. Princeton: Princeton University Press.

Mosse, George L. 1978. *Toward the Final Solution: A History of European Racism*. New York: Howard Fertig.

Mulkay, Michael. 1979. *Science and the Sociology of Knowledge*. London: Allen and Unwin.

Müller, Fritz. 1869. *Facts and Arguments for Darwin*. London. Reprint, Farnborough: Gregg International, 1968.

Muller, Hermann. 1949. "The Darwinian and Modern Conceptions of Natural Selection." *Proc. Am. Phil. Soc.* 90: 459–70.

———. 1959. "One Hundred Years without Darwin Are Enough." *The Humanist* 19: 139–49.

Murphy, Terence D. 1976. "Jean-Baptiste Robinet: The Career of a Man of Letters." *Stud. Voltaire & 18th Cent.* 150: 183–250.

Muschinske, David. 1977. "The Nonwhite as Child: G. Stanley Hall on the Education of Nonwhite Peoples." *J. Hist. Behavioral Sci.* 13: 328–336.

Nägeli, Carl von. 1898. *A Mechanico-Physiological Theory of Organic Evolution*. Chicago.

Needham, John Turberville. 1748. "A Summary of Some Late Observations upon the Generation, Composition, and Decomposition of Animal and Vegetable Substances." *Phil. Trans. Roy. Soc.* 45: 615–666.

Nelkin, Dorothy. 1977. *Science Textbook Controversies and the Politics of Equal Time*. Cambridge, Mass.: MIT Press.

———. 1983. *The Creation Controversy: Science or Scripture in Public Schools*. New York: Norton.

Nelson, Gareth. 1978. "From Candolle to Croizat: Comments on the History of Biogeography." *J. Hist. Biology* 11: 269–305.

Nelson, Gareth, and Don E. Rosen, eds. 1981. *Vicariance Biogeography: A Critique*. New York: Columbia University Press.

Nicholson, A. J. 1960. "The Role of Population Dynamics in Natural Selection."

In *Evolution after Darwin,* edited by Sol Tax, 1:477–522. Chicago: University of Chicago Press.

Nicolson, Malcolm. 1987. "Alexander von Humboldt, Humboldtian Science, and the Origins of the Study of Vegetation." *History of Science* 25: 167–194.

———. 1990. "Alexander von Humboldt and the Geography of Vegetation." In *Romanticism and the Sciences,* edited by Andrew Cunningham and Nicholas Jardine, pp. 169–185. Cambridge: Cambridge University Press.

Nisbet, Robert. 1980. *History of the Idea of Progress.* London: Heinemann.

Nitecki, Matthew H., ed. 1988. *Evolutionary Progress.* Chicago: University of Chicago Press.

Nitecki, Mathew H., and Doris V. Nitecki, eds. 1992. *History and Evolution.* Albany: State University of New York Press.

Nordau, Max. 1895. *Degeneration.* 2d ed. London.

Nordenskiöld, Erik. 1946. *The History of Biology.* Reprint, New York: Tudor Publishing.

Norton, Bernard J. 1973. "The Biometric Defense of Darwinism." *J. Hist. Biology* 6: 283–316.

———. 1975a. "Biology and Philosophy: The Methodological Foundations of Biometry." *J. Hist. Biology* 8: 85–93.

———. 1975b. "Metaphysics and Population Genetics: Karl Pearson and the Background to Fisher's Multi-Factorial Theory of Inheritance." *Annals of Science* 32: 537–553.

———. 1983. "Fisher's Entrance into Evolutionary Science: The Role of Eugenics." In *Dimensions of Darwinism: Themes and Counterthemes in Twentieth-Century Evolutionary Theory,* edited by Marjorie Grene, pp. 19–30. Cambridge: Cambridge University Press.

Norton, Bernard J., and E. S. Pearson. 1976. "A Note on the Background to, and Refereeing of, R. A. Fisher's Paper 'On the Correlation {...}' " *Notes and Records of Roy. Soc. London* 31: 151–162.

Numbers, Ronald L. 1977. *Creation by Natural Law: Laplace's Nebular Hypothesis in American Thought.* Seattle: University of Washington Press.

———. 1992. *The Creationists.* New York: Alfred A. Knopf.

———. 1998. *Darwinism Comes to America.* Cambridge, Mass.: Harvard University Press.

———, ed. 1994–95. *Creationism in Twentieth-Century America.* 10 vols. New York: Goddard.

Numbers, Ronald L., and John Stenhouse, eds. 1999. *Disseminating Darwinism: The Role of Place, Race, Religion, and Gender.* Cambridge: Cambridge University Press.

Nye, R. 1976. "Heredity or Milieu: The Foundations of European Criminological Theory." *Isis* 67: 335–355.

Nyhart, Lynn K. 1995. *Biology Takes Form: Animal Morphology and the German Universities, 1800–1900.* Chicago: University of Chicago Press.

Oberg, Barbara Bowen. 1976. "David Harley and the Association of Ideas." *J. Hist. Ideas* 37: 441–454.

O'Brien, Charles F. 1970. "*Eozoön canadense:* The Dawn Animal of Canada." *Isis* 61: 209–223.

———. 1971. *Sir William Dawson: A Life in Science and Religion. Mem. Am. Phil. Soc.* (New York) 84.

Ogilvie, M. B. 1975. "Robert Chambers and the Nebular Hypothesis." *Brit. J. Hist. Sci.* 7: 214–232.

Oken, Lorenz. 1847. *Elements of Physico-Philosophy.* Translated by Alfred Tulk. London.

Olby, Robert C. 1966. *The Origins of Mendelism.* London: Constable.

———. 1974. *The Path to the Double Helix.* Foreword by Francis Crick. Seattle: University of Washington Press.

———. 1979. "Mendel No Mendelian." *History of Science* 17: 53–72.

———. 1985. *The Origins of Mendelism.* 2d ed. Chicago: University of Chicago Press.

———. 1989. "Dimensions of Scientific Controversy: The Biometric-Mendelian Debate." *Brit. J. Hist. Sci.* 22: 299–320.

Olby, Robert C., and Peter Gautry. 1968. "Eleven References to Mendel before 1900." *Annals of Science* 24: 7–20.

Oldroyd, D. R. 1980. *Darwinian Impacts: An Introduction to the Darwinian Revolution.* Milton Keynes, U.K.: Open University Press.

———. 1984. "How Did Darwin Arrive at His Theory?" *History of Science* 22: 325–374.

———. 1996. *Thinking about the Earth: A History of Geological Ideas.* London: Athlone.

Oldroyd, D. R., and Ian Langham, eds. 1983. *The Wider Domain of Evolutionary Thought.* Dordrecht: D. Reidel.

Oleson, Alexandra, and Sanborn C. Brown, eds. 1976. *The Pursuit of Knowledge in the Early American Republic: American Scientific and Learned Societies from Colonial Times to the Civil War.* Baltimore: Johns Hopkins University Press.

Oparin, A. I. 1938. *The Origin of Life.* Translated by Sergius Morgulis. 2d ed., New York: Dover, 1953.

Oppenheimer, Jane. 1967. *Essays in the History of Embryology.* Cambridge, Mass.: MIT Press.

Orel, Viteslav. 1984. *Mendel.* Oxford: Oxford University Press.

———. 1995. *Gregor Mendel: The First Geneticist.* Oxford: Oxford University Press.

Osborn, Henry Fairfield. 1908. "The Four Inseparable Factors of Evolution." *Science* 27: 148–150.

———. 1912. "The Continuous Origin of Certain Unit Characters as Observed by a Paleontologist." *Am. Naturalist* 46: 185–206, 249–278.

———. 1917. *The Origin and Evolution of Life on the Theory of Action, Reaction, and Interaction of Energy.* New York.

———. 1929. *The Titanotheres of Ancient Wyoming, Dakota, and Nebraska.* 2 vols. U.S. Geological Survey Monograph No. 55. Washington, D.C.

———. 1931. *Cope, Master Naturalist: The Life and Writings of Edward Drinker Cope.* Princeton: Princeton University Press.

———. 1934. "Aristogenesis: The Creative Principle in the Origin of Species." *Am. Naturalist* 68: 193–235.

Osler, M., ed. 1985. *Science, Religion, and World View.* Cambridge: Cambridge University Press.

Ospovat, Alexander M. 1969. "Reflections on A. C. Werner's 'Kurze Klassification.' " In *Toward a History of Geology,* edited by Cecil J. Schneer, pp. 242–256. Cambridge, Mass.: MIT Press.

Ospovat, Dov. 1976. "The Influence of Karl Ernst von Baer's Embryology, 1828–1859: A Reappraisal in Light of Richard Owen and William B. Carpenter's 'Paleontological Application of von Baer's Law.' " *J. Hist. Biology* 9: 1–28.

———. 1977. "Lyell's Theory of Climate." *J. Hist. Biology* 10: 317–339.

———. 1978. "Perfect Adaptation and Teleological Explanation: Approaches to the Problem of the History of Life in the Mid-Nineteenth Century." *Studies in the History of Biology* 2: 33–56.

———. 1979. "Darwin after Malthus." *J. Hist. Biology* 12: 211–230.

———. 1981. *The Development of Darwin's Theory: Natural History, Natural Theology, and Natural Selection, 1838–59.* Cambridge and New York: Cambridge University Press.

Outram, Dorinda. 1984. *Georges Cuvier: Vocation, Science, and Authority in Post-Revolutionary France.* Manchester: Manchester University Press.

———. 1986. "Uncertain Legislator: Georges Cuvier's Laws of Nature and Their Intellectual Context." *J. Hist. Biology* 19: 323–368.

———. 1995. *The Enlightenment.* Cambridge: Cambridge University Press.

Owen, Richard. 1848. *On the Archetype and Homologies of the Vertebrate Skeleton.* London. Reprint, New York: AMS Press.

———. 1849. *On the Nature of Limbs.* London.

———. 1851. "Lyell on Life and Its Successive Development." *Quarterly Rev.* 89: 412–451.

———. 1860. *Palaeontology: Or a Systematic Study of Extinct Animals and Their Geological Relations.* Edinburgh.

———. 1866–68. *The Anatomy of the Vertebrates.* 3 vols. London. Reprint, New York: AMS Press, 1973.

Oyama, Susan. 1985. *The Ontogeny of Information: Developmental Systems and Evolution.* Cambridge: Cambridge University Press.

Packard, Alpheus. 1889. "The Cave Fauna of North America." *Mem. Nat. Acad. Sci.* 4: 1–156.

———. 1901. *Lamarck, the Founder of Evolution: His Life and Work. With Translations of His Writings on Organic Evolution.* New York.

Page, Leroy E. 1969. "Diluvialism and Its Critics." In *Toward a History of Geology,* edited by Cecil J. Schneer, pp. 257–271. Cambridge, Mass.: MIT Press.

Paley, William. 1802. *Natural Theology: Or Evidences of the Existence and*

Attributes of the Deity Collected from the Appearances of Nature. London. Reprint, Farnborough: Gregg International, 1970.

Pancaldi, Giuliano. 1994. "The Technology of Nature: Marx's Thoughts on Darwin." In *The Natural Sciences and the Social Sciences*, edited by I. B. Cohen, pp. 257–274. Dordrecht: Kluwer.

Paradis, James G. 1978. *T. H. Huxley: Man's Place in Nature.* Lincoln: University of Nebraska Press.

Parker, G. E. 1980. *Creation: The Facts of Life.* San Diego: C.L.P. Publishers.

Pastore, Nicholas. 1949. *The Nature-Nurture Controversy.* New York: King's Crown Press.

Patterson, Colin. 1980. "Cladistics." *Biologist* 27: 234–240.

———. 1982. "Cladistics and Classification." *New Scientist* 94: 303–306.

Paul, Diane. 1984. "Eugenics and the Left." *J. Hist. Ideas* 45: 567–590.

———. 1988. "The Selection of the 'Survival of the Fittest.' " *J. Hist. Biology* 21: 411–424.

———. 1998. *The Politics of Heredity: Essays on Eugenics, Biomedicine, and the Nature-Nurture Debate.* Albany: State University of New York Press.

Pauly, Philip J. 1982. "Samuel Butler and His Darwinian Critics." *Victorian Studies.* 25: 161–180.

Peacocke, A. R. 1980. *Creation and the World of Science.* Oxford: Oxford University Press.

Pearson, Karl. 1894. "Socialism and Natural Selection." *Fortnightly Revue*, n.s., 56: 1–21.

———. 1896. "Regression, Heredity, and Panmixia." *Phil. Trans. Roy. Soc.*, ser. A, 197: 253–318.

———. 1898. "Mathematical Contributions to the Theory of Evolution: On the Law of Ancestral Heredity." *Proc. Roy. Soc.* 57: 386–412.

———. 1900. *The Grammar of Science.* 2d ed. London.

———. 1914–30. *The Life, Letters, and Labours of Francis Galton.* 3 vols. Cambridge: Cambridge University Press.

Peckham, Morse. 1959. "Darwinism and Darwinisticism." Reprinted in *The Triumph of Romanticism: Collected Essays*, by Peckham, pp. 176–201. Columbia: University of South Carolina Press, 1970.

Peel, J. D. Y. 1971. *Herbert Spencer: The Evolution of a Sociologist.* London: Heinemann.

Pennock, Robert T. 1999. *Tower of Babel: The Evidence against the New Creationism.* Cambridge, Mass.: MIT Press.

Persell, Stuart M. 1999. *Neo-Lamarckism in France, 1870–1920.* Lewiston, N.Y.: Mellen.

Persons, Stow, ed. 1956. *Evolutionary Thought in America.* New York: George Braziller.

Petersen, William. 1979. *Malthus.* Cambridge, Mass.: Harvard University Press.

Pfeifer, Edward J. 1965. "The Genesis of American Neo-Lamarckism." *Isis* 56: 156–167.

————. 1974. "United States." In *The Comparative Reception of Darwinism*, edited by Thomas F. Glick, pp. 168–206. Austin: University of Texas Press. New ed., Chicago: University of Chicago Press, 1988.

Pick, Daniel. 1989. *Faces of Degeneration: Aspects of European Cultural Disorder, 1848–1918*. Cambridge: Cambridge University Press.

Pickens, D. K. 1968. *Eugenics and the Progressives*. Nashville, Tenn.: Vanderbilt University Press.

Pickering, Mary. 1994. *Auguste Comte: An Intellectual Biography*. Cambridge: Cambridge University Press.

Pinto-Correia, C. 1997. *The Ovary of Eve: Egg and Sperm in Preformation*. Chicago: University of Chicago Press.

Pittenger, Mark. 1993. *American Socialists and Evolutionary Thought, 1870–1920*. Madison: University of Wisconsin Press.

Plate, Robert. 1964. *The Dinosaur Hunters: Othniel C. Marsh and Edward D. Cope*. New York: D. McKay.

Playfair, John. 1802. *Illustrations of the Huttonian Theory of the Earth*. Edinburgh. Reprint, New York: Dover, 1964.

Poliakov, L. 1970. *The Aryan Myth: A History of Racist and Nationalist Ideas in Europe*. New York: Basic Books.

Pollard, Sydney. 1968. *The Idea of Progress: History and Society*. Reprint, Harmondsworth, Middlesex: Penguin Books, 1971.

Pope, Alexander. 1751. *The Works of Alexander Pope*. 8 vols. London.

Popkin, R. H. 1987. *Isaac La Peyrère (1596–1676): His Life, Work, and Influence*. Leiden: Brill.

Popper, Karl. 1959. *The Logic of Scientific Discovery*. London: Hutchinson.

————. 1962. *The Open Society and Its Enemies*. 4th ed. 2 vols. Princeton: Princeton University Press.

————. 1974. *The Philosophy of Karl Popper*. Edited by Paul A. Schfipp. 2 vols. La Salle, Ill.: Open Court.

————. 1978. "Natural Selection and the Emergence of Mind." *Dialectica* 32: 339–355.

Porter, Roy. 1977. *The Making of Geology: Earth Sciences in Britain, 1660–1815*. Cambridge: Cambridge University Press.

————. 1989. "Erasmus Darwin: Doctor of Evolution?" In *History, Humanity, and Evolution: Essays for John C. Greene*, edited by James R. Moore, pp. 39–70. Cambridge: Cambridge University Press.

————. 1990. *The Enlightenment*. 2d ed., Basingstoke: Macmillan, 2001.

Posner, E., and J. Skutil. 1968. "The Great Neglect: The Fate of Mendel's Classic Paper between 1865 and 1900." *Medical History* 12: 122–136.

Poulton, Edward Bagnall. 1890. *The Colors of Animals: Their Meaning and Use, Especially Considered in the Case of Insects*. New York.

————. 1908. *Essays in Evolution: 1889–1907*. Oxford.

Powell, Baden. 1855. *Essays on the Spirit of the Inductive Philosophy, the Unity of Worlds, and the Philosophy of Creation*. London. Reprint, Farnborough: Gregg International, 1969.

Powell, James L. 1998. *Night Comes to the Cretaceous: Dinosaur Extinction and the Transformation of Modern Geology.* New York: W. H. Freeman.

Priest, Josiah. 1843. *Slavery: As It Relates to the Negro or African Race.* Albany, N.Y. Reprint, New York: Arno, 1977.

Provine, William B. 1971. *The Origins of Theoretical Population Genetics.* Chicago: University of Chicago Press.

———. 1973. "Geneticists and the Biology of Race Crossing." *Science* 182: 790–796.

———. 1978. "The Role of Mathematical Population Genetics in the Evolutionary Synthesis of the 1930s and 1940s." *Stud. Hist. Biology* 2: 167–192.

———. 1979. "Francis B. Sumner and the Evolutionary Synthesis." *Stud. Hist. Biology* 3: 211–240.

———. 1986. *Sewall Wright and Evolutionary Biology.* Chicago: University of Chicago Press.

———. 1988. "Progress in Evolution and Meaning in Life." In *Evolutionary Progress,* edited by Matthew H. Nitecki, pp. 49–74. Chicago: University of Chicago Press. Reprinted in *Julian Huxley: Biologist and Statesman of Science,* edited by C. Kenneth Waters and Albert Van Helden, pp. 165–180. Houston: Rice University Press.

Punnett, R. C. 1915. *Mimicry in Butterflies.* Cambridge: Cambridge University Press.

Quammen, David. 1996. *The Song of the Dodo: Island Biogeography in an Age of Extinction.* New York: Scribner.

Quetelet, Lambert. 1842. *A Treatise on Man and the Development of His Faculties.* London. Reprint, Delmar, N.Y.: Scholars' Facsimiles and Reprints.

Raby, Peter. 2001. *Alfred Russel Wallace: A Life.* London: Chatto and Windus.

Radick, Gregory. 2000. "Morgan's Vanon, Garner's Phonograph, and the Evolutionary Origins of Language and Reason." *Brit. J. Hist. Sci.* 32: 3–23.

Radl, Emmanuel. 1930. *The History of Biological Theories.* Translated by E. I. Hatfield. Oxford: Oxford University Press; London: Humphrey Milford.

Raff, Rudolph A. 1996. *The Shape of Life: Genes, Development, and the Evolution of Animal Form.* Chicago: University of Chicago Press.

Rainger, Ronald. 1981. "The Continuation of the Morphological Tradition in American Paleontology, 1880–1910." *J. Hist. Biology* 14: 129–158.

———. 1991. *An Agenda for Antiquity: Henry Fairfield Osborn and Vertebrate Paleontology at the American Museum of Natural History.* Tuscaloosa: University of Alabama Press.

Rainger, Ronald, Keith R. Benson, and Jane Maienschein, eds. 1988. *The American Development of Biology.* Philadelphia: University of Pennsylvania Press.

Rappaport, Rhoda. 1997. *When Geologists Were Historians, 1665–1750.* Ithaca, N.Y.: Cornell University Press.

Rasmussen, Nicolas. 1991. "The Decline of Recapitulationism in Early Twentieth-Century Biology." *J. Hist. Biology* 24: 51–89.

Raup, David. 1991. *Extinction: Bad Genes or Bad Luck?* New York: Norton.

Raven, C. E. 1942. *John Ray, Naturalist: His Life and Work.* Cambridge: Cambridge University Press.

Ray, John. 1692. *Miscellaneous Discourses Concerning the Changes of the World.* London. Reprint, Hildesheim: Georg Olms, 1968.

———. 1713. *Three Physico-Theological Discourses.* 3d ed. London. Reprint, New York: Arno Press, 1977.

———. 1717. *The Wisdom of God Manifested in the Works of Creation.* 7th ed., reprint, New York: Arno Press, 1977.

———. 1724. *Synopsis Methodica Stirpum Britannicum.* 3d ed. London. Reprinted with Linnaeus, *Flora Anglica.* London: Ray Society, 1973.

Reader, J. 1981. *Missing Links: The Hunt for Earliest Man.* London: Collins.

Rehbock, Philip F. 1975. "Huxley, Haeckel, and the Oceanographers: The Case of *Bathybius Haeckelii.*" *Isis* 66: 504–533.

———. 1983. *The Philosophical Naturalists: Themes in Early Nineteenth-Century British Biology.* Madison: University of Wisconsin Press.

Reif, Wolf-Ernst. 1983. "Evolutionary Theory in German Paleontology." In *Dimensions of Darwinism: Themes and Counterthemes in Twentieth-Century Evolutionary Theory,* edited by Marjorie Grene, pp. 71–93. Cambridge: Cambridge University Press.

———. 1986. "The Search for a Macroevolutionary Theory in German Paleontology." *J. Hist. Biology* 19: 79–130.

Reif, Wolf-Ernst, Thomas Junker, and Uwe Hossfeld. 2000. "The Synthetic Theory of Evolution: General Problems and the German Contributions to the Synthesis." *Theory in Biosciences* 119: 41–91.

Rensch, Bernhard. 1983. "The Abandonment of Lamarckian Explanations: The Case of Climatic Parallelism of Animal Characteristics." In *Dimensions of Darwinism: Themes and Counterthemes in Twentieth-Century Evolutionary Theory,* edited by Marjorie Grene, pp. 31–42. Cambridge: Cambridge University Press.

Richards, Evelleen. 1983. "Darwin and the Descent of Woman." In *The Wider Domain of Evolutionary Thought,* edited by D. R. Oldroyd and Ian Langham, pp. 57–111. Dordrecht: D. Reidel.

———. 1987. "A Question of Property Rights: Richard Owen's Evolutionism Reassessed." *Brit. J. Hist. Sci.* 20: 129–172.

———. 1989a. "The 'Moral Anatomy' of Robert Knox: The Interplay between Biological and Social Thought in Victorian Scientific Naturalism." *J. Hist. Biology* 22: 373–436.

———. 1989b. "Huxley Finds Man, Loses Woman: The 'Woman Question' and the Control of Victorian Anthropology." In *History, Humanity, and Evolution: Essays for John C. Greene,* edited by James R. Moore, pp. 253–284. Cambridge: Cambridge University Press.

Richards, Robert J. 1987. *Darwin and the Emergence of Evolutionary Theories of Mind and Behavior.* Chicago: University of Chicago Press.

———. 1991. *The Meaning of Evolution: The Morphological Construction and*

Ideological Reconstruction of Darwin's Theory. Chicago: University of Chicago Press.

———. 1997. "Darwin and the Inefficiency of Artificial Selection." *Stud. Hist. & Phil. Sci.* 28: 75–97.

———. 2002. *The Romantic Conception of Life: Science and Philosophy in the Age of Goethe*. Chicago: University of Chicago Press.

Richardson, R. Alan. 1981. "Biogeography and the Genesis of Darwin's Ideas on Transmutation." *J. Hist. Biology* 14: 1–41.

Ridley, Mark. 1982. "Coadaptation and the Inadequacy of Natural Selection." *Brit. J. Hist. Sci.* 15: 45–68.

———. 1993. *The Red Queen: Sex and the Evolution of Human Nature*. London: Viking.

Rieppel, Olivier. 1988. "The Reception of Leibniz's Philosophy in the Writings of Charles Bonnet (1720–1793)." *J. Hist. Biology* 21: 119–145.

Rinard, R. G. 1988. "Neo-Lamarckism and Technique: Hans Spemann and the Development of Experimental Embryology." *J. Hist. Biology* 21: 95–118.

Ritterbush, Phillip C. 1964. *Overtures to Biology: The Speculation of the Eighteenth-Century Naturalists*. New Haven: Yale University Press.

Ritvo, Lucille B. 1990. *Darwin's Influence on Freud*. New Haven: Yale University Press.

Roberts, H. F. 1929. *Plant Hybridization before Mendel*. Princeton: Princeton University Press.

Roberts, Jon H. 1988. *Darwin and the Divine in America: Protestant Intellectuals and Organic Evolution, 1859–1900*. Madison: University of Wisconsin Press.

Robinet, J. B. 1761–1766. *De la nature*. 4 vols. Amsterdam.

Robinson, Gloria. 1979. *A Prelude to Genetics: Theories of a Material Substance of Heredity, Darwin to Weismann*. Lawrence, Kans.: Coronado Press.

Roe, Shirley A. 1981. *Matter, Life, and Generation: Eighteenth-Century Embryology and the Haller-Wolff Debate*. Cambridge: Cambridge University Press.

———. 1983. "John Turberville Needham and the Generation of Living Organisms." *Isis* 74: 159–184.

———. 1985. "Voltaire versus Needham: Atheism, Materialism, and the Generation of Life." *J. Hist. Ideas* 46: 65–87.

Roger, Jacques. 1997. *Buffon: A Life in Natural History*. Translated by Sarah L. Bonnefoi. Edited by L. Pierce Williams. Ithaca, N.Y.: Cornell University Press.

———. 1998. *The Life Sciences in Eighteenth-Century French Thought*. Translated by Robert Ellrich. Edited by Keith R. Benson. Stanford: Stanford University Press.

Rogers, James Allen. 1972. "Darwinism and Social Darwinism." *J. Hist. Ideas* 33: 265–280.

———. 1974. "The Reception of Darwin's *Origin of Species* by Russian Scientists." *Isis* 64: 484–503.

Roll-Hansen, Nils. 1985. "A New Perspective on Lysenko?" *Annals of Science* 42: 261–276.

———. 1988. "The Progress of Eugenics: Growth of Knowledge and Change in Ideology." *History of Science* 26: 295–331.

Roller, Duane H. D., ed. 1971. *Perspectives in the History of Science and Technology.* Norman: University of Oklahoma Press.

Romanes, George John. 1886. "Physiological Selection: An Additional Suggestion on the Origin of Species." *J. Linn. Soc. (Zool.)* 19: 337–411. Abstracted *Nature* 34: 314–316, 336–440, and 362–365.

———. 1888. *Mental Evolution in Man.* London.

———. 1892–97. *Darwin and after Darwin: An Exposition of the Darwinian Theory and a Discussion of Post-Darwinian Problems.* 3 vols. London.

———. 1899. *An Examination of Weismannism.* 2d ed. Chicago.

Romer, Alfred Sherwood. 1933. *Vertebrate Paleontology.* Chicago: University of Chicago Press.

Rose, Michael R. 1998. *Darwin's Specter: Evolutionary Biology in the Modern World.* Princeton: Princeton University Press.

Rose, Steven, Leon Kamin Jr., and R. C. Lewontin. 1984. *Not in Our Genes: Biology, Ideology, and Human Nature.* New York: Pantheon.

Rosenfield, Leonora C. 1968. *From Beast-Machine to Man-Machine: Animal Soul in French Letters from Descartes to La Mettrie.* New ed. New York: Octagon Books.

Ross, Edward Alsworth. 1927. *Standing Room Only?* New York and London. Reprint, New York: Arno, 1977.

Rossi, Paolo. 1984. *The Dark Abyss of Time: The History of the Earth and the History of Nations from Hooke to Vico.* Chicago: University of Chicago Press.

Rousseau, G. S., and Roy Porter, eds. 1980. *The Ferment of Knowledge: Studies in the Historiography of Eighteenth-Century Science.* Cambridge: Cambridge University Press.

Rousseau, Jean-Jacques. 1994. *Discourse on the Origin and Foundations of Inequality among Men.* Translated by Franklin Phillips. Oxford: Oxford University Press.

Rudberg, Daniel. 1752. "Peloria." In *Amoenitates Academicae, seu Dissertationes variae physicae, medicae et botanicae,* edited by Carolus Linnaeus, 2:280–298. Leiden and Amsterdam, 1749–90.

Rudge, David Wyss. 1999. "Taking the Peppered Moth with a Pinch of Salt." *Biology and Philosophy* 14: 9–37.

Rudwick, Martin J. S. 1970. "The Strategy of Lyell's *Principles of Geology.*" *Isis* 61: 5–33.

———. 1971. "Uniformity and Progression: Reflections on the Structure of Geological Theory in the Age of Lyell." In *Perspectives in the History of Science and Technology,* edited by Duane H. D. Roller, pp. 209–227. Norman: University of Oklahoma Press.

———. 1972. *The Meaning of Fossils: Episodes in the History of Paleontology.* 2d ed., New York: Science History Publications, 1976.

———. 1974a. "Poulett Scrope on the Volcanoes of Auvergne: Lyellian Time and Political Economy." *Brit. J. Hist. Sci.* 7: 205–242.

———. 1974b. "Darwin and Glen Roy: A 'Great Failure' in Scientific Method?" *Stud. Hist. & Phil. Sci.* 5: 97–185.

———. 1975. "Caricature as a Source for the History of Science: De la Beche's Anti-Lyellian Sketches of 1831." *Isis* 66: 534–560.

———. 1982. "Charles Darwin in London: The Integration of Public and Private Science." *Isis* 73: 186–206.

———. 1985. *The Great Devonian Controversy: The Shaping of Scientific Knowledge among Gentlemanly Specialists.* Chicago: University of Chicago Press.

———. 1997. *Georges Cuvier, Fossil Bones, and Geological Catastrophes.* Chicago: University of Chicago Press.

Rupke, Nicolaas A. 1983. *The Great Chain of History: William Buckland and the English School of Geology (1814–1849).* Oxford: Oxford University Press.

———. 1994. *Richard Owen: Victorian Naturalist.* New Haven: Yale University Press.

Ruse, Michael. 1971. "Natural Selection in the *Origin of Species.*" *Stud. Hist. & Phil. Sci.* 1: 311–351.

———. 1974. "The Darwin Industry: A Critical Evaluation." *History of Science* 7: 43–58.

———. 1975a. "Charles Darwin and Artificial Selection." *J. Hist. Ideas* 36: 339–350.

———. 1975b. "Darwin's Debt to Philosophy: An Examination of the Influence of the Philosophical Ideas of John F. W. Herschel and William Whewell on the Development of Charles Darwin's Theory of Evolution." *Stud. Hist. & Phil. Sci.* 6: 159–181.

———. 1975c. "The Relationship between Science and Religion in Britain, 1830–1870." *Church History* 34: 505–522.

———. 1977. "Karl Popper's Philosophy of Biology." *Philosophy of Science* 44: 638–661.

———. 1979a. *The Darwinian Revolution: Science Red in Tooth and Claw.* Chicago: University of Chicago Press. 2d ed., Chicago: University of Chicago Press, 1999.

———. 1979b. *Sociobiology: Sense or Nonsense?* Dordrecht: D. Reidel.

———. 1982. *Darwinism Defended: A Guide to the Evolution Controversies.* Reading, Mass.: Addison-Wesley.

———. 1996. *Monad to Man: The Concept of Progress in Evolutionary Biology.* Cambridge, Mass.: Harvard University Press.

———, ed. 1994. Special Issue on Ernst Mayr at Ninety. *Biology and Philosophy* 9: 263–427.

Russell, Bertrand. 1961. *The Basic Writings of Bertrand Russell.* Edited by Lester E. Dennon and Robert E. Egner. London: Allen and Unwin.

Russell, E. S. 1916. *Form and Function: A Contribution to the History of Animal Morphology.* London: Murray. Reprint, Farnborough: Gregg International, 1972.

Russett, Cynthia Eagle. 1976. *Darwin in America: The Intellectual Response.* San Francisco: W. H. Freeman.

———. 1989. *Sexual Science: The Victorian Contribution to Womanhood.* Cambridge, Mass.: Harvard University Press.

Ryan, Frank X., ed. 2001. *Darwin's Impact: Social Evolution in America, 1880–1920.* 3 vols. Bristol: Thoemmes Press.

———. 2002. *Darwinism and Theology.* 4 vols. Bristol: Thoemmes Press.

Saint-Simon, Claude-Henri de Rouvray, Comte de. 1952. *Selected Writings.* Translated by F. M. H. Markham. Oxford: Oxford University Press.

Sandler, Iris. 1983. "Pierre Louis Moreau de Maupertuis—a Precursor of Mendel?" *J. Hist. Biology* 16: 101–136.

Santurri, Edmund N. 1982. "Theodicy and Social Policy in Malthus' Thought." *J. Hist. Ideas* 43: 315–330.

Sapp, Jan. 1983. "The Struggle for Authority in the Field of Heredity, 1900–1932." *J. Hist. Biology* 16: 311–342.

———. 1987. *Beyond the Gene: Cytoplasmic Inheritance and the Struggle for Authority in Genetics.* Oxford: Oxford University Press.

Sarker, Sahotra, ed. 1992. *The Founders of Evolutionary Genetics.* Dordrecht: Kluwer.

Schaffer, Simon. 1989. "The Nebular Hypothesis and the Science of Progress." In *History, Humanity, and Evolution: Essays for John C. Greene,* edited by James R. Moore, pp. 131–164. Cambridge: Cambridge University Press.

Schiller, Joseph. 1971. "L'échelle des êtres et la série chez Lamarck." In *Colloque international "Lamarck" tenue au Muséum National d'Histoire Naturelle,* edited by Joseph Schiller, pp. 87–103. Paris: Blanchard.

———. 1974. "Queries, Answers, and Unsolved Problems in Eighteenth-Century Biology." *History of Science* 12: 184–199.

———, ed. 1971. *Colloque international "Lamarck" tenue au Muséum National d'Histoire Naturelle.* Paris: Blanchard.

Schindewolf, Otto. 1993. *Basic Questions in Paleontology.* Translated by Judith Schaefer. Introduction by S. J. Gould. Chicago: University of Chicago Press.

Schneer, Cecil J., ed. 1969. *Toward a History of Geology.* Cambridge, Mass.: MIT Press.

Schuchert, Charles, and Clara Mae Levene. 1940. *O. C. Marsh: Pioneer in Paleontology.* New Haven: Yale University Press.

Schwartz, Joel S. 1974. "Charles Darwin's Debt to Malthus and Edward Blyth." *J. Hist. Biology* 7: 301–318.

———. 1995. "George John Romanes's Defence of Darwinism: The Correspondence of Charles Darwin and His Chief Disciple." *J. Hist. Biology* 28: 281–316.

Schweber, Sylvan S. 1977. "The Origin of the *Origin* Revisited." *J. Hist. Biology* 10: 229–316.

———. 1979. "Essay Review: The Young Charles Darwin." *J. Hist. Biology* 13: 175–192.

———. 1980. "Darwin and the Political Economists: Divergence of Character." *J. Hist. Biology* 13: 195–289.

———. 1989. "John Herschel and Charles Darwin: A Study in Parallel Lives." *J. Hist. Biology* 22: 1–71.

Scopes, John Thomas. 1967. *Center of the Storm*. New York: Holt, Rinehart, and Winston.

Scott, Clifford H. 1976. *Lester Frank Ward*. Boston: Twayne Publishers.

Scott-Ram, N. R. 1990. *Transformed Cladistics, Taxonomy, and Evolution*. Cambridge: Cambridge University Press.

Scrope, George Poulett. 1827. *Memoir on the Geology of Central France*. London.

Searle, C. R. 1976. *Eugenics and Politics in Britain: 1900–1914*. Leiden: Noordhoff International Publishing.

———. 1979. "Eugenics and Politics in Britain in the 1930s." *Annals of Science* 36: 159–169.

Secord, James A. 1981. "Nature's Fancy: Charles Darwin and the Breeding of Pigeons." *Isis* 72: 163–186.

———. 1986. *Controversy in Victorian Geology: The Cambrian-Silurian Debate*. Princeton: Princeton University Press.

———. 1989. "Behind the Veil: Robert Chambers and *Vestiges*." In *History, Humanity, and Evolution: Essays for John C. Greene*, edited by James R. Moore, pp. 165–194. Cambridge: Cambridge University Press.

———. 1991. "Edinburgh Lamarckians: Robert Jameson and Robert E. Grant." *J. Hist. Biology*. 24: 1–18.

———. 2000. *Victorian Sensation: The Extraordinary Publication, Reception, and Secret Authorship of* Vestiges of the Natural History of Creation. Chicago: University of Chicago Press.

Sedgwick, Adam. 1845. "Vestiges of the Natural History of Creation." *Edinburgh Review* 82: 1–85.

Segerstrale, Ullica. 2000. *Defenders of the Truth: The Battle for Science in the Sociobiology Debate and Beyond*. Oxford: Oxford University Press.

Sellars, Roy Wood. 1922. *Evolutionary Naturalism*. Chicago. Reprint, New York: Russell and Russell, 1969.

Semmel, Bernard. 1960. *Imperialism and Social Reform: English Socio-Imperial Thought, 1895–1914*. London: Allen and Unwin.

Settle, M. L. 1972. *The Scopes Trial*. New York: Franklin Watts.

Shapin, Steven. 1979. "Homo Phrenologicus: Anthropological Perspectives on a Historical Problem." In *Natural Order: Historical Studies of Scientific Culture*, edited by Barry Barnes and Stephen Shapin, pp. 41–71. Beverly Hills: Sage Publications.

———. 1982. "History of Science and Its Sociological Reconstructions." *History of Science* 20: 157–211.

Shaw, George Bernard. 1921. *Back to Methuselah: A Metabiological Pentateuch*. London.

Sheets-Johnstone, Maxine. 1982. "Why Lamarck Did Not Discover the Principle of Natural Selection." *J. Hist. Biology* 15: 443–65.

Sheets-Pyenson, Susan. 1989. *Cathedrals of Science: The Development of Colonial Natural History Museums.* Montreal: McGill-Queen's University Press.

———. 1996. *John William Dawson: Faith, Hope, Science.* Montreal: McGill-Queen's University Press.

Shine, Ian B., and Sylvia Wrobel. 1976. *Thomas Hunt Morgan: Pioneer of Genetics.* Lexington: University of Kentucky Press.

Shor, Elizabeth Noble. 1974. *The Fossil Feud between E. D. Cope and O. C. Marsh.* Hicksville, N.Y.: Exposition Press.

Simpson, George Gaylord. 1944. *Tempo and Mode in Evolution.* New York: Columbia University Press.

———. 1953a. "The Baldwin Effect." *Evolution* 7: 110–117.

———. 1953b. *The Major Features of Evolution.* New York: Columbia University Press.

———. 1963. *This View of Life: The World of an Evolutionist.* New York: Harcourt, Brace, and World.

Sloan, Phillip R. 1972. "John Locke, John Ray, and the Problem of the Natural System." *J. Hist. Biology* 5: 1–53.

———. 1976. "The Buffon-Linnaeus Controversy." *Isis* 67: 356–375.

———. 1985. "Darwin's Invertebrate Program, 1826–1836." In *The Darwinian Heritage: A Centennial Retrospect,* edited by David Kohn, pp. 71–120. Princeton: Princeton University Press.

———. 1986. "Darwin, Vital Matter, and the Transformism of Species." *J. Hist. Biology* 19: 369–445.

Smedley, Audrey. 1993. *Race in North America: Origins and Evolution of a World View.* Oxford: Westview Press.

Smiles, Samuel. 1859. *Self-Help.* London.

Smith, Adam. 1910. *The Wealth of Nations.* 2 vols. Reprint, London: Everyman.

———. 1978. *Lectures on Jurisprudence.* Edited by R. L. Meek. Oxford: Clarendon Press.

Smith, Crosbie W., and M. Norton Wise. 1989. *Energy and Empire: A Biographical Study of Lord Kelvin.* Cambridge: Cambridge University Press.

Smith, Grafton Elliot. 1924. *The Evolution of Man: Essays.* Oxford: Oxford University Press.

Smith, J. Maynard. 1982. *Evolution and the Theory of Games.* Cambridge: Cambridge University Press.

Smith, R. 1972. "Alfred Russel Wallace: Philosophy of Nature and Man." *Brit. J. Hist. Sci.* 6: 177–199.

Smith, William. 1815. *A Memoir to the Map and Delineation of the Strata of England . . .* London.

Smocovitis, Vassiliki Betty. 1996. *Unifying Biology: The Evolutionary Synthesis and Evolutionary Biology.* Princeton: Princeton University Press.

Smout, Kary Doyle. 1998. *The Creation/Evolution Controversy: A Battle for Cultural Power.* Westport, Conn.: Praeger.

Snyder, Louis L. 1962. *The Idea of Racialism: Its Meaning and History.* Princeton: Princeton University Press.

Sober, Elliot, and David S. Wilson. 1998. *Unto Others: The Evolution and Psychology of Unselfish Behavior.* Cambridge, Mass.: Harvard University Press.

Sokal, Alan, and Jean Bricmont. 1998. *Intellectual Impostures: Postmodern Philosophers' Abuse of Science.* London: Profile Books.

Soyfer, Valery N. 1994. *Lysenko and the Tragedy of Soviet Science.* Translated by Leo and Rebecca Gruliow. New Brunswick, N.J.: Rutgers University Press.

Spallanzani, Lazzaro. 1769. *Nouvelles recherches sur les découvertes microscopiques, et la génération des corps organizées.* 2 vols. London and Paris.

Spary, Elizabeth C. 2000. *Utopia's Garden: French Natural History from Old Regime to Revolution.* Chicago: University of Chicago Press.

Spencer, Frank. 1990. *Piltdown: A Scientific Forgery.* London: Oxford University Press.

Spencer, Herbert. 1851. *Social Statics: Or the Conditions Essential to Human Happiness Specified, and One of Them Adopted.* London. Reprint, Farnborough: Gregg International.

———. 1855. *Principles of Psychology.* London. Reprint, Farnborough: Gregg International. 1881 ed. reprinted (2 vols.) Boston: Longwood Press.

———. 1862. *First Principles of a New Philosophy.* London.

———. 1864. *Principles of Biology.* 2 vols. London.

———. 1883. *Essays Scientific, Political, and Speculative.* 3 vols. London.

———. 1893. "The Inadequacy of Natural Selection." *Contemporary Review* 63: 153–166, 439–456.

———. 1904. *An Autobiography.* 2 vols. New York.

———. 1969. *The Man versus the State.* Edited by Donald Macrae. Harmondsworth: Penguin Books.

Stafleu, F. 1971. *Linnaeus and the Linnaeans: The Spreading of Their Ideas in Systematic Botany.* Utrecht: Oosthoek.

Stanton, William. 1960. *The Leopard's Spots: Scientific Attitudes toward Race in America, 1815–1859.* Chicago: Phoenix Books.

Staum, Martin S. 1980. *Cabanis: Enlightenment and Medical Philosophy in the French Revolution.* Princeton: Princeton University Press.

Stebbins, Robert E. 1974. "France." In *The Comparative Reception of Darwinism,* edited by Thomas F. Glick, pp. 117–167. Austin: University of Texas Press. New ed., Chicago: University of Chicago Press, 1988.

Steele, Edward J. 1979. *Somatic Selection and Adaptive Evolution: On the Inheritance of Acquired Characters.* Toronto: Williams and Wallace International. Reprint, Chicago: University of Chicago Press, 1982.

Steele, Edward J., Robyn A. Lindley, and Robert V. Blanden. 1998. *Lamarck's Signature: How Retrogens Are Changing the Natural Selection Paradigm.* St. Leonards, NSW: Allen and Unwin; Reading, Mass.: Perseus Books.

Steno, Nicholas. 1916. *The Prodromus of Nicholas Steno's Dissertation*

Concerning a Solid Body Enclosed by Process of Nature within a Solid. Translated by J. G. Winter. Vol. 1, pt. 2. University of Michigan Humanistic Studies. Reprint, New York: Hafner, 1968.

Stepan, Nancy. 1982. *The Idea of Race in Science: Great Britain, 1800–1960.* London: Macmillan.

Stephens, Lester G. 1982. *Joseph LeConte: Gentle Prophet of Evolution.* Baton Rouge: Louisiana State University Press.

Stern, Curt, and E. R. Sherwood, eds. 1966. *The Origin of Genetics: A Mendel Sourcebook.* San Francisco: W. H. Freeman.

Stevens, Peter F. 1994. *The Development of Biological Systematics: Antoine-Laurent de Jussieu, Nature, and the Natural System.* New York: Columbia University Press.

Stocking, George W. 1962. "Lamarckianism in American Social Science." *J. Hist. Ideas* 23: 239–256.

———. 1968. *Race, Culture, and Evolution.* New York: Free Press.

———. 1987. *Victorian Anthropology.* New York: Free Press.

———. 1996. *After Tylor: British Social Anthropology, 1888–1951.* London: Athlone.

Strahler, Arthur N. 1988. *Science and Earth History: The Evolution/Creation Controversy.* New York: Prometheus Books.

Strick, James. 1999. "Darwinism and the Origin of Life: The Role of H. C. Bastian in the British Spontaneous Generation Debates, 1868–1873." *J. Hist. Biology* 32: 51–92.

———. 2000. *Sparks of Life: Darwinism and the Victorian Debates over Spontaneous Generation.* Cambridge, Mass.: Harvard University Press.

———, ed. 2001. *Evolution and the Spontaneous Generation Debates.* Bristol: Thoemmes Press.

Strickberger, Monroe W. 2000. *Evolution.* 3d ed. Sudbury, Mass.: James and Bartlett.

Stubbe, Hans. 1972. *History of Genetics: From Prehistoric Times to the Rediscovery of Mendel's Laws.* Translated by T. H. Waters. Cambridge, Mass.: MIT Press.

Sturtevant, A. H. 1965. *A History of Genetics.* New York: Harper and Row.

Sulloway, Frank J. 1979a. "Geographic Isolation in Darwin's Thinking: The Vicissitudes of a Crucial Idea." *Stud. Hist. Biology* 3: 23–65.

———. 1979b. *Freud: Biologist of the Mind: Beyond the Psychoanalytic Legend.* London: Burnett Books.

———. 1982a. "Darwin and His Finches: The Evolution of a Legend." *J. Hist. Biology* 15: 1–54.

———. 1982b. "Darwin's Conversion: The *Beagle* Voyage and Its Aftermath." *J. Hist. Biology* 15: 323–396.

———. 1996. *Born to Rebel: Birth Order, Family Dynamics, and Creative Lives.* New York: Pantheon.

Sumner, F. B. 1924. "The Stability of Subspecific Characters under Changed Conditions of Environment." *Am. Naturalist* 58: 481–505.

Swetlitz, Marc. 1995. "Julian Huxley and the End of Evolution." *J. Hist. Biology* 28: 181–217.

Swinburne, R. G. 1965. "Galton's Law—Formulation and Development." *Annals of Science* 21: 15–31.

Symondson, Anthony, ed. 1970. *The Victorian Crisis of Faith.* London: SPCK.

Tammone, William. 1995. "Competition, the Division of Labor, and Darwin's Principle of Divergence." *J. Hist. Biology* 28: 109–131.

Taylor, M. W. 1992. *Men Versus the State: Herbert Spencer and Late Victorian Individualism.* Oxford: Clarendon Press.

Teich, M., and R. M. Young, eds. 1973. *Changing Perspectives in the History of Science.* London: Heinemann.

Teilhard de Chardin, Pierre. 1959. *The Phenomenon of Man.* Introduction by Julian Huxley. London: Collins.

Temkin, Owsei. 1950. "German Concepts of Ontogeny and Development around 1800." *Bull. Hist. Medicine* 24: 227–246.

———. 1959. "The Idea of Descent in Post-Romantic German Biology." In *Forerunners of Darwin, 1745–1859,* edited by Bentley Glass, Owsei Temkin, and William Strauss Jr., pp. 323–355. Baltimore: Johns Hopkins University Press.

Tennenbaum, J. 1956. *Race and Reich: The Story of an Epoch.* Boston: Twayne Publishers.

Tennyson, Alfred. 1973. *In Memoriam.* Edited by Robert H. Ross. New York: Norton.

Theunissen, Bert. 1986. "The Relevance of Cuvier's *Lois zoologiques* for His Paleontological Work." *Annals of Science* 43: 543–556.

———. 1989. *Eugene Dubois and the Ape-Man from Java: The History of the First 'Missing Link' and Its Discoverer.* Dordrecht: Kluwer.

Todes, Daniel P. 1989. *Darwin without Malthus: The Struggle for Existence in Russian Evolutionary Thought.* New York: Oxford University Press.

Topham, Jonathan R. 1998. "Beyond the 'Common Context': The Production and Reading of the Bridgewater Treatises." *Isis* 89: 233–262.

Topsell, Edward. 1608. *The Historie of Foure-Footed Beastes.* Reprint, Amsterdam: Da Capo, 1973.

Torrey, Norman L. 1930. *Voltaire and the English Deists.* New Haven: Yale University Press.

Toulmin, Stephen, and June Goodfield. 1965. *The Discovery of Time.* Chicago: University of Chicago Press.

Trautmann, Thomas R. 1987. *Lewis Henry Morgan and the Invention of Kinship.* Berkeley and Los Angeles: University of California Press.

Trembley, Abraham. 1973. *Memoirs on the Natural History of the Freshwater Polyp.* Translated by H. F. Ewer. New York: Johnson Reprint Corporation.

Trigger, Bruce G. 1989. *A History of Archaeological Thought.* Cambridge: Cambridge University Press.

Trinkaus, Eric, and Pat Shipman. 1993. *The Neanderthals: Of Skeletons, Scientists, and Scandal.* New York: Knopf.

Turgot, Anne Robert Jacques. 1973. *Turgot on Progress, Sociology, and Economics.* Translated by Robert L. Meek. Cambridge: Cambridge University Press.

Turner, Frank Miller. 1974. *Between Science and Religion: The Reaction to Scientific Naturalism in Late Victorian England.* New Haven: Yale University Press.

———. 1978. "The Victorian Conflict between Science and Religion: A Professional Dimension." *Isis* 69: 356–376.

Turrill, William Bertram. 1963. *Joseph Dalton Hooker: Botanist, Explorer, and Administrator.* London: Scientific Book Guild.

Tylor, Edward B. 1865. *Researches into the Early History of Mankind.* London.

Tyndall, John. 1902. *Fragments of Science.* New ed. 2 vols. New York.

Van Doren, Charles L. 1967. *The Idea of Progress.* New York: F. A. Praeger.

Van Riper, A. Bowdoin. 1993. *Men among the Mammoths: Victorian Science and the Discovery of Human Prehistory.* Chicago: University of Chicago Press.

Van Valen, L. 1973. "A New Evolutionary Law." *Evolution Theory* 1: 1–30.

Vartanian, Aram. 1950. "Trembley's Polyp, La Mettrie, and Eighteenth-Century French Materialism." *J. Hist. Ideas* 11: 259–286.

———. 1953. *Diderot and Descartes: A Study of Scientific Naturalism in the Enlightenment.* Princeton: Princeton University Press.

Velikovsky, Immanuel. 1950. *Worlds in Collision.* Garden City, N.Y.: Doubleday.

Vico, Giambattista. 1948. *The New Science of Giambattista Vico.* Translated by Thomas Goddard Bergin and Max Harold Fisch. Ithaca, N.Y.: Cornell University Press.

Vorzimmer, Peter J. 1963. "Charles Darwin and Blending Inheritance." *Isis* 54: 371–390.

———. 1965. "Darwin's Ecology and Its Influence upon His Theory." *Isis* 56: 148–155.

———. 1968. "Darwin and Mendel: The Historical Connection." *Isis* 59: 72–82.

———. 1969a. "Darwin, Malthus, and the Theory of Natural Selection." *J. Hist. Ideas* 30: 527–542.

———. 1969b. "Charles Darwin's 'Questions on the Breeding of Animals.' " *J. Hist. Biology* 2: 269–281.

———. 1970. *Charles Darwin, the Years of Controversy: The Origin of Species and Its Critics, 1859–82.* Philadelphia: Temple University Press.

———. 1977. "The Darwin Reading Notebooks (1838–1860)." *J. Hist. Biology* 10: 107–152.

Vucinich, Alexander. 1988. *Darwin in Russian Thought.* Berkeley and Los Angeles: University of California Press.

Vyverberg, Henry. 1989. *Human Nature, Cultural Diversity, and the French Enlightenment.* New York: Oxford University Press.

Waddington, C. H. 1957. *The Strategy of the Gene.* London: Allen and Unwin.

———. 1975. *The Evolution of an Evolutionist*. Edinburgh: Edinburgh University Press.

Wade, Ira O. 1971. *The Intellectual Origins of the French Enlightenment*. Princeton: Princeton University Press.

Waerden, B. L. van der. 1968. "Mendel's Experiments." *Centaurus* 12: 275–288.

Wagner, Moritz. 1873. *The Darwinian Theory and the Law of the Migration of Organisms*. Translated by I. L. Laird. London.

Waisbren, Steven James. 1988. "The Importance of Morphology in the Evolutionary Synthesis." *J. Hist. Biology* 21: 291–330.

Wallace, Alfred Russel. 1855. "On the Law Which Has Regulated the Introduction of New Species." *Ann. & Mag. of Nat. Hist.* 26: 184–196.

———. 1869. *The Malay Archipelago: The Land of the Orang-Utan and the Bird of Paradise: A Narrative of Travel with Studies of Man and Nature*. London. Rev. ed. 1890. Reprint, New York: Dover, 1962.

———. 1870. *Contributions to the Theory of Natural Selection*. London. Reprint, New York: AMS Press, 1973.

———. 1876. *The Geographical Distribution of Animals*. 2 vols. London. Reprint, New York: Hafner, 1962.

———. 1889. *Darwinism: An Exposition of the Theory of Natural Selection*. London. Reprint, New York: AMS Press, 1975.

———. 1891. *Natural Selection and Tropical Nature*. London. Reprint, Farnborough: Gregg International, 1969.

———. 1905. *My Life: A Record of Events and Opinions*. London. Reprint, Farnborough: Gregg International, 1969.

Wallace, David Rains. 1999. *The Bonehunters' Revenge: Dinosaurs, Greed, and the Greatest Scientific Fraud of the Gilded Age*. Boston: Houghton Mifflin.

Waller, John C. 2001a. "Gentlemanly Men of Science: Sir Francis Galton and the Professionalization of the British Life Sciences." *J. Hist. Biology* 34: 83–114.

———. 2001b. "Ideas of Heredity, Reproduction, and Eugenics in Britain, 1800–1875." *Stud. Hist. & Phil. Sci. (Biomed.)*.

———. 2002. "Putting Method First: Re-appraising the Extreme Determinism and Hard Hereditarianism of Sir Francis Galton." *History of Science* 40: 35–62.

Ward, Lester. 1883. *Dynamic Sociology: Or Applied Social Science as Based upon Statistical Sociology and the Less Complex Sciences*. Reprint, with an introduction by David W. Noble, New York: Johnson Reprint Corporation, 1968.

———. 1906. *Applied Sociology: A Treatise on the Conscious Improvement of Society by Society*. Reprint, New York: Johnson Reprint Corporation, 1968.

Wasserman, Gerhard D. 1978. "Testability and the Role of Natural Selection within Theories of Population Genetics and Evolution." *Brit. J. Phil. Sci.* 29: 223–242.

Waters, C. Kenneth, and Albert Van Helden, eds. 1992. *Julian Huxley: Biologist and Statesman of Science*. Houston: Rice University Press.

Weikart, Richard. 1999. *Socialist Darwinism: Evolution in German Socialist Thought from Marx to Bernstein*. San Francisco: International Scholars Publications.

Weindling, Paul. 1989a. "Ernst Haeckel, Darwinismus, and the Secularization of Nature." In *History, Humanity, and Evolution: Essays for John C. Greene*, edited by James R. Moore, pp. 311–328. Cambridge: Cambridge University Press.

——. 1989b. *Health, Race, and German Politics between National Unification and Nazism, 1870–1945*. Cambridge: Cambridge University Press.

Weiner, J. S. 1955. *The Piltdown Hoax*. Oxford: Oxford University Press.

Weiner, Jonathan. 1994. *The Beak of the Finch: A Story of Evolution in Our Time*. New York: Alfred A. Knopf.

Weingart, Peter. 1989. "German Eugenics between Science and Politics." *Osiris*, n.s., 5: 260–282.

Weinstein, A. 1977. "How Unknown Was Mendel's Paper?" *J. Hist. Biology* 10: 341–364.

Weismann, August. 1882. *Studies in the Theory of Descent*. Translated by Raphael Meldola. 2 vols. London. Reprint, New York: AMS Press.

——. 1891–92. *Essays upon Heredity and Kindred Biological Problems*. Edited by E. B. Poulton et al. 2 vols. Oxford.

——. 1893a. *The Germ Plasm: A Theory of Heredity*. Translated by W. Newton Parker and Harriet Ronfeldt. London.

——. 1893b. "The All-Sufficiency of Natural Selection." *Contemporary Review* 64: 309–338, 596–610.

——. 1896. *On Germinal Selection*. Chicago. Also printed in *The Monist* 6: 250–293.

——. 1904. *The Evolution Theory*. Translated by J. Arthur Thomson and M. R. Thomson. 2 vols. London.

——. 1999. *Ausgewahlte Briefe und Dokumente/Selected Letters and Documents*. Edited by F. Churchill and H. Risler. 2 vols. Freiburg i. Br.: Universitatsbibliothek.

Weiss, Sheila Faith. 1986. "Wilhelm Schallmayer and the Logic of German Eugenics." *Isis* 77: 33–46.

——. 1988. *Race Hygiene and National Efficiency: The Eugenics of Wilhelm Schallmayer*. Berkeley and Los Angeles: University of California Press.

Weldon, W. F. R. 1894–95. "An Attempt to Measure the Death Rate Due to the Selective Destruction of *Carcinas moenas* with Respect to Particular Dimensions." *Proc. Roy. Soc.* 57: 360–379.

——. 1898. President's Address, Zoological Section. *Report of the British Association for the Advancement of Science*, 887–902.

——. 1901. "A First Study of Natural Selection in *Clausilia laminata*." *Biometrika* 1: 109–124.

Wells, George A. 1967. "Goethe and Evolution." *J. Hist. Ideas* 28: 537–550.

Wells, H. G., Julian Huxley, and G. P. Wells. 1930. *The Science of Life.* New York: Doran.

Wells, Kentwood D. 1973a. "The Historical Context of Natural Selection: The Case of Patrick Matthew." *J. Hist. Biology* 6: 223–258.

———. 1973b. "William Charles Wells and the Races of Man." *Isis* 64: 213–225.

Werner, Abraham Gottlob. 1971. *Short Classification and Description of the Various Rocks.* Translated by Alexander M. Ospovat. New York: Hafner.

Werskey, Gary. 1978. *The Visible College: The Collective Biography of British Scientific Socialists of the 1930s.* New York: Holt, Rinehart, and Winston; London: Allan Lane.

Westfall, Richard S. 1958. *Science and Religion in Seventeenth-Century England.* New Haven: Yale University Press.

Whewell, William. 1847a. *History of the Inductive Sciences.* New ed. 3 vols. London. Reprint, Hildesheim: Georg Olms.

———. 1847b. *Philosophy of the Inductive Sciences.* New ed. 2 vols. London. Reprint, with an introduction by J. W. Herivel, New York: Johnson Reprint Corporation.

———. 1989. *Whewell's Theory of Scientific Method.* Edited by Robert E. Butts. 2d ed. Indianapolis: Hackett.

Whiston, William. 1696. *A New Theory of the Earth from Its Original to the Consummation of All Things.* London. Reprint, New York: Arno Press, 1977.

Whitcomb, J. C., and H. M. Morris. 1961. *The Genesis Flood.* Nutley, N.J.: Presbyterian and Reformed Publishing Company.

White, Andrew Dickson. 1896. *A History of the Warfare of Science with Theology.* 2 vols. Reprint, New York: Dover, 1969.

Whitehead, Alfred North. 1929. *Process and Reality: An Essay in Cosmology.* Cambridge: Cambridge University Press.

Wiener, Philip P. 1949. *Evolution and the Founders of Pragmatism.* Cambridge, Mass.: Harvard University Press.

Wilkie, J. S. 1955. "Galton's Contribution to the Theory of Evolution, with Special Reference to His Use of Models and Metaphors." *Annals of Science* 11: 194–205.

———. 1956. "The Idea of Evolution in the Writings of Buffon." *Annals of Science* 12: 48–62, 212–247, 255–266.

———. 1962. "Some Reasons for the Rediscovery and Appreciation of Mendel's Work in the First Years of the Present Century." *Brit. J. Hist. Sci.* 1: 5–17.

Wilkins, Adam S. 2001. *The Evolution of Developmental Pathways.* Sunderland, Mass.: Sinauer Associates.

Willey, Basil. 1940. *The Eighteenth-Century Background: The Idea of Nature in the Thought of the Period.* London: Chatto and Windus.

———. 1949. *Nineteenth-Century Studies: Coleridge to Matthew Arnold.* London: Chatto and Windus.

———. 1956. *More Nineteenth-Century Studies: A Group of Honest Doubters.* London: Chatto and Windus.

————. 1960. *Darwin and Butler: Two Versions of Evolution.* London: Chatto and Windus.

Williams, G. C. 1966. *Adaptation and Natural Selection.* Princeton: Princeton University Press. Reprint, 1996.

Williams-Ellis, Amabel. 1966. *Darwin's Moon: A Biography of Alfred Russel Wallace.* London: Blackie.

Wilson, Catherine. 1994. *The Invisible World: Early Modern Philosophy and the Invention of the Microscope.* Princeton: Princeton University Press.

Wilson, David B. 1977. "Victorian Science and Religion." *History of Science* 15: 52–67.

Wilson, Edward O. 1975. *Sociobiology: The New Synthesis.* Cambridge, Mass.: Harvard University Press.

————. 1978. *On Human Nature.* Cambridge, Mass.: Harvard University Press.

Wilson, Leonard G. 1967. "The Origins of Charles Lyell's Uniformitarianism." *Geol. Soc. Am., Special Paper,* 89: 35–62.

————. 1969. "The Intellectual Background in Charles Lyell's *Principles of Geology, 1830–33.*" In *Toward a History of Geology,* edited by Cecil J. Schneer, pp. 426–443. Cambridge, Mass.: MIT Press.

————. 1972. *Charles Lyell: The Years to 1841: The Revolution in Geology.* New Haven: Yale University Press.

————. 1980. "Geology on the Eve of Charles Lyell's First Visit to America, 1841." *Proc. Am. Phil. Soc.* 124: 168–202.

————. 1996. "The Gorilla and the Question of Human Origins: The Brain Controversy." *J. Hist. Medicine* 51: 184–207.

Winchester, Simon. 2001. *The Map That Changed the World: The Tale of William Smith and the Birth of a Science.* London: Viking.

Winslow, John. 1971. "Darwin's Victorian Malady: Evidence for Its Medically Induced Origins." *Mem. Am. Phil. Soc.* 88.

Winsor, Mary P. 1976. *Starfish, Jellyfish, and the Order of Life.* New Haven: Yale University Press.

————. 1979. "Louis Agassiz and the Species Question." *Stud. Hist. Biology* 3: 89–117.

————. 1991. *Reading the Shape of Nature: Comparative Zoology at the Agassiz Museum.* Chicago: University of Chicago Press.

Wohl, R. 1960. "Buffon and the Project for a New Science." *Isis* 51: 186–199.

Wood, Robert Muir. 1985. *The Dark Side of the Earth.* London: Allen and Unwin.

Woodcock, George. 1969. *Henry Walter Bates: Naturalist of the Amazons.* London: Faber and Faber.

Woodward, John. 1695. *An Essay toward a Natural History of the Earth and Terrestrial Bodyes.* London. Reprint, New York: Arno Press, 1977.

Wright, Sewall. 1930. "The Genetical Theory of Natural Selection: A Review." *J. Heredity* 21: 349–356.

————. 1931. "Evolution in Mendelian Populations." *Genetics* 16: 97–159.

————. 1966. "Mendel's Ratios." In *The Origin of Genetics: A Mendel*

Sourcebook, edited by Curt Stern and E. R. Sherwood, pp. 173–175. San Francisco: W. H. Freeman.

————. 1968–78. *Evolution and the Genetics of Populations.* 4 vols. Chicago: University of Chicago Press.

————. 1986. *Evolution: Selected Papers.* Chicago: University of Chicago Press.

Wyllie, Irvine. 1959. "Social Darwinism and the Businessman." *Proc. Am. Phil. Soc.* 103: 629–635.

Wynne-Edwards, V. C. 1962. *Animal Dispersion and the Relation to Social Behaviour.* Edinburgh: Oliver and Boyd.

Yeo, Richard. 1984. "Science and Intellectual Authority in Mid-Nineteenth-Century Britain: Robert Chambers and *Vestiges of the Natural History of Creation.*" *Victorian Studies* 28: 5–31.

————. 1993. *Defining Science: William Whewell, Natural Knowledge, and Public Debate in Early Victorian Britain.* Cambridge: Cambridge University Press.

Yolton, John W. 1983. *Thinking Matter: Materialism in Eighteenth-Century Britain.* Minneapolis: University of Minnesota Press.

Young, Robert M. 1969. "Malthus and the Evolutionists: The Common Context of Biological and Social Theory." *Past and Present* 43: 109–145.

————. 1970a. *Mind, Brain, and Adaptation in the Nineteenth Century: Cerebral Localization and Its Biological Context from Gall to Ferrier.* Oxford: Clarendon Press.

————. 1970b. "The Impact of Darwin on Conventional Thought." In *The Victorian Crisis of Faith*, edited by Anthony Symondson, pp. 13–36. London: SPCK.

————. 1971a. "Darwin's Metaphor: Does Nature Select?" *Monist* 55: 442–503.

————. 1971b. "Evolutionary Biology and Ideology: Then and Now." *Science Studies* 1: 177–206.

————. 1973. "The Historiographical and Ideological Context of the Nineteenth-Century Debate on Man's Place in Nature." In *Changing Perspectives in the History of Science,* edited by M. Teich and R. M. Young, pp. 344–438. London: Heinemann.

————. 1985. *Darwin's Metaphor: Nature's Place in Victorian Culture.* Cambridge: Cambridge University Press.

Yule, G. Udney. 1902. "Mendel's Laws and Their Probable Relations to Intraracial Heredity." *New Phytologist* 1: 193–207, 222–238.

Ziadat, Adel A. 1986. *Western Science in the Arab World: The Impact of Darwinism, 1860–1930.* London: Macmillan.

Zimmer, Carl. 2001. *Evolution: The Triumph of an Idea.* New York: HarperCollins.

Zirkle, Conway. 1949. *Death of a Science in Russia: The Fate of Genetics as Described in Pravda and Elsewhere.* Philadelphia: University of Pennsylvania Press.

————. 1951. "Gregor Mendel and His Precursors." *Isis* 42: 97–104.

————. 1959. *Evolution, Marxian Biology, and the Social Scene.* Philadelphia: University of Pennsylvania Press.

————. 1968. "The Role of Liberty Hyde Bailey and Hugo De Vries in the Rediscovery of Mendelism." *J. Hist. Biology* 1: 205–218.

Zmarzlik, Gunter. 1972. "Social Darwinism in Germany." In *Republic to Reich: The Making of the Nazi Revolution,* edited by H. Holborn, pp. 435–474. New York: Pantheon.

Index

Acquired characteristics, inheritance of. *See* Lamarckism

Actualism, 130

Adanson, Michel, 68

Adaption, 10–12, 40, 67, 77–79, 104, 118–120, 125–126, 134; and Darwinism, 149, 151, 154–155, 161–162, 166, 170–172, 175, 181–182, 189, 194–195, 205–206, 213, 228–229, 231, 252, 258–259, 328–334, 338, 357–358; and Lamarckism, 86, 90–93, 229, 237–238, 239–240, 243, 367; rejection of, 11, 198–199, 203–205, 241–242, 247–250, 267–270, 350, 363–364, 368, 369–371

Adaptionism, 363–364

Adaptive radiation, 233–234, 249

Agassiz, Louis, 108, 122–123, 136, 186–187, 203, 237, 241, 248, 295

Aggression, 345, 358

Agnosticism, 219, 318

Alexander, Samuel, 321

Allen, Garland E., 230

Allometry, 250

Altruism, 316, 322, 359–60

Alvarez, L. W., 352

America, 189, 194. *See also* South America; United States

American school of neo-Lamarckism, 237, 240–244, 248–249, 294

Ancestral inheritance, law of, 257

Ancon sheep, 200

Animal machine doctrine, 45, 54, 71–72, 75, 81–82

Anthropology, 211, 285–289, 297, 345. *See also* Human race

Apes, 30, 50–53, 208–209, 278, 280–282, 293, 296, 354–355. *See also* Primate studies

Archaeology, 210–211, 278, 285–287

Archaeopteryx, 194

Archetype, 121, 125, 190–191, 227–228, 364, 368

Ardrey, Robert, 345

Argument from design. *See* Design, argument from

Argyll, Duke of, 206–207, 215

Aristotle, 19

Artificial selection, 147, 160–161, 166, 168, 170, 176, 184, 200, 308, 332

Artificial system (in taxonomy), 68

Association of ideas, 53, 220

Atheism. *See* Agnosticism; Materialism

Australopithecus, 284, 341–342, 353–354

Babbage, Charles, 137

Bacon, Francis, 35

Baer, K. E. von, 122, 124, 135, 169

Bagehot, Walter, 304

Bain, Alexander, 220

Balance of nature, 40, 67, 151. *See also* Ecological relationships
Balancing selection, 329, 358
Baldwin, J. M., 243
Baldwin effect, 243, 367
Balfour, Arthur, 318
Bannister, Robert C., 299–300
Barnacles, 168–169
Barnes, Bishop E. W., 323
Barzun, Jacques, 142
Bastian, Charles, 231
Bates, H. W., 174, 189
Bateson, William, 230, 232, 245, 260, 267–268, 272
Bauplan, 364, 368. *See also* Archetype; Idealism
Beagle, voyage of, 149–155
Behe, Michael, 378
Belloc, Hilaire, 323, 331
Bentham, Jeremy, 99, 102–103
Berg, Leo S., 249
Bergson, Henri, 320–321
Bernhardi, Friedrich von, 305
Big-bang cosmology, 8
Bilberg, Isaac, 67
Binomial nomenclature, 68–69
Biogeography. *See* Geographical distribution
Biometry, 256–260, 268, 271–272, 310, 329–330
Bipedalism, 213, 280, 341
Birds, 194, 232, 349, 372–373
Biston betularia, 332, 338
Black, Davidson, 283
Blending heredity. *See* Heredity: blending
Blumenbach, J. F., 52–53
Blyth, Edward, 158
Boas, Franz, 288–289, 297
Bonnet, Charles, 63–66, 72, 89
Borelli, Giovanni, 45
Botany, 43–44, 67–68, 70, 115, 139, 165–166, 189, 239–240
Boucher des Perthes, 210
Boule, Marcellin, 280
Bowlby, John, 148
Boyle, Robert, 39

Brady, R. H., 370
Brain, evolution of, 138, 208–210, 280–281, 341, 355. *See also* Mind; Phrenology
Breeders, influence of, on Darwin, 147, 160–1
Brewster, David, 162
Bridgewater Treatises, 116–117, 120
Brinton, Daniel, 287
Britain, 96–97, 99, 102–103, 106–108, 115–120, 127–129, 135, 162, 266, 310, 318, 322–324; reception of Darwinism in, 184–186, 202–203, 217–222, 301
British Association for the Advancement of Science, 107, 185
Brongniart, Adolphe, 115
Brongniart, Alexander, 113
Bronn, H. G., 139
Brooks, W. K., 267
Broom, Robert, 233, 249, 250, 284, 341
Brown-Sequard, C. E., 244
Bryan, William J., 324
Buckland, William, 116–117, 118, 119–120, 130
Buffon, Georges, Comte de: on evolution, 19, 75–80, 82–83; on geology, 57–59; on humans, 51, 56–57
Burnet, Thomas, 32–34
Butler, Samuel, 20, 75, 85, 238–239, 367
Butler Act, 324

Cabanis, Pierre, 99
Calculating engine, 137
Cambrian system, 194, 350
Camerarius, R. J., 68
Camper, Petrus, 52–53
Candolle, Alphonse de, 151
Capitalism. *See* Laissez-faire economics
Carbon dioxide, declining level of, 115
Carnegie, Andrew, 302

Carpenter, W. B., 124, 170, 204–205
Cartesianism, 31–32, 45, 54
Castle, W. E., 332
Catastrophism, 32, 37, 61, 66, 112–117,
    130, 233, 235; modern, 352
Catholicism. *See* Roman Catholicism,
    and evolution
Causation, 100, 131–132, 134, 145,
    158, 197–198, 218, 250, 318,
    320–321. *See also* Naturalism
Cell theory. *See* Cytology
Celts, 293, 294
Cesalpino, Andreas, 43–44
Chain of being, 1, 12, 50, 52–53, 62–66,
    68, 89, 109, 191
Chambers, Robert, 20, 24, 98–99,
    134–140, 164, 174, 179–180
Chance, 82–83, 160, 163, 186, 197–198,
    204, 225, 272, 320–321, 350, 353
Chemical evolution, 339–340
Chemistry, in eighteenth century, 88
Chesterton, G. K., 323
Chetverikov, S. S., 334–335
Chomsky, Noam, 355
Christianity, 1–4, 27; Darwin and, 146,
    149, 170, 177; and early geology,
    29–30, 31–32, 36, 38–39; and
    Enlightenment, 48, 54–55, 58,
    81–82; and evolution, 185, 202–203,
    215, 322–324, 345, 375–380; and
    early nineteenth-century science,
    103–104, 115–117, 136–137, 139.
    See also Design, argument from;
    Human race: origin of; Roman
    Catholicism, and evolution
Chromosomes, 253, 263, 271–272
Churchill, Frederick, 253
Circular system of classification, 126,
    136
Cladism, 371–374
Classification. *See* Taxonomy
Clausius, Rudolph, 235
Climate change, 32, 59, 80, 113–115,
    118–120, 131–132, 151, 202,
    232–233, 235, 352
Coleridge, Samuel Taylor, 99, 102
Combe, George, 128

Comparative anatomy, 77, 93,
    109–111, 125, 127, 184, 190–191,
    225–230, 261, 334
Comte, August, 100, 162
Condillac, E. B. de, 53
Condorcet, M. J. A., 50, 56, 97
Continental drift, 352, 369
Continuity, 9–10, 54, 63, 65, 76, 89,
    130–132, 136; in evolution, 144, 150,
    166–168, 182, 189, 190, 194,
    199–200, 232–233, 257–258, 265,
    337–338; between humans and apes,
    208–210, 213–214, 278–279. *See
    also* Catastrophism; Saltations
Convergent evolution, 229–230
Cooling-earth theory, 34, 59, 79, 115,
    120, 131–132, 201–202, 234–235
Cooperation. *See* Mutual aid
Cope, Edward D., 196, 241–243, 294
Copernicus, Nicholas, 1, 27
Coral reefs, 150
Correlation of growth, 198
Correns, Carl, 267
Cosmology, 8, 31–34, 58–60
Creation, 3–6, 27, 37–40, 48, 112,
    116–117, 120, 134, 186, 194–195,
    203; Darwin and, 154–155, 163, 170,
    183. *See also* Law of creation
Creation, biblical story of. *See* Genesis,
    book of
Creationism, modern, 2, 27, 375–379
Creative evolution, 317, 320–321
Crick, Francis, 367
Croizat, Leon, 368–369
Cuénot, Lucien, 240
Culture, evolution of, 101, 211, 215,
    284–286, 344–346, 360. *See also*
    Society, evolution of
Curie, Pierre, 235
Cuvier, Georges, 93–94, 108–115, 127,
    187
Cytology, 253–254, 271–272, 350
Cytoplasm, 272

D'Alembert, Jean, 83
Daniken, Erich von, 379
Darrow, Clarence, 324

Dart, Raymond, 284
Darwin, Charles, 22, 84, 96, 142–148, 177, 235; on *Beagle*, 149–155; and design, 146, 151, 170, 183, 203–204, 206; and divergent evolution, 12, 167–168, 170–173, 181–182; early life, 148–149; on heredity, 24, 92, 147, 159–160, 199–202, 254; on humans, 163, 207, 211–214, 215, 289–290, 300, 303, 314; and Lyell, 150, 182; and Malthus, 22, 145, 161–162; and *Origin of Species*, 176, 178, 180–184; on selection, 156–164, 166–168, 170–172, 181, 190, 196–199; and Wallace, 173–176, 196–197
Darwin, Erasmus, 85–86, 103, 148, 157, 239
Darwin, Francis, 157, 239
Darwinian revolution, 24–25, 139–140, 142, 177–179
Darwinism, 7–13, 179, 183–187; arguments for, 152–155, 160–164, 166–173, 180–183, 188–195, 226–227, 231–233, 249, 251–259; arguments against, 186, 196–202, 224–225, 229, 234–236, 237–239, 248–249, 267–270, 362–374; and design, 5–6, 151–155, 163, 170, 186, 196, 202–207, 323; modern, 325–339, 356–361; and progress, 12, 146, 163–164, 169, 191–193, 216–222, 301, 323, 344; social relations of, 144–146, 152, 161–164, 172, 216–222, 297–302, 309, 340–345, 360–361
Daubenton, J. L. M., 77, 109
Davenport, C. B., 311
Dawkins, Richard, 361, 370, 371
Dawson, Charles, 282, 283
Dawson, Sir J. W., 194–196, 203
De Beer, Sir Gavin, 143, 335
Degeneration, 77, 229, 231, 235, 240, 249, 276, 307
De Maillet, Benoît, 34–35, 58, 72–73
De Mortillet, Gabriel, 285–286
Dennett, Daniel, 361

Descartes, Rene, 31–32, 54. *See also* Cartesianism
Design, argument from, 5–6, 38–40, 46, 116–120, 378; and Darwinism, 5–6, 151–155, 163, 170, 186, 196, 202–207, 323, 344; elimination of, 31–32, 37, 57–58, 71–73, 75–76, 80–84, 127–128, 318–320; Lamarckism and, 237–239, 242–243; laws of nature and, 86, 134, 136–140, 196, 204–205, 233, 250, 321–323; and taxonomy, 41–42, 62–65, 67, 123, 125–126
Desmond, Adrian, 143, 146, 349
Determinism, 249, 319–320, 331, 380. *See also* Chance; Evolution: predetermined; Materialism; Naturalism
Determinism, biological. *See* Heredity, as determinant of human character
Development, as model for evolution, 24, 121–125, 135–138, 169, 191–193, 228–229, 241–242, 249, 294. *See also* Evolution: predetermined; Ontogeny; Recapitulation theory
De Vries, Hugo, 224, 267, 268–269, 270, 327, 328
Dewey, John, 320
D'Holbach, baron, 83–84, 100
Dialectical materialism, 101, 339–340. *See also* Marxism
Diderot, Denis, 82–83
Diluvialism, 115–116. *See also* Flood, biblical
Dinosaurs, 118–119, 123, 194, 349
Directed evolution. *See* Evolution: predetermined; Orthogenesis
Discontinuity. *See* Catastrophism; Saltations
Divergence. *See* Evolution: divergent
Division of labor, 172
DNA, 46, 367. *See also* Germ plasm
Dobzhansky, Theodosius, 334, 335, 336–337, 338, 345
Dohrn, Anton, 199, 229, 232
Döllinger, Ignatius, 122
Donald, Mervin, 355

Dorlodot, Henri de, 324
Draper, J. W., 3–4
Drosophila, 271–272
Drummond, Henry, 322
Dualism, cartesian, 81
Dubois, Eugene, 280
Dunbar, Robin, 355
Durkheim, Émile, 288

Earth: age of, 132, 202, 234–236;
   cooling of, 59, 79, 115, 120, 132,
   201–202, 234–235; origin of, 31–34,
   58–60, 132
Earthquakes, 37, 115, 131, 150
East, Edward, 330
Ecological relationships, 40, 67, 120,
   151, 172
Eimer, Theodor, 240, 247, 248
Eiseley, Loren, 200, 224
Eldredge, Niles, 357, 362–365
Electricity, 71, 88, 135
Elie de Beaumont, Léonce, 115
Ellegard, Alvar, 178, 204
Emblematic view of nature, 43
Embryo. *See* Development, as model
   for evolution; Ontogeny; Parallel-
   ism, law of; Recapitulation theory
Emergent evolution, 291, 321, 323
Emotions, expression of, 212, 289
Empiricism, 29, 35, 197
Enlightenment, philosophy of, 48–50,
   53–57, 62–63, 71, 75–76
Environment. *See* Adaptation; Eco-
   logical relationships
Eozoön canadense, 192–194
Epigenetic evolution, 366
Equal-time movement, 377
Essentialism. *See* Species: reality of
Ethics. *See* Morality, and evolution
Eugenics, 256–259, 304, 308–313,
   331–332, 361
Evening primrose, 269
"Evo-devo," 366
Evolution, 3–13, 183, 220; divergent,
   12, 74, 77–78, 125; in early nineteenth century, 93–94, 126–128,
135–140; in eighteenth century, 74,
   77–78, 84–93; of humans, 51, 127,
   133, 138, 163, 183, 207–216,
   278–285, 341–343, 353–356; opposition to, 94, 110–112, 134, 138–139,
   196, 375–380; predetermined, 11,
   136, 196, 229, 233, 247–250, 272,
   337–338, 350, 366; and religion, 5–6,
   151–156, 163, 170, 185–186, 196,
   202–208, 237, 242, 327, 343–345,
   375–380; and society, 30–31, 54–57,
   97, 99–106, 134, 211, 214–222,
   275–277, 284–291, 297–302. *See
   also* Creative evolution; Darwinism;
   Design, argument from;
   Development, as model for evolution; Emergent evolution;
   Lamarckism; Progress
Experimental biology, 226, 230,
   244–245, 266–268. *See also*
   Morphology: revolt against
Extinction, 37–38, 80, 112–114, 117,
   133, 350, 351–353; and Darwinism,
   150, 172; early rejection of, 37–38,
   63, 87
Extraterrestrial origin of life, 72, 339,
   379

Facial angle, 52–53
Falsification of theories, 14–15,
   370–371, 374
Farley, John, 339
Ferguson, Adam, 56
Fichte, J. G., 100
Field studies, 149–155, 174, 189–190,
   251–252, 259, 325–326, 334–337,
   357
Fish, 122, 123, 232
Fisher, R. A., 309, 311, 323, 328,
   330–331, 357
Fiske, John, 222
Fitzroy, Robert, 149
Flood, biblical, 32–34, 36, 58, 115–116,
   376–378
Forbes, Edward, 126
Ford, E. B., 335
Fossey, Dian, 346

Fossil record: discontinuity of, 113–114, 117, 123, 190, 376; early studies of, 35–38, 58, 80, 111–114, 118–120; as evidence for evolution, 135–136, 139, 150, 182, 194–196, 228, 232–234, 337–338, 349–350; of human origins, 209, 278–283, 341–342, 353–354; imperfection of, 113–114, 182; and Lamarckism, 241–243; and orthogenesis, 196, 249–250; specialization in, 125, 170, 194–195. See also Living fossils

Foucault, Michel, 49, 93, 95

France, 56, 76, 82, 87, 93–94, 107, 127, 187, 272, 305. See also French revolution

Frazer, J. G., 286

French revolution, 36, 109

Freud, Sigmund, 277, 291–292

Frontier hypothesis, 316

Function, 109–110; change of, 199. See also Adaptation; Design, argument from; Lamarckism

Galápagos islands, 153–155, 170

Galileo, 2, 28

Gall, F. J., 128

Galton, Francis, 256–258, 265, 309–310

Gasman, Daniel, 305

Gaupp, Ernst, 233

Geddes, Patrick, 314

Gegenbaur, Carl, 190–191

Gender, science and, 313–315, 342, 345–346

Generation, early theories of, 45–47, 65, 76–77, 79, 82–83, 86; Darwinism and, 159–160. See also Development; Ontogeny

Genesis, book of, 1–3, 27–28, 29–34, 36, 57–58, 61–62, 115–116, 294, 376–378

Genetic determinism. See Eugenics; Heredity: as determinant of human character

Genetics, 47, 246, 261–273, 327–333, 361, 367–368; and Darwinism, 266–267, 269, 272, 328–329, 330–333, 334–336, 361; and eugenics, 311, 312, 331. See also Heredity

Genotype, 271

Geoffroy Saint-Hilaire, E., 121, 127, 261

Geographical distribution, 80, 126, 139, 189–190, 233–234, 252, 366–367, 368–369; Darwin and, 150–151, 153–155, 172–173

Geology: in eighteenth century, 57–62, 87; modern, 352, 368; in nineteenth century, 10, 112–118, 123, 129–134, 149–150, 202, 234–235; origins of, 31–38

Germany, 100–102, 107, 120–121, 187, 227, 272–273, 294–296, 305, 311–312, 333, 338–339

Germ plasm, theory of, 253–256. See also DNA; Genetics

Gesner, Conrad, 43

Ghiselin, Michael T., 143

Giard, Alfred, 240

Gillispie, Charles C., 87

Gish, Duane T., 377

Godwin, William, 55

Goethe, J. W. von, 100, 121, 187

Goldschmidt, Richard, 273, 337, 364

Goodall, Jane, 346, 354–355

Gore, Charles, 322

Gould, John, 155

Gould, Stephen Jay, 12, 326, 338; and modern evolutionism, 350, 357, 362–366

Graaf, Regnier de, 46

Gradualism. See Continuity

Grant, Robert, 128, 134, 148

Gray, Asa, 176, 184, 187, 189, 205–206

Greene, John C., 142–143

Group selection, 214, 359

Growth.. See Development; Ontogeny; Recapitulation theory

Gulick, J. T., 252

Habit, change of, 237, 239, 242–243, 289

Haeckel, Ernst, 184, 187, 191–193,

227–228, 230–232, 240; on humans, 210, 280–281, 294, 305

Haldane, J. B. S., 312, 328, 331–332, 339, 358, 359

Hall, G. Stanley, 316

Hamilton, W. D., 359

Hardy, Sir Alistair, 367

Hardy, J. Keir, 306

Hartley, David, 53, 86

Harwood, Jonathan, 333

Hegel, G. W. F., 101

Helvétius, C. A., 53

Hennig, Willi, 371–372

Henslow, George, 239–240

Henslow, John, 149

Herder, J. G. von, 101

Heredity: blending, 199–202, 257–259; Darwinism and, 159–160, 199–202, 253–259; as determinant of human character, 307–313. 360–361; early views of, 45–47, 65, 73–75, 76–77; "soft," 91–92, 236–240, 244–247, 298, 367–368. *See also* Genetics

Herschel, Sir J. F. W., 134, 158, 186, 197, 204

Herschel, William, 59

Higher criticism, 203

Hill, James J., 302

Himmelfarb, Gertrude, 142

History, theories of, 29–31, 54–57, 99–104. *See also* Progress; Society, evolution of

Hobbes, Thomas, 30

Hodge, M. J. S., 88, 159

Hofstadter, Richard, 299, 316

Holmes, Arthur, 236

Homo erectus, 280, 342, 355

Homology, 125, 190–191, 230

Hooke, Robert, 37–38

Hooker, Sir J. D., 165–167, 176, 184, 185, 189

Hopeful monster, 5, 200, 364. *See also* Saltations

Horse, evolution of, 194–195

Hull, David L., 178, 185, 374

Human race: antiquity of, 29, 210–211, 284–285; origin of, 29–31, 52–54,

56–57, 138, 163, 183, 207–216, 222, 278–284, 323, 341–343, 353–356; relation to apes of, 30, 51–53, 208–209, 323, 341, 354–355; relation to nature of, 1, 7, 163, 208, 219, 308, 319–320. *See also* Anthropology; Races, human; Society, evolution of

Humboldt, Alexander von, 120–121, 149, 151

Hume, David, 100

Hutton, James, 61–62, 115, 132

Huxley, Aldous, 317

Huxley, Julian S., 224, 250, 323, 343–344, 371

Huxley, T. H., 107–108, 139, 192, 194, 197, 217, 227, 232–233, 235, 265, 339; and Darwinism, 23, 179, 184–185, 218; on humans, 208–209, 217–219, 313–314, 318–319

Hyatt, Alpheus, 196, 241–243, 249

Hybridization, 70–71, 78–79, 262–264. *See also* Sterility, between species

Hydra, freshwater. *See* Polyp

Hypothetico-deductive method, 14, 147, 158, 197

Ice ages, 115, 123, 126

Idealism, 100–102, 318; in early nineteenth-century biology, 121–123, 125; in post-Darwinian biology, 186, 204–205, 241, 248, 368, 250. See also *Bauplan;* Species: reality of

Individualism. *See* Laissez-faire economics

Induction, 14, 197

Industrial revolution, 97, 99, 106, 275

Inheritance of acquired characteristics. *See* Lamarckism

Instincts, 163, 212, 214, 220, 239, 289–290, 317, 335, 345, 360

Intelligence, evolution of, 138, 211–214, 221–222, 280–281, 290, 355. *See also* Brain, evolution of

Intelligence testing, 311

Intelligent design, 378. *See also* Design, argument from

Invertebrates, 110–111; origin of, 229, 231–232
Ireland, 293, 294, 312
Irish elk, 248, 250
Islam, evolutionism and, 322, 375, 379
Isolating mechanisms, 252, 337
Isolation, geographical, 79, 153–155, 168, 172–173, 189–190, 251–252, 334–335

James, William, 320
Jameson, Robert, 128
Java man. See Pithecanthropus
Jenkin, Fleeming, 200–201, 266, 329
Johannsen, Wilhelm, 270–271
Johanson, Donald, 354
Johnson, Philip E., 378
Jordan, Karl, 252
Jussieu, A. L. de, 70

Kammerer, Paul, 245–246, 317
Kant, Immanuel, 58, 100, 129
Keith, Sir Arthur, 282–283, 303, 345
Kellogg, Vernon L., 259, 305
Kelvin, Lord (William Thomson), 132, 201–202, 234–235, 269
Kettlewell, H. B. D., 338
Kielmeyer, C. F., 122
Kingsley, Charles, 102, 203
Kin selection, 359–360
Kipling, Rudyard, 371
Kirwan, Richard, 61–62
Knox, Robert, 128, 293, 295
Koestler, Arthur, 245, 317, 367
Kovalevskii, Alexandr, 232
Kroeber, A. L., 288
Kropotkin, Peter, 316
Kuhn, Thomas S., 15–16, 24, 185, 347

Lack, David, 154
Laissez-faire economics, 55–56, 103–105, 145, 161–162, 217–218, 299–300; and Spencer, 221, 301–302
Lamarck, J. B., 11, 19, 50, 86–94, 110, 134, 187, 236, 239; on human origins, 51

Lamarckism, 11, 94, 127; Darwin and, 160, 236; opposition to, 255–256, 266, 272, 308; post-Darwinian, 191, 229, 232, 236–247, 323, 367–368; social implications of, 127, 315–317
La Mettrie, J. O. de, 54, 81–82
Landau, Misia, 282
Language, evolution of, 51, 167–168, 212, 344, 354–355
Lankester, E. Ray, 229, 307, 319, 339
La Peyrère, Isaac, 31
Laplace, P. S., 58–59, 129
Lavoisier, Antoine, 88
Law of creation, 125–126, 134–138, 186, 203–206, 242
Leakey, Louis, 346, 354
Leakey, Richard, 354
Le Conte, Joseph, 316
Leeuwenhoek, A. van, 46
Leibniz, G. W. von, 34
Lewontin, R. C., 11, 363–364
Lhwyd, Edward, 37
Life, origin of. See Extraterrestrial origin of life; Spontaneous generation
Linnaeus, Carolus (Karl von Linné), 49–50, 51, 67–71, 262
Literature, Darwinism and, 238–239, 274, 282, 307, 317
Living fossils, 182, 191, 228–229
Locke, John, 30, 44, 53
Lombroso, C., 284
Lorenz, Konrad, 345
Lubbock, Sir John, 211, 215, 285–286
Lyell, Sir Charles, 9–10, 21, 129–134, 174; and Darwin, 21, 97–98, 134, 150, 152, 176, 182, 202; on evolution, 134; on humans, 133, 210, 215
Lysenko, T. D., 246–247, 339

MacBride, E. W., 229, 232, 245, 294, 312, 317
McDougall, William, 245, 290
Mackenzie, Donald, 311
MacLeay, W. S., 126, 136
Magnello, Eileen, 311

Malinowski, Bronislaw, 288

Malthus, T. R., 22, 104–106, 146, 161–162, 175

Mammals, 125, 194–195, 233

Mammoth, 59, 112

Man. *See* Human race

Mantell, Gideon, 118–119

Marsh, O. C., 187, 194–195, 241

Marx, Karl, 17, 101, 145, 306

Marxism, 101, 246–247, 306, 312, 339–340

Mass extinctions, 352. *See also* Catastrophism; Extinction

Materialism, 38, 98, 104, 126–129; of Darwinism, 163, 205, 212, 219, 320; in Enlightenment, 75–76, 81–84. *See also* Marxism; Mechanical philosophy; Naturalism

Matthew, Patrick, 158

Maupertuis, P. L., 52, 73–75

Mayr, Ernst, 5, 123, 143, 145, 155–156, 181, 224; on human origins, 342; and modern synthesis, 326, 328, 333–334, 336–337, 363, 371–372

Mead, Margaret, 288–289

Mechanical philosophy, 28, 31, 38–39, 45–46

Meckel, J. F., 122

Melanism, industrial, 332, 338

Mendel, Gregor, 226, 261–264

Mendelism. *See* Genetics

Mental testing, 311

Merz, J. T., 145

Messenger, Ernest, 324

Methodology. *See* Scientific method

Midwife toad, case of, 245–246

Migration, 80, 114, 126, 133–134, 153, 168, 189, 233–234

Mill, J. S., 99, 197, 220, 315

Miller, Hugh, 118, 123, 138–139

Miller, Stanley, 340

Milne-Edwards, Henri, 172

Mimicry, 189, 248, 340

Mind: and body, 53–54, 82, 98, 128, 163, 219–220; evolution of, 138, 163, 212–216, 221, 239, 243, 280–282, 289–292, 316, 321, 344–345,

354–355. *See also* Materialism; Psychology

Miracles. *See* Creation

"Missing links," 188, 194, 208. *See also* Human race, evolution of

Mitochondrial DNA, 356

Mivart, St. G. J., 186, 196, 198–199, 204, 225, 239

Modernism, 319; in theology, 323

Modern synthesis, 272, 325–328, 333–339, 340–348

Molecular clock, 348, 351

Monboddo, Lord James, 51

Monism, 305

Monkey trial (of John T. Scopes), 324, 376

Monstrosities, 82–83, 127. *See also* Saltations

Montesquieu, baron de, 55

Moore, James R., 3–4, 143, 146, 222

Morality, and evolution, 163, 214, 219, 221–222, 297–305, 316–317, 360–361

Moral sense, 56, 100, 104, 133, 215, 293; evolution of, 163, 207, 212–214, 285, 290, 301, 316, 318–319, 322–323, 344–345

Morgan, C. Lloyd, 243, 290–291, 321, 323

Morgan, Lewis H., 286–287

Morgan, Thomas Hunt, 269–270, 271–272, 327, 328–329

Morphology, 121–122, 125, 190–191, 225–226, 227–230; revolt against, 226, 229–230, 250, 261, 265. *See also* Comparative anatomy

Morris, Desmond, 345

Morris, Simon Conway, 350

Muller, Fritz, 191–192

Muller, Herman J., 329

Museum d'histoire naturelle, Paris, 87, 107, 109

Museums, 108–109, 233

Mutation, genetic, 268, 271, 327–329, 332, 351

Mutation theory, of De Vries, 268–270, 327, 328. *See also* Saltations

Mutual aid, 316

Nägeli, Carl von, 198, 248, 253, 264
Narrative structure of evolution theories, 282
Nationalism, 304–305
Naturalism, 6, 163, 170, 202, 217–220, 318–320. *See also* Materialism
Natural selection, 10–11, 155–164, 166–173, 175–178, 180–181, 189–190, 205–206, 243, 250–260; arguments against, 180–181, 186, 196–202, 204, 206–207, 225, 234, 237, 239, 248, 266–270, 361–370; modern theory of, 272, 325, 327–333, 335–338, 356–361. *See also* Darwinism; Social Darwinism
Natural theology, 38–41, 103, 116, 149, 274. *See also* Design, argument from
Nature, balance of. *See* Balance of nature; Ecological relationships
Nature (journal), 185
Nazism, 101, 296, 305
Neanderthals, 209–210, 278–283, 303, 342, 355–356
Nebular hypothesis, 58–60, 129, 135
Needham, J. T., 79, 84
Neo-Darwinism, 251–60. *See also* Darwinism, modern
Neo-Lamarckism. *See* Lamarckism: post-Darwinian
Neptunism, 34, 61, 72, 113, 115
Neutralism, 351
Newton, Sir Isaac, 28, 58
Newtonianism, 32, 74, 76
Nichol, J. P., 129
Nicholson, A. J., 175
Nietzsche, Friedrich, 305
Nilsson-Ehle, H., 329–330
Nomenclature, binomial, 68–69
Nonadaptive characters, 198–199, 204–207, 247–250, 266–269, 338, 363–364
Nordau, Max, 307
Norton, H. T. J., 330
Nucleus, of cell, 253, 271–272

Oenothera lamarckiana, 269
Oken, Lorenz, 121

Ontogeny, 121—11, lution, 135—13, 228–229, 241
Oparin, Alexand...
Orangutan, 51, 2...
Organic selectio... effect
Origin of life. *See* origin of life; ... tion
Origin of Species, 180–183
Orthogenesis, 19, 247–251, 272,
Osborn, H. F., 22, 249–250, 296
Ospovat, Dov, 17...
Owen, Sir Richa..., 139, 170, 196, ... on humans, 200
Oyama, Susan, 3...

Packard, A. S., 23...
Paleoanthropolo..., 341–342, 353 race
Paleontology. *See*...
Paley, William, 3..., 163, 378
Pallas, P. S., 61
Panbiogeography...
Pangenesis, 159—, 253–254, 256,
Parallel evolutio... evolution; Evo... mined; Ortho...
Parallelism, law...
Parasites, 40, 203...
Pasteur, Louis, 33...
Patterson, Colin, ...
Pearson, Karl, 22..., 309–310, 330
Peirce, Charles, 3...
Peking man, 283
Peppered moth (..., 338
Phillips, John, 11...

Philosophy, 53–55, 81–84, 99–103, 218–222, 317–322

Phrenology, 98, 128, 138, 163, 220

Phylogenies, 191–196, 226, 230–234, 349–351

Physiological selection, 251–252

Piaget, Jean, 291

Piltdown man, 282–283

Pithecanthropus, 278, 280. See also *Homo erectus*

Plants: classification of, 43; evolution of, 239–240; sexuality of, 68, 70

Plato, 5

Playfair, John, 62, 115

Polydactyly, 73

Polygenism, 294–297

Polyp (freshwater hydra), 54, 81

Pope, Alexander, 63

Popper, Sir Karl, 14–15, 101, 370–371, 374

Populational concept of species, 5, 11, 156, 160–163, 166–167, 170–172, 189–190, 256–260. *See also* Population genetics

Population genetics, 327–333, 357–361

Population, principle of, 104–105, 145, 161–162

Positivism, 100, 162

Poulton, E. B., 252

Powell, Baden, 139, 204

Pragmatism, 320

Preadamites, 31

Preadaptation, 199, 232

Preformation theory, 46–47, 65–66, 73, 77

Priestly, Joseph, 53, 55

Primate studies, 346, 354–355

Professionalization in science, 19, 106–108, 185, 217–218, 225–227, 266, 272–273, 326, 338

Progress, 8–9, 12, 86–90, 97, 99, 101–102, 118–124, 135–139, 216, 219–221, 274–277, 321–323; and chain of being, 66; and Darwinism, 143–146, 163–164, 169–172, 183, 191–194, 231–234, 290, 298–302, 334, 343–344; rejection of, 104, 133,

215, 229, 244, 248–250, 288, 307, 319–320; in society, 51, 55–57, 101, 103, 106–107, 126–128, 210–211, 213–214, 216–223, 284–288, 297–301

Provine, William B., 326

Psychology, 220, 288, 290–292. *See also* Mind

Punctuated equilibrium, 326–325

Punnett, R. C., 330

Pure lines, 263, 270

Purpose. *See* Design, argument from; Teleology

Quetelet, Lambert, 162, 256

Races, human, 51–52, 211, 284–289, 292–297, 303–304, 309, 343, 355–356; origin of, 30–31, 52, 215, 294–296

Radicalism, 82–83, 94–95, 126–129, 134–135, 187. *See also* Marxism; Socialism

Radioactivity, 235–236

Random variation. *See* Variation, random

Raven, Charles, 323

Ray, John, 37–38, 39–40, 41, 42–45, 67, 323

Rayleigh, R. L. Strutt, Lord, 235–236

Recapitulation theory, 135–138, 191–193, 228–229, 241–242, 253, 335; and human races, 294, 316. *See also* Parallelism, law of

Redi, Francesco, 45

Religion. *See* Christianity; Design, argument from; Islam, evolutionism and

Rensch, Bernhard, 334, 336

Reproduction. *See* Generation, early theories of

Reptiles, 123, 194, 232–233, 372–373

Retreating-ocean theory. *See* Neptunism

Revolutions. See Catastrophism; French revolution; Industrial revolution; Scientific revolutions

Revolutions in science. *See* Scientific revolutions
Richards, Robert J., 146, 164, 169, 183, 301
Rivers, W. H. R., 288
Robinet, J. B., 65–66
Rockefeller, John D., 302
Roman Catholicism, and evolution, 27, 198, 323–324
Romanes, G. J., 236, 251–252, 290
Romanticism, 87, 100–101, 120, 218. *See also* Idealism
Romer, A. S., 232–233, 250, 334
Rosse, Lord, 129
Rossi, Paolo, 29
Rousseau, J. J., 56
Rudimentary organs. *See* Vestigial organs
Ruse, Michael, 22, 141, 143, 146, 158, 179, 230; on modern evolutionism, 327, 360, 371
Russell, Bertrand, 319–320
Russia, 246–247, 334–335, 339, 345

Saint-Simon, Marquis de, 99–100
Saltations, 5, 10, 127, 136, 184, 242, 265–270, 273, 364; Darwin and, 200–201; and human origins, 215. *See also* Mutation theory, of De Vries
Saunderson, Nicholas, 82
Schelling, F. W. J. von, 100–101, 102
Schindewolf, Otto, 339
Schwalbe, Gustav, 280
Schweber, Sylvan S., 145, 158
Science wars, 379
Scientific community, 15–16, 23, 106–108, 272; and Darwinism, 178–179, 181, 184–187, 217, 338
Scientific method, 13–16, 131–132, 370–371, 380–381; of Darwin, 147, 158, 197–198
Scientific revolution, of seventeenth century, 27–29
Scientific revolutions, 15, 24–25, 139–140, 142, 177–179
Scopes, John T., 324, 376

Scopes monkey trial, 324, 376
Scrope, G. P., 130
Secord, James A., 99, 135, 140, 178
Sedgwick, Adam, 118, 138–139, 149
Selection. *See* Artificial selection; Group selection; Kin selection; Natural selection; Sexual selection
Sellars, R. W., 321
Senility, racial, 244, 249
Sensationalist philosophy, 44, 53, 100, 102–103, 197
Sex, evolution of, 314. *See also* Gender, science and; Generation, early theories of; Heredity
Sexual selection, 196–197, 207, 314, 358
Shaw, George Bernard, 244, 317, 232, 367
Shelley, Mary, 88
Simpson, G. G., 250, 337–338, 343, 344–345, 350, 362
Sloan, Philip R., 76, 159
Smiles, Samuel, 301
Smith, Adam, 55, 56, 103, 145, 172
Smith, J. Maynard, 358
Smith, Sir G. Elliot, 279–280, 283
Smith, William, 117–118
Social Darwinism, 2, 144–145, 161–164, 221–222, 297–305, 309, 360
Socialism, 306, 309, 316. *See also* Marxism; Radicalism
Social relations of science, 15–17, 23, 142–143; pre-Darwinian, 48–49, 63, 71, 81–84, 93–99, 106–108, 116–117, 126–130, 134–135; post-Darwinian, 145–146, 152–153, 161–164, 172, 177–179, 185–187, 202–204, 211, 216–223, 276, 286–288. *See also* Darwinism: social relations of; Lamarckism: social relations of
Society, evolution of, 30–31, 54–57, 97, 99–106, 134, 211, 214, 216–222, 275–277, 284–289, 297–302
Sociobiology, 359–360
Sociology, 100, 288, 360. *See also* Anthropology; Society, evolution of

South America, 150–153

"Space gods" theory, 379

Spallanzani, Lazzaro, 79

Specialization, in development of life, 124–126, 170–172, 194–195

Speciation, 189–190, 251–252, 269, 334, 337, 363; Darwin on, 153–155, 166–167, 172–173

Species, 5, 41–42, 63–65, 89; Darwinian view of, 153–155, 166–167, 172–173, 184, 189–190, 251–252; reality of, 5, 10, 41–42, 63, 67, 76–78, 110–111, 123–124. *See also* Populational concept of species; Speciation

Species selection, 364–365

Spencer, Herbert: on Lamarckism, 139, 221, 238, 255–256, 316; on morality, 221, 301, 324; on progress, 8, 100, 106, 144, 276; on social evolution, 214, 220–222, 287, 301, 304, 314, 316

Spontaneous generation, 45, 79–80, 81–84, 87–88, 135, 183, 192, 231, 339–340

Sports of nature. *See* Saltations

Stalin, Joseph, 246

State, role in history of, 101, 102, 304–305

State of nature, 30, 56

Steady-state world view, 9, 132

Steele, Ted, 367–368

Steno, Nicholas, 36–37

Sterility, between species, 166–167, 251–252

Stone age, 210–211, 285–286

Stratigraphy, 36–37, 61, 112–114, 117–118

Straus, D. F., 203

Stromatolites, 350

Struggle, for existence, 85, 104–105, 145, 175, 270, 275, 297, 301; Darwin and, 143, 151–152, 161–162, 166, 170–172, 275, 300.

Struggle, in society, 30, 55, 103–105, 143, 145, 152, 161–162, 172, 175, 218, 221–222, 298–305. *See also* Social Darwinism

Subspecies. *See* Varieties

Succession of types, 150, 153

Sulloway, Frank J., 147, 148, 154

Sumner, Francis B., 336

Sumner, W. G., 302

Survival of the fittest, 221, 300–302; tautologous, 369–370. *See also* Natural selection; Struggle, for existence

Swammerdam, Jan, 46

Swamping of variations, 200–201, 266, 329

Symbiosis, 350

Synthetic philosophy, of Spencer, 220–223, 301–302

Synthetic theory of evolution. *See* Modern synthesis

Tautology, natural selection as, 369–370

Taxonomy, 41–44, 49–50, 62–64, 66–70, 76, 87–88, 89–90, 109–111, 121–126, 189, 242; Darwin and, 155, 167–168, 169; modern, 336–338, 371–374

Teilhard de Chardin, P., 7, 343–344

Teleology, 6, 38–40, 62, 67, 103–104, 109–110, 116–117, 322–323, 343; Darwinism and, 149, 152, 163, 170, 198–199, 202–207, 378; Lamarckism and, 237, 239–240, 242–243, 316–317, 323. *See also* Design, argument from

Temple, Frederick, 203

Tennyson, Alfred, 152

Theistic evolutionism, 139, 186, 196, 203–207, 242, 250, 343, 378

Thermodynamics, 201, 234–235

Thomson, J. Arthur, 314

Thomson, William (Lord Kelvin), 132, 201–202, 234–235, 269

Timofeef-Ressovsky, N. W., 335

Tournefort, Joseph de, 44

Trembley, Abraham, 54, 81

Tschermak, E. V., 267

Turgot, A. R. J., 56

Turner, F. J., 316

Tylor, E. B., 286, 293
Tyndall, John, 219
Type concept, of Cuvier, 110–111, 124
Typological view of species. *See*
    Species: reality of
Typostrophe theory, 339
Tyson, Edward, 30, 51

Unger, Franz, 262
Uniformitarian geology, 9, 61, 87,
    129–134; and Darwinism, 150, 202,
    234–235
United States, 108, 186–187, 222, 266,
    291, 296–297, 302, 310–311, 316,
    320, 324, 332–333, 336, 338, 341,
    375–379
Upright posture, of humans. *See* Bi-
    pedalism
Use-inheritance. *See* Lamarckism
Ussher, Archbishop J., 4, 29
Utilitarian argument from design,
    39–40, 103–104. *See also* Design, ar-
    gument from
Utility, principle of, 102–104

Van Valen, L., 358
Variation: directed, 11, 186, 196, 199,
    204–206, 225, 237, 241–242,
    247–250, 366, 369; genetic, 255, 267,
    269–272, 325, 328–331, 336; limit to,
    201, 270; random, 10–11, 83,
    159–160, 186, 201, 206, 255,
    257–259, 325. *See also* Saltations
Varieties, 42, 78, 155, 166–167, 175,
    252, 262–263, 269
Velikovsky, Immanuel, 352
Vera causa, 197–198
Vertebrates, 110–111, 125, 229; origin
    of, 232, 351. *See also* Birds; Fish;
    Mammals; Reptiles
*Vestiges of the Natural History of
    Creation*, 98–99, 134–139, 164, 174,
    179–180. *See also* Chambers, Robert
Vestigial organs, 168, 206, 238, 256
Vico, Giambattista, 55
Virchow, Rudolph, 240

Vitalism, 81, 242, 320–321
Vogt, Carl, 296
Voltaire, 55, 58
Vorzimmer, Peter J., 161, 200
Vulcanism, 61, 130–131

Waddington, C. H., 367
Wagner, Moritz, 190, 251
Wallace, A. R., 144, 173–176, 189,
    196–197, 201, 233–234, 235, 251; on
    humans, 213, 215–216, 296, 306
Ward, Lester F., 316
Washburn, Sherwood L., 342, 345
Wasserman, G. D., 371
Watson, H. C., 139
Watson, James, 367
Watson, J. B., 291
Weber, Max, 288
Weismann, August, 225, 238, 251,
    253–256
Weldon, W. F. R., 226, 258–260
Wells, H. G., 307, 319, 323, 335
Wells, William C., 158
Werner, A. G., 61
Whewell, William, 158, 198
Whig history, 18
Whiston, William, 32
White, A. D., 3–4
Whitehead, Alfred N., 321
Wilberforce, Bishop S., 185, 208
Williams, G. C., 359
Willoughby, Francis, 44
Wilson, E. O., 360
Winsor, Mary P., 124
Women, and biology. *See* Gender,
    science and
Woodward, A. Smith, 282
Woodward, John, 36
Wright, Sewall, 332–333, 334, 336, 338
Wyman, Jeffries, 186
Wynne-Edwards, V. C., 359

X-club, 185

Young, Robert M., 105, 143, 144, 162
Yule, G. Udney, 329

Compositor: Binghamton Valley Composition, LLC
Text: 10/13 Aldus
Display: Aldus
Printer and Binder: Sheridan Books, Inc.